FUNDAMENTAL PRINCIPLES OF POLYMERIC MATERIALS

FUNDAMENTAL PRINCIPLES OF POLYMERIC MATERIALS

Third Edition

CHRISTOPHER S. BRAZEL
Department of Chemical and Biological Engineering
The University of Alabama
Tuscaloosa, Alabama

STEPHEN L. ROSEN
Department of Chemical Engineering
University of Missouri-Rolla
Rolla, Missouri

A JOHN WILEY & SONS, INC., PUBLICATION

Published by John Wiley & Sons, Inc., Hoboken, New Jersey
Published simultaneously in Canada

For general information on our other products and services or for technical support, please contact our Customer Care Department within the United States at (800) 762-2974, outside the United States at (317) 572-3993 or fax (317) 572-4002.

Wiley also publishes its books in a variety of electronic formats. Some content that appears in print may not be available in electronic formats. For more information about Wiley products, visit our web site at www.wiley.com.

Library of Congress Cataloging-in-Publication Data:

Brazel, Christopher S., 1970-
  Fundamental principles of polymeric materials / Christopher S. Brazel, Stephen L. Rosen. --
3rd ed.
       pages cm
  Revised edition of: Fundamental principles of polymeric materials / Stephen L. Rosen. 2nd
ed. c1993.
       Includes bibliographical references and index.
  ISBN 978-0-470-50542-7
  1. Polymers. I. Rosen, Stephen L., 1937- II. Rosen, Stephen L., 1937- Fundamental
principles of polymeric materials. III. Title.
  TA455.P58R63 2012
  668.9--dc23

                                                                                    2011052328

Printed in the United States of America

ISBN: 9780470505427

V10016055_120419

CONTENTS

PREFACE   xiii

PREFACE TO THE SECOND EDITION   xv

ACKNOWLEDGMENTS   xvii

1 INTRODUCTION   1

   Problems   7
   References   7

PART I. POLYMER FUNDAMENTALS   9

2 TYPES OF POLYMERS   11

   2.1 Reaction to Temperature   11
   2.2 Chemistry of Synthesis   12
   2.3 Structure   19
   2.4 Conclusions   30
   Problems   30
   Reference   34

3 MOLECULAR STRUCTURE OF POLYMERS   35

   3.1 Types of Bonds   35
   3.2 Bond Distances and Strengths   35
   3.3 Bonding and Response to Temperature   37
   3.4 Action of Solvents   38

3.5   Bonding and Molecular Structure   39

3.6   Stereoisomerism in Vinyl Polymers   40

3.7   Stereoisomerism in Diene Polymers   42

3.8   Summary   44

Problems   44

References   45

4   **POLYMER MORPHOLOGY**   **46**

4.1   Amorphous and Crystalline Polymers   47

4.2   The Effect of Polymer Structure, Temperature, and Solvent
      on Crystallinity   48

4.3   The Effect of Crystallinity on Polymer Density   49

4.4   The Effect of Crystallinity on Mechanical Properties   50

4.5   The Effect of Crystallinity on Optical Properties   51

4.6   Models for the Crystalline Structure of Polymers   53

4.7   Extended Chain Crystals   56

4.8   Liquid Crystal Polymers   57

Problems   59

References   60

5   **CHARACTERIZATION OF MOLECULAR WEIGHT**   **61**

5.1   Introduction   61

5.2   Average Molecular Weights   62

5.3   Determination of Average Molecular Weights   66

5.4   Molecular Weight Distributions   75

5.5   Gel Permeation (or Size-Exclusion) Chromatography (GPC, SEC)   79

5.6   Summary   85

Problems   86

References   89

6   **THERMAL TRANSITIONS IN POLYMERS**   **91**

6.1   Introduction   91

6.2   The Glass Transition   91

6.3   Molecular Motions in an Amorphous Polymer   92

6.4   Determination of T_g 92

6.5   Factors that Influence T_g 95

6.6   The Effect of Copolymerization on T_g 97

6.7   The Thermodynamics of Melting   97

6.8   The Metastable Amorphous State   100

6.9   The Influence of Copolymerization on Thermal Properties   101

6.10 Effect of Additives on Thermal Properties   102

6.11   General Observations about T_g and T_m 103

6.12   Effects of Crosslinking   103

6.13   Thermal Degradation of Polymers   103

6.14   Other Thermal Transitions   104

Problems   104

References   106

7   POLYMER SOLUBILITY AND SOLUTIONS   107

7.1   Introduction   107

7.2   General Rules for Polymer Solubility   107

7.3   Typical Phase Behavior in Polymer–Solvent Systems   109

7.4   The Thermodynamic Basis of Polymer Solubility   110

7.5   The Solubility Parameter   112

7.6   Hansen's Three-Dimensional Solubility Parameter   114

7.7   The Flory–Huggins Theory   116

7.8   Properties of Dilute Solutions   118

7.9   Polymer–Polmyer-Common Solvent Systems   121

7.10   Polymer Solutions, Suspensions, and Emulsions   121

7.11   Concentrated Solutions: Plasticizers   122

Problems   124

References   126

PART II. POLYMER SYNTHESIS   129

8   STEP-GROWTH (CONDENSATION) POLYMERIZATION   131

8.1   Introduction   131

8.2   Statistics of Linear Step-Growth Polymerization   132

8.3   Number-Average Chain Lengths   133

8.4   Chain Lengths on a Weight Basis   136

8.5   Gel Formation   137

8.6   Kinetics of Polycondensation   142

Problems   143

References   145

9   FREE-RADICAL ADDITION (CHAIN-GROWTH) POLYMERIZATION   146

9.1   Introduction   146

9.2   Mechanism of Polymerization   147

9.3   Gelation in Addition Polymerization   148

9.4   Kinetics of Homogeneous Polymerization   149

9.5   Instantaneous Average Chain Lengths   153

9.6   Temperature Dependence of Rate and Chain Length   155
9.7   Chain Transfer and Reaction Inhibitors   157
9.8   Instantaneous Distributions in Free-Radical Addition
        Polymerization   160
9.9   Instantaneous Quantities   165
9.10 Cumulative Quantities   166
9.11 Relations Between Instantaneous and Cumulative Average
        Chain Lengths for a Batch Reactor   169
9.12 Emulsion Polymerization   173
9.13 Kinetics of Emulsion Polymerization in Stage II, Case 2   176
9.14 Summary   180
Problems   180
References   183

10   ADVANCED POLYMERIZATION METHODS   185

10.1 Introduction   185
10.2 Cationic Polymerization   185
10.3 Anionic Polymerization   186
10.4 Kinetics of Anionic Polymerization   192
10.5 Group-Transfer Polymerization   194
10.6 Atom Transfer Radical Polymerization   195
10.7 Heterogeneous Stereospecific Polymerization   196
10.8 Grafted Polymer Surfaces   202
10.9 Summary   203
Problems   203
References   205

11   COPOLYMERIZATION   207

11.1 Introduction   207
11.2 Mechanism   207
11.3 Significance of Reactivity Ratios   209
11.4 Variation of Composition with Conversion   210
11.5 Copolymerization Kinetics   216
11.6 Penultimate Effects and Charge-Transfer Complexes   216
11.7 Summary   217
Problems   217
References   219

12   POLYMERIZATION PRACTICE   220

12.1 Introduction   220
12.2 Bulk Polymerization   220

12.3 Gas-Phase Olefin Polymerization 225

12.4 Solution Polymerization 226

12.5 Interfacial Polycondensation 228

12.6 Suspension Polymerization 229

12.7 Emulsion Polymerization 232

12.8 Summary 234

Problems 234

References 235

PART III. POLYMER PROPERTIES 237

13 RUBBER ELASTICITY 239

13.1 Introduction 239

13.2 Thermodynamics of Elasticity 239

13.3 Statistics of Ideal Rubber Elasticity 246

13.4 Summary 248

Problems 248

References 249

14 INTRODUCTION TO VISCOUS FLOW AND THE RHEOLOGICAL BEHAVIOR OF POLYMERS 250

14.1 Introduction 250

14.2 Basic Definitions 251

14.3 Relations Between Shear Force and Shear Rate: Flow Curves 252

14.4 Time-Dependent Flow Behavior 254

14.5 Polymer Melts and Solutions 255

14.6 Quantitative Representation of Flow Behavior 256

14.7 Temperature Dependence of Flow Properties 259

14.8 Influence of Molecular Weight on Flow Properties 262

14.9 The Effects of Pressure on Viscosity 263

14.10 Viscous Energy Dissipation 264

14.11 Poiseuille Flow 265

14.12 Turbulent Flow 268

14.13 Drag Reduction 269

14.14 Summary 271

Problems 271

References 274

15 LINEAR VISCOELASTICITY 276

15.1 Introduction 276

15.2 Mechanical Models for Linear Viscoelastic Response 276

15.3   The Four-Parameter Model and Molecular Response   285

15.4   Viscous or Elastic Response? The Deborah Number   288

15.5   Quantitative Approaches to Model Viscoelasticity   289

15.6   The Boltzmann Superposition Principle   293

15.7   Dynamic Mechanical Testing   297

15.8   Summary   304

Problems   304

References   307

16   POLYMER MECHANICAL PROPERTIES   **308**

16.1   Introduction   308

16.2   Mechanical Properties of Polymers   308

16.3   Axial Tensiometers   309

16.4   Viscosity Measurement   311

16.5   Dynamic Mechanical Analysis: Techniques   316

16.6   Time–Temperature Superposition   323

16.7   Summary   329

Problems   329

References   332

PART IV. POLYMER PROCESSING AND PERFORMANCE   **335**

17   PROCESSING   **337**

17.1   Introduction   337

17.2   Molding   337

17.3   Extrusion   344

17.4   Blow Molding   347

17.5   Rotational, Fluidized-Bed, and Slush Molding   348

17.6   Calendering   349

17.7   Sheet Forming (Thermoforming)   350

17.8   Stamping   351

17.9   Solution Casting   351

17.10   Casting   351

17.11   Reinforced Thermoset Molding   352

17.12   Fiber Spinning   353

17.13   Compounding   355

17.14   Lithography   358

17.15   Three-Dimensional (Rapid) Prototyping   358

17.16   Summary   359

Problems   359

References   360

18 POLYMER APPLICATIONS: PLASTICS AND PLASTIC ADDITIVES   **361**

    18.1   Introduction   361
    18.2   Plastics   361
    18.3   Mechanical Properties of Plastics   362
    18.4   Contents of Plastic Compounds   363
    18.5   Sheet Molding Compound for Plastics   371
    18.6   Plastics Recycling   373
    Problems   374
    References   374

19 POLYMER APPLICATIONS: RUBBERS AND THERMOPLASTIC ELASTOMERS   **375**

    19.1   Introduction   375
    19.2   Thermoplastic Elastomers   375
    19.3   Contents of Rubber Compounds   376
    19.4   Rubber Compounding   379
    References   379

20 POLYMER APPLICATIONS: SYNTHETIC FIBERS   **380**

    20.1   Synthetic Fibers   380
    20.2   Fiber Processing   380
    20.3   Fiber Dyeing   381
    20.4   Other Fiber Additives and Treatments   381
    20.5   Effects of Heat and Moisture on Polymer Fibers   381

21 POLYMER APPLICATIONS: SURFACE FINISHES AND COATINGS   **383**

    21.1   Surface Finishes   383
    21.2   Solventless Coatings   385
    21.3   Electrodeposition   387
    21.4   Microencapsulation   387
    Problem   389
    References   389

22 POLYMER APPLICATIONS: ADHESIVES   **390**

    22.1   Adhesives   390
    References   394

INDEX   **395**

PREFACE

This work has been edited and organized to provide a solid understanding of the main concepts of polymeric materials at an introductory level, suitable for undergraduate and beginning graduate students in disciplines ranging from chemistry and chemical engineering to materials science, polymer engineering, and mechanical engineering. The second edition of the textbook was organized in a way that flowed naturally from molecular-level considerations to bulk properties, mechanical behavior, and processing methods. I have kept that organization intact with this third edition. I have used this book in teaching a polymer materials engineering course over the past several years, and find that enough information is presented without overwhelming students in detail (i.e., for more detailed courses beyond the introductory polymer class).

One of the big challenges in updating a textbook is to include some of the newer materials, methods, and issues surrounding polymer science while editing and refining the original material so that the end product remains fairly streamlined and provides a balance between describing theories and methodologies while treating each subject with an appropriate weighting. (Of course, instructors are certainly invited to pick and choose topics for their classes, and add material to that covered in the text, but I hope that this provides a good, solid read for students learning the material for the first time without a significant need to supplement the book on the instructor's behalf and without providing so much information that significant portions of the book must be passed over in a one-semester course.)

The text should be suitable for advanced undergraduates and beginning graduate students in disciplines ranging from chemical engineering and chemistry to materials science and mechanical engineering. I have taught mixed classes with just this background using the second edition, and usually found that some introductory information (such as organic chemical structures) was needed—both for students who had not been formally trained in organic chemistry and for those who needed a refresher. This edition now includes a short section in Chapters 1 and 2 on organic functional groups, with an emphasis on some of the structures found in condensation polymerizations. Several instances of natural polymers are included in structures and examples, including starches and poly-peptides, to make the important connection that may building blocks of biology are also polymers.

Some reorganization and combination was done in the third edition, with Chapters 3 and 4 from the previous edition combined, and some of the detailed information on polymer rheology and transport was shortened so that students could be introduced to the material without being overwhelmed. Only small sections were removed, and at many instances, new materials were added, such as the addition of techniques for polymer analysis, processing techniques (including three-dimensional prototyping), and the inclusion of microencapsulation with the coatings section. Updates to advanced polymerization techniques includes some of the emerging techniques to make well-defined polymers, such as atom-transfer radical polymerization, although these methods are treated in a rather brief sense, so that students can understand the basics of the technique improvements and what advantages are achieved compared to other techniques. (In most cases, references are given for those seeking more detail.)

Some of the things that I liked best about this book for teaching an introductory polymers course have been retained. These areas include the description of processes to formulate different products, along with sketches of the processes, the arrangement of the book in going from molecular to macromolecular to physical structures, and the general tone of the book that attempts to connect with the reader through examples that may be familiar to them.

New homework problems have been introduced throughout, primarily those that I have found useful in teaching.

CHRISTOPHER S. BRAZEL

Tuscaloosa, AL
August 2011

PREFACE TO THE SECOND EDITION

This work was written to provide an appreciation of those fundamental principles of polymer science and engineering that are currently of practical relevance. I hope the reader will obtain both a broad, unified introduction to the subject matter that will be of immediate practical value and a foundation for more advanced study.

A decade has passed since the publication of the first Wiley edition of this book. New developments in the polymer area during that decade justify an update. Having used the book in class during the period, I've thought of better ways of explaining some of the material, and these have been incorporated in this edition.

But the biggest change with this edition is the addition of end-of-chapter problems at the suggestion of some academic colleagues. This should make the book more suitable as an academic text. Most of these problems are old homework problems or exam questions. I don't know what I'm going to do for new exam questions, but I'll think of something. Any suggestions for additional problems will be gratefully accepted.

The first Wiley edition of this book in 1982 was preceded by a little paperback intended primarily as a self-study guide for practicing engineers and scientists. I sincerely hope that by adding material aimed at an academic audience I have not made the book less useful to that original audience. To this end, I have retained the worked-out problems in the chapters and added some new ones. I have tried to emphasize a qualitative understanding of the underlying principles before tackling the mathematical details, so that the former may be appreciated independently of the latter (I don't recommend trying it the other way around, however), and I have tried to include practical illustrations of the material whenever possible.

In this edition, previous material has been generally updated. In view of commercial developments over the decade, the discussion of extended-chain crystals has been increased and a section on liquid-crystal polymers has been added. The discussion of phase behavior in polymer-solvent systems has been expanded and the Flory–Huggins theory is introduced. All kinetic expressions are now written in terms of conversion (rather than monomer concentration) for greater generality and ease of application. Also, in deference to the ready availability of numerical-solution software, kinetic expressions now incorporate the possibility of a variable-volume reaction mass, and the effects of variable volume are illustrated in several examples. A section on group-transfer polymerization has been added and a quantitative treatment of Ziegler–Natta polymerization has been attempted for the first time, including three new worked-out examples. Processes based on these catalysts are presented in greater detail. The "modified Cross" model, giving viscosity as a function of both shear rate and temperature, is introduced and its utility is illustrated. A section on scaleup calculations for the laminar flow of non-Newtonian fluids has been added, including two worked-out examples. The discussion of three-dimensional stress and strain has been expanded and includes two new worked-out examples. Tobolsky's "Procedure X" for extracting discrete relaxation times and moduli from data is introduced.

Obviously, the choice of material to be covered involves subjective judgment on the part of the author. This, together with space limitations and the rapid expansion of knowledge in the field, has resulted in the omission or shallow treatment of many interesting subjects. I apologize to friends and colleagues who have suggested incorporation of their work but don't find it here. Generally, it's fine work, but too specialized for a book of this nature. The end-of-chapter references are chosen to aid the reader who wishes to pursue a subject in greater detail.

I have used the previous edition to introduce the macromolecular gospel to a variety of audiences. Parts 1, 2 and most of 3 were covered in a one-semester course with chemistry and chemical engineering seniors and graduate students at Carnegie-Mellon. At Toledo, Parts 1 and 2 were covered in a one-quarter course with chemists and chemical engineers. A second quarter covered Part 3 with additional quantitative material on processing added. The audience for this included chemical and mechanical engineers (we didn't mention chemical reactions). Finally, I covered Parts 1 and 3 in one quarter with a diverse audience of graduate engineers at the NASA–Lewis Research Labs.

A word to the student: To derive maximum benefit from the worked-out examples, make an honest effort to answer them before looking at the solutions. If you can't do one, you've missed some important points in the preceding material, and you ought to go back over it.

STEPHEN L. ROSEN

Rolla, Missouri
November 1992

ACKNOWLEDGMENTS

The most important person to acknowledge is Dr. Stephen Rosen, who penned the first and second editions of this book with a great vision for organizing the wealth of information on polymers into a textbook covering the fundamentals that provided an excellent tool for classroom learning. The guinea pigs (or students) who helped do a trial run of this edition in my polymeric materials classes in 2011 provided corrections and suggestions throughout the semester.

I greatly appreciate my departmental colleagues and university for allowing me a sabbatical from my normal professor duties to expand my research and write several papers as well as updating this book. I am also grateful to the U.S.–U.K. Fulbright Commission, which partially funded my stay in the United Kingdom during which I began writing this third edition.

CHAPTER 1

INTRODUCTION

Although relatively new to the scene of materials science, polymers have become ubiquitous over the past century. In fact, since the Second World War, polymeric materials represent the fastest growing segment of the U.S.' chemical industry. It has been estimated that more than a third of the chemical research dollar is spent on polymers, with a correspondingly large proportion of technical personnel working in the area. From the beginning, the study of polymers was an interdisciplinary science, with chemists, chemical engineers, mechanical engineers, and materials scientists working to understand the chemical structure and synthesis of polymers, develop methods to scale up and process polymers, and evaluate the wide range of mechanical properties existing within the realm of polymeric materials. The molecular structure of polymers is far more complex than the molecules you may have studied in a general chemistry course: just compare the molecular weights, H_2O is 18, NaCl is about 58, but polymers have molecular weights from 10,000 to tens of millions (or possibly much higher for cross-linked polymers). Many of the structures you might have seen in a general cell biology course are made of polymers—proteins, polysaccharides, and DNA are all notable biological polymers. In a material science course, you may have studied crystal structures in metals to understand the mechanical behavior of different alloys (polymers can form crystals, too, but imagine the difficulty of trying to line up a huge polymer molecule into a crystal structure). Polymers are a unique class of materials having wide ranging applications.

A modern automobile contains over 300 lb (150 kg) of plastics, and this does not include paints, the rubber in tires, or the fibers in tires and upholstery. Newer aircraft incorporate increasing amounts of polymers and polymer-based composites. With the need to save fuel and therefore weight, polymers will continue to replace traditional materials in the automotive and aircraft industries. Similarly, the applications of polymers in the building construction industry (piping, resilient flooring, siding, thermal and electrical insulation, paints, decorative laminates) are already impressive and will become even more so in the

Fundamental Principles of Polymeric Materials, Third Edition. Christopher S. Brazel and Stephen L. Rosen.
© 2012 John Wiley & Sons, Inc. Published 2012 by John Wiley & Sons, Inc.

future. A trip through your local supercenter will quickly convince anyone of the importance of polymers in the packaging (bottles, films, trays), clothing (even cotton is a polymer), and electronics industries. Many other examples from pharmaceutical coatings to playground equipment could be cited, but to make a long story short, the use of polymers now outstrips that of metals not just on a volume basis but also on a *mass* basis.

People have objected to synthetic polymers because they are not "natural." Well, botulism is natural, but it is not particularly desirable. Seriously, if all the polyester and nylon fibers in use today were to be replaced by cotton and wool, their closest natural counterparts, calculations show that there would not be enough arable land left to feed the populace, and we would be overrun by sheep. The fact is that there simply are no practical natural substitutes for many of the synthetic polymers used in modern society.

Since most modern polymers have their origins in petroleum, it has been argued that this increased reliance on polymers constitutes an unnecessary drain on energy resources. However, the raw materials for polymers account for less than 2% of total petroleum and natural gas consumption, so even the total elimination of synthetic polymers would not contribute significantly to the conservation of hydrocarbon resources. Furthermore, when *total* energy costs (raw materials plus energy to manufacture and ship) are compared, the polymeric item often comes out well ahead of its traditional counterpart, for example, glass versus plastic beverage bottles. In addition, the manufacturing processes used to produce polymers often generate considerably less environmental pollution than the processes used to produce the traditional counterparts, for example, polyethylene film versus brown kraft paper for packaging.

Ironically, one of the most valuable properties of polymers, their chemical inertness, causes problems because polymers do not normally degrade in the environment. As a result, they increasingly contribute to litter and the consumption of scarce landfill space. One of the challenges in using polymers in materials is developing suitable methods for recycle or effective methods to improve the degradation of disposable items.

Environmentally degradable polymers are being developed, although this is basically a wasteful approach and we are not yet sure of the impact of the degradation products. Burning polymer waste for its fuel value makes more sense, because the polymers retain essentially the same heating value as the raw hydrocarbons from which they were made. Still, the polymers must be collected and this approach wastes the value added in manufacturing the polymers.

This ultimate solution is recycling. If waste polymers are to be recycled, they must first be collected. Unfortunately, there are literally dozens (maybe hundreds) of different polymers in the waste mix, and mixed polymers have mechanical properties similar to Cheddar cheese. Thus, for anything but the least- demanding applications (e.g., parking bumpers, flower pots), the waste mix must be separated prior to recycling. To this end, several automobile manufacturers have standardized plastics used in cars that can be easily removed, remolded, and reused in newer models. Another identifier helpful in recycling plastics is obvious if you have ever looked at the bottom of a plastic soda bottle; there are molded-in numbers on most of the large volume commodity plastics, allowing hand sorting of different materials.

Processes have been developed to separate the mixed plastics in the waste. The simplest of these is a sink–float scheme that takes advantage of density differences among various plastics. Unfortunately, many plastic items are foamed, plated, or filled (mixed with nonpolymer components), which complicates density-based separations. Other separation processes are based on solubility differences between various polymers. An intermediate approach chemically degrades the waste polymer to the starting materials from which new polymer can be made. Other efforts related to polymeric waste have focused on reducing the seemingly infinite lifetime of many plastics in the environment by developing biodegradable commodity polymers.

There are five major areas of application for polymers: (1) plastics, (2) rubbers or elastomers, (3) fibers, (4) surface finishes and protective coatings, and (5) adhesives. Despite the fact that all five applications are based on polymers, and in many cases the same polymer is used in two or more, the industries pretty much grew up separately. It was only after Dr. Harmann Staudinger [1,2] proposed the "macromolecular hypothesis" in the 1920s explaining the common molecular makeup of these materials (for which he won the 1953 Nobel Prize in chemistry in belated recognition of the importance of his work) that polymer science began to evolve from the independent technologies. Thus, a sound fundamental basis was established for continued technological advances. The history of polymer science is treated in detail elsewhere [3,4].

Economic considerations alone would be sufficient to justify the impressive scientific and technological efforts expended on polymers in the past several decades. In addition, however, this class of materials possesses many interesting and useful properties completely different from those of the more traditional engineering materials and that cannot be explained or handled in design situations by the traditional approaches. A description of three simple experiments should make this obvious.

1. Silly putty, a silicone polymer, bounces like rubber when rolled into a ball and dropped. On the other hand, if the ball is placed on a table, it will gradually spread to a puddle. The material behaves as an elastic solid under certain conditions and as a viscous liquid under others.

2. If a weight is suspended from a rubber band, and the band is then heated (taking care not to burn it), the rubber band will contract appreciably. All materials other than polymers will undergo thermal expansion upon heating (assuming that no phase transformation has occurred over the temperature range).

3. When a rotating rod is immersed in a molten polymer or a fairly concentrated polymer solution, the liquid will actually climb up the rod. This phenomenon, the Weissenberg effect, is contrary to what is observed in nonpolymer liquids, which develop a curved surface profile with a lowest point at the rod, as the material is flung outward by centrifugal force.

Although such behavior is unusual in terms of the more familiar materials, it is a perfectly logical consequence of the *molecular structure* of polymers. This molecular structure is the key to an understanding of the science and technology of polymers and will underlie the chapters to follow.

Figure 1.1 illustrates the followings questions to be considered:

1. How is the desired molecular structure obtained?
2. How do the polymer's processing (i.e., formability) properties depend on its molecular structure?
3. How do its material properties (mechanical, chemical, optical, etc.) depend on molecular structure?
4. How do material properties depend on a polymer's processing history?
5. How do its applications depend on its material properties?

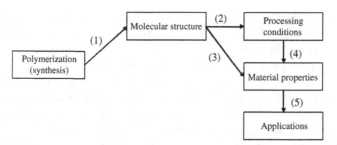

FIGURE 1.1   The key role of molecular structure in polymer science and technology.

The word polymer comes from the Greek word meaning "many- membered." Strictly speaking, it could be applied to any large molecule formed from a relatively large number of smaller units or "mers," for example, a sodium chloride crystal, but it is most commonly (and exclusively, here) restricted to materials in which the mers are held together by covalent bonding, that is, shared electrons. For our purposes, only a few bond valences need be remembered:

$$-\overset{|}{\underset{|}{C}}- \quad \overset{|}{\underset{/\backslash}{N}} \quad -O- \quad Cl- \quad F- \quad H- \quad -\overset{|}{\underset{|}{Si}}-$$

It is always a good idea to "count the bonds" in any structure written to make sure they conform to the above.

Example 1.1   Carbon is the most common element in polymers. Why?

Solution. To make a polymer, the atoms that make up the repeat units must be capable of forming at least two bonds, so H, F, and Cl would terminate a polymer, and can be only on the ends or in functional groups on the sides of the polymer. Two bonds are enough for bonding to occur to continue a polymer structure, but using only oxygen ($O-O-O$...) would not provide any functionality. C and Si both can make four bonds, and since most polymers are derived from petroleum sources, carbon is found in the vast majority of polymers. The four bonds allow two to be used to form the backbone ($C-C-C$...), but the remaining two can be used to include different functional groups, which give polymers their unique characteristics.

The basic structure of a polymer consists of a backbone and pendant (or side) groups (Figure 1.2). Atoms that are covalently linked and stretch from one end of a polymer to the other make up the polymer backbone (which is often not only carbon, but can also contain other atoms such as N, O, or Si). All other atoms are part of the side groups (H is the simplest, methyl ($-CH_3$) or alcohol ($-OH$) groups are among many possibilities, making for a wide variety of structures that make up polymers). Polymers are named based on the repeat unit since that can describe a rather lengthy molecule in a succinct way.

Because carbon is so versatile and can form the backbone of a polymer while being covalently attached to two other side groups, some basic organic chemistry will help to understand the molecular structure of polymers. While organic chemistry is not a prerequisite for learning about polymers, some basic terminology related to organic functional groups will help in understanding molecular structures and polymerization reactions.

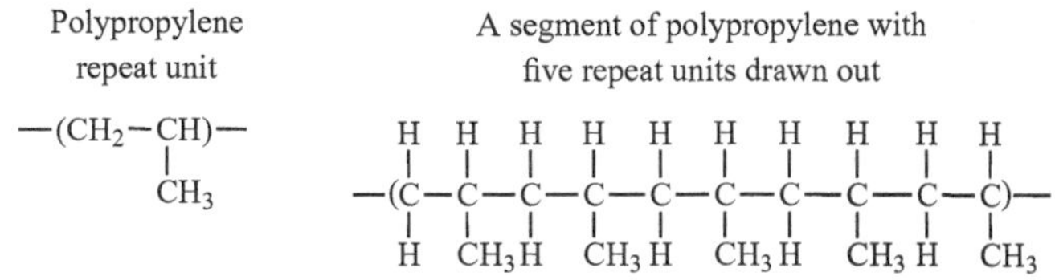

Polypropylene
repeat unit

A segment of polypropylene with
five repeat units drawn out

$-(CH_2-CH)-$
 $\quad\quad CH_3$

H H H H H H H H H H
| | | | | | | | | |
$-(C-C-C-C-C-C-C-C-C-C)-$
| | | | | | | | | |
H CH₃H CH₃H CH₃H CH₃H CH₃

FIGURE 1.2 Structure of a typical polymer polypropylene is shown here, with an all-carbon backbone and side groups of −H or −CH₃. Note that fixed bond angles make the three-dimensional structure more complex than shown here.

Organic chemicals can be categorized based on common functional groups (types of bonds) that are found within a structure (Figure 1.3). These functionalities are helpful in determining if a molecule might be reactive, whether it is likely to be hydrophilic (or hydrophobic), and how (or if) a molecule can participate in a polymerization reaction. Alcohols (−OH) are perhaps one of the more familiar functional groups, and they can be reacted with carboxylic acids (−COOH) to form an ester bond. As long as each molecule has at least two reactive functional groups (e.g., a dialcohol (more commonly referred to as a diol) and a dicarboxylic acid (a diacid) can react to form a continuous polymer molecule with multiple ester linkages along the polymer backbone, yielding a polyester!). Alcohols can also be found in the side groups of polymers, as in poly(vinyl alcohol), where the −OH group stays intact after the reaction.

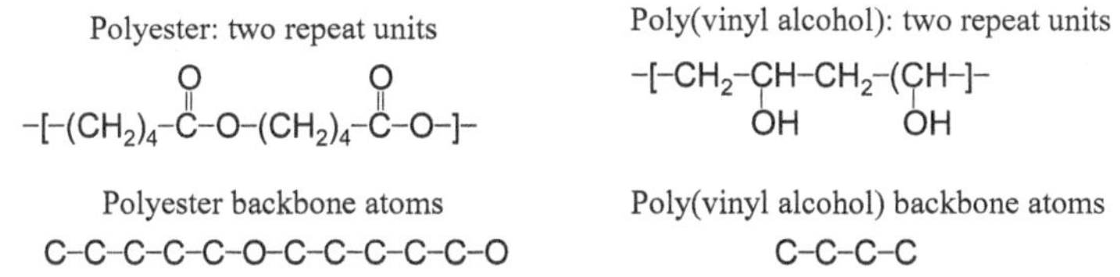

Polyester: two repeat units

$-[-(CH_2)_4-\overset{O}{\overset{||}{C}}-O-(CH_2)_4-\overset{O}{\overset{||}{C}}-O-]-$

Poly(vinyl alcohol): two repeat units

$-[-CH_2-CH-CH_2-(CH-]-$
 $\quad\quad OH \quad\quad\quad OH$

Polyester backbone atoms

C–C–C–C–C–O–C–C–C–C–C–O

Poly(vinyl alcohol) backbone atoms

C–C–C–C

Note that the hydrogens written alongside carbons are not part of the backbone of a polymer molecule. They can bond only to one other atom (the carbon), so a more accurate drawing would show the hydrogens off to the side of the backbone, just as the −OH group is in the poly(vinyl alcohol).

Some of the organic functional groups that are reactive to form polymers include alcohols, amines, carboxylic acids, and alkenes. Although alkanes (C−C bonds) are common in polymers, both in the backbone and in the side groups, they are not reactive. Other bonds commonly found in polymers include amides (for nylons), esters, carbonates, imides, and urethanes. Of this last set, even those who have had organic chemistry probably have not run across the last three, as they are more common in polymers than in traditional organic chemistry. Further reviews of organic chemistry functional groups are also available, and many others can be found online [5,6].

While there is no simple rule about the structures found in a polymer, the most common polymers have only a few organic functional groups, that repeat over and over to make the macromolecules. Of the groups shown in Figure 1.3, alkanes, esters, urethanes, carbonates, aryls, silanes, amides, and imides are commonly found in the backbone of polymers. Alcohols, carboxylic acids, and amines are either reactive parts of monomers or can be found in polymer side groups.

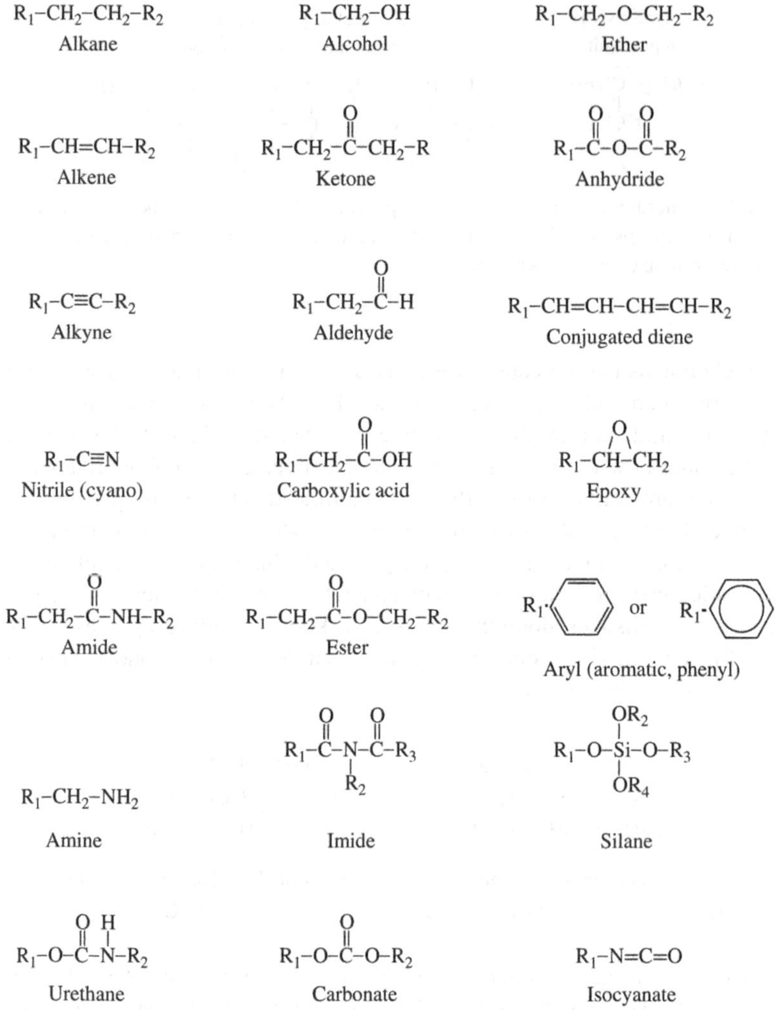

R₁–CH₂–CH₂–R₂ → $R_1\text{–}CH_2\text{–}CH_2\text{–}R_2$
Alkane

$R_1\text{–}CH_2\text{–}OH$
Alcohol

$R_1\text{–}CH_2\text{–}O\text{–}CH_2\text{–}R_2$
Ether

$R_1\text{–}CH{=}CH\text{–}R_2$
Alkene

$$R_1\text{–}CH_2\overset{\overset{\displaystyle O}{\|}}{\text{–}C}\text{–}CH_2\text{–}R$$
Ketone

$$R_1\overset{\overset{\displaystyle O}{\|}}{\text{–}C}\text{–}O\overset{\overset{\displaystyle O}{\|}}{\text{–}C}\text{–}R_2$$
Anhydride

$R_1\text{–}C{\equiv}C\text{–}R_2$
Alkyne

$$R_1\text{–}CH_2\overset{\overset{\displaystyle O}{\|}}{\text{–}C}\text{–}H$$
Aldehyde

$R_1\text{–}CH{=}CH\text{–}CH{=}CH\text{–}R_2$
Conjugated diene

$R_1\text{–}C{\equiv}N$
Nitrile (cyano)

$$R_1\text{–}CH_2\overset{\overset{\displaystyle O}{\|}}{\text{–}C}\text{–}OH$$
Carboxylic acid

$$R_1\text{–}CH\overset{\displaystyle O}{\overset{\diagup\ \diagdown}{\text{–}CH_2}}$$
Epoxy

$$R_1\text{–}CH_2\overset{\overset{\displaystyle O}{\|}}{\text{–}C}\text{–}NH\text{–}R_2$$
Amide

$$R_1\text{–}CH_2\overset{\overset{\displaystyle O}{\|}}{\text{–}C}\text{–}O\text{–}CH_2\text{–}R_2$$
Ester

$R_1\cdot$⟨benzene⟩ or $R_1\cdot$⟨benzene⟩
Aryl (aromatic, phenyl)

$R_1\text{–}CH_2\text{–}NH_2$
Amine

$$R_1\overset{\overset{\displaystyle O}{\|}}{\text{–}C}\text{–}\underset{\underset{\displaystyle R_2}{|}}{N}\overset{\overset{\displaystyle O}{\|}}{\text{–}C}\text{–}R_3$$
Imide

$$R_1\text{–}O\text{–}\underset{\underset{\displaystyle OR_4}{|}}{\overset{\overset{\displaystyle OR_2}{|}}{Si}}\text{–}O\text{–}R_3$$
Silane

$$R_1\text{–}O\overset{\overset{\displaystyle O}{\|}}{\text{–}C}\text{–}\underset{}{\overset{\overset{\displaystyle H}{|}}{N}}\text{–}R_2$$
Urethane

$$R_1\text{–}O\overset{\overset{\displaystyle O}{\|}}{\text{–}C}\text{–}O\text{–}R_2$$
Carbonate

$R_1\text{–}N{=}C{=}O$
Isocyanate

FIGURE 1.3 Common organic functional groups (note: R_1 and R_2 stand for general organic molecules that continue the molecule away from the functional group).

Also keep in mind that when the covalently bound atoms differ in electronegativity, the electrons are not shared evenly between them. In gaseous HCl, for example, the electrons cluster around the electronegative chlorine atom, giving rise to a molecular dipole:

$$
\begin{array}{cc}
\text{H} & \text{:}\ddot{\text{C}}\text{l:} \\
\delta^+ & \delta^-
\end{array}
$$

Electrostatic forces between such dipoles can play an important role in determining polymer properties, as can be imagined for polymers containing the electronegative oxygen, nitrogen, chlorine, or fluorine atoms (which are generally δ^-). These atoms pull the electrons away from other atoms they are covalently attached to (often carbons), making the carbon atom somewhat positive (δ^+).

The most important constituents of living organisms, cellulose and proteins, are naturally occurring polymers (as is DNA), but most commercially important polymers are synthetic or modified natural polymers.

PROBLEMS

1 Consider the room you are in while doing your homework (dorm room, library, cafeteria, pub., and so on).

    (a) Identify five items in it that are made of polymers.

    (b) What would you make those items of if there were no polymers?

    (c) Why do you suppose polymers were chosen over competing materials (if any) for each particular application?

2 Consider the plastic used in the construction of an automobile. List four different parts of a car that may be made of a polymer (consider the exterior, under the hood, and interior of the car—there are many more than four!). For each part, list two physical or chemical properties important for that part. For example, for plastic cup holders, the polymer must be moldable to form into correct shape and should not turn rubbery when heated to $\sim 150°F$ (if exposed to the sun or a hot cup of coffee).

3 You wish to develop a polymer to replace glass in window glazing. What properties must a polymer have for that application?

4 Vinyl chloride is the monomer from which the commercially important polymer poly (vinyl chloride), PVC, is made. It has the chemical formula C_2H_3Cl. Show the structure of vinyl chloride and identify any dipoles present.

5 Acrylonitrile monomer, C_3H_3N, is an important constituent of acrylic fibers and nitrile rubber. It does not have a cyclic structure and it has only one double bond.

    (a) Show the structure of acrylonitrile and identify any dipoles present.

    (b) Draw the repeating structure of polyacrylonitrile, ignore end groups.

6 Poly(octyl cyanoacrylate) is an important adhesive. With the help of the Internet, draw the repeating structure for this polymer, show any dipoles present, and find one trademarked product made from poly(octyl cyanoacrylate).

7 Propylene (C_3H_6) is the monomer from which the fastest growing plastic, polypropylene, is made. It contains one double bond. Show its structure and identify any dipoles present.

REFERENCES

[1] Staudinger, H., *Ber. Dtsch. Chem. Ges.* **53**, 1073 (1920).

[2] Staudinger, H. and J. Fritsch, *Helv. Chim. Acta* **5**, 778 (1922).

[3] Morawetz, H., *Polymers: The Origins and Growth of a Science*, Wiley-Interscience, New York, 1985.

[4] Stahl, G.A., *CHEMTECH*, August 1984, 492.

[5] Richardson, P.N. and R.C. Kierstead, *SPE J.* **25**(9), 54 (1969).

[6] Leach, M.R., *Organic Functional Groups*, on website Chemistry Tutorials and Drills, www.chemistry-drills.com/functional-groups.php?q=simple, 2009.

PART I

POLYMER FUNDAMENTALS

This part covers the fundamental building blocks, basic structures, and nomenclature for polymers. To start, we need to know a little about the molecular organization of polymers, how they are named, and some of the important techniques to characterize polymeric materials. Part I also covers some of the unique thermal, solution, and optical properties of polymers.

Some of the key points to learn from this part of the book are the following:

Polymers are huge—molecular weights of 1,000,000 are not uncommon.

Building polymers requires reactive organic functional groups.

Polymers twist and turn, leave space open like a sponge, but can also crystallize.

Determining a polymer's molecular weight is complicated, but important.

Polymer structure and organization determines whether the material can crystallize.

Why some polymers are transparent but others are opaque.

Besides melting points, polymers have another important thermal transition: the glass transition temperature.

In a good solvent, a polymer can dissolve, but even at low concentrations, the solution quickly becomes viscous.

Fundamental Principles of Polymeric Materials, Third Edition. Christopher S. Brazel and Stephen L. Rosen.
© 2012 John Wiley & Sons, Inc. Published 2012 by John Wiley & Sons, Inc.

CHAPTER 2

TYPES OF POLYMERS

The large number of natural and synthetic polymers has been classified in several ways: thermal behavior, route of chemical synthesis, and structural organization. These will be outlined below, and in the process many terms important in polymer science and technology will be introduced.

2.1 REACTION TO TEMPERATURE

The earliest distinction between types of polymers was made long before any concrete knowledge of their molecular structure. It was a purely phenomenological distinction based on their reaction to heating and cooling.

2.1.1 Thermoplastics

It was noted that certain polymers would soften upon heating and could then be made to flow when a stress was applied. When cooled again, they would reversibly regain their solid or rubbery nature. These polymers are known as *thermoplastics*. By analogy, ice and solder, though not polymers, behave as thermoplastics.

Some of the most commercially important thermoplastics include polyethylene (PE), polypropylene (PP), poly(vinyl chloride) (PVC), and polystyrene (PS). PE is used in products ranging from plastic bags to detergent bottles and has the simplest possible repeat structure of any polymer, since all the pendant groups are hydrogens. PP is also found in a wide range of products, such as plastic storage containers, and competes with other polymers in making plastic bags and pipes. PVC is commonly found in materials as diverse as rigid drain pipes, shower curtains, and raincoats, while PS has been used in foam coffee cups and disposable cutlery. Because these materials are thermoplastics, they are typically

Fundamental Principles of Polymeric Materials, Third Edition. Christopher S. Brazel and Stephen L. Rosen.
© 2012 John Wiley & Sons, Inc. Published 2012 by John Wiley & Sons, Inc.

made into pellets after polymerization, and then can be melted and extruded or shaped into final products.

2.1.2 Thermosets

Other polymers, although they might be heated to the point where they would soften and could be made to flow under stress *once*, would not do so reversibly; that is, heating caused them to undergo a "curing" reaction. Sometimes these materials emerge from the synthesis reaction in a cured state. Further heating of these *thermosetting* polymers ultimately leads only to degradation (as is sometimes attested to by the smell of a short-circuited electrical appliance) and not softening and flow. Again by analogy, eggs and concrete behave as thermosets. Continued heating of thermoplastics will also ultimately lead to degradation, but they will generally soften at temperatures below their degradation point.

Commercially important thermosets include epoxies, polyesters, and phenolic resins. Each of these materials starts out as (often viscous) liquids that set by curing into a final shape. Because these materials set the first time they are made, they cannot be reheated after the polymer is formed without degrading the structure.

Natural rubber is a classic example of the difference between a thermoplastic and a thermoset. Introduced to Europe by Columbus, natural rubber did not achieve commercial significance for centuries; because it was a thermoplastic, articles made of it would become soft and sticky on hot days. In 1839, Charles Goodyear discovered the curing reaction with sulfur (which he called *vulcanization* in honor of the Roman god of fire) that converted the polymer to a thermoset. This allowed the rubber to maintain its useful properties to much higher temperatures, which ultimately led to its great commercial importance.

2.2 CHEMISTRY OF SYNTHESIS

Pioneering workers in the field of polymer chemistry soon observed that they could produce polymers by two familiar types of organic reactions: condensation and addition. As monomers react, new structures are created during the polymerization:

- dimers (or 2-mers): the combination of two monomers,
- trimers (or 3-mers): three monomers strung together, and other #-mers, with the number related to the degree of polymerization, and
- oligomers (or a few mers): 10–20 or so repeat units along a backbone.

An oligomer is simply a small polymer, but it has unique properties since it has a small number of repeat units. Eventually, enough reactions take place to form a large polymer, where it is not uncommon to have thousands or tens of thousands of repeat units connected along a single polymer backbone. In the formation from monomer to polymer, the basic repeat unit remains the same, but the material changes from a monomer that is liquid (or even a gas) to oligomers that may be highly viscous liquids to solid polymers.

Several elementary organic functional groups are worth reviewing, as they play important roles in the synthesis of polymers (Figure 1.3). Because the chemical structures resulting from condensation and addition polymerizations are a bit different, the method of polymerization is one of the major distinctions used in describing polymers. Note that certain functional groups are reactive for forming polymers (common ones include alkenes, alcohols, carboxylic acids, and amines), while others are more commonly found in the resulting structures (alkanes, amides, esters, imides, and urethanes). Other functional groups may not participate in reactions, but be present as side groups or within the polymer backbone (e.g., aryl groups, and ethers).

2.2.1 Condensation Polymerization

Polymers formed from a typical organic condensation reaction, in which a small molecule (most often water) is split out, are known, logically enough, as *condensation polymers*. The common esterification reaction of an organic acid and an organic base (an alcohol in this case) illustrates the simple "lasso chemistry" involved:

$$R\text{–}O\text{(H)} + \text{(HO)}\text{–}\overset{\overset{\displaystyle O}{\|}}{C}\text{–}R' \rightarrow R\text{–}O\text{–}\overset{\overset{\displaystyle O}{\|}}{C}\text{–}R' + H_2O$$

Alcohol + Acid $\rightarrow$ Ester

The $-OH$ group on the alcohol and the $\overset{\overset{\displaystyle O}{\|}}{HO\text{–}C\text{–}}$ on the acid are known as *functional* groups, those parts of a molecule that participate in a reaction, while R and R' are abbreviations that represent the remainder of the molecule—they are generic organic groups and will commonly be used in organic chemistry. Of course, the ester formed in the preceding reaction is not a polymer because we have hooked up only two small molecules, and the reaction is finished far short of anything that might be considered "many membered."

At this point, it is useful to introduce the concept of monomer functionality. Functionality is the number of bonds a mer can form with the other mers in a reaction. In condensation polymerization, it is equal to the number of (organic) functional groups on the mer. These groups must be reactive (such as carboxylic acids, alcohols, aldehydes, or amines).

In the above example, the reactants are monofunctional (there is one alcohol group on one reactant and one carboxylic acid group on the other); thus, if there are no other functional groups present in the rest of the molecule (the R group), the reaction will form only one new bond (an ester). Consider, though, what happens if both reactants are difunctional and the reaction progresses at each end.

$$x\,HO\text{–}R\text{–}OH + x\,HO\text{–}\overset{\overset{\displaystyle O}{\|}}{C}\text{–}R'\text{–}\overset{\overset{\displaystyle O}{\|}}{C}\text{–}OH \rightarrow H\!\left\{O\text{–}R\text{–}O\text{–}\overset{\overset{\displaystyle O}{\|}}{C}\text{–}R'\text{–}\overset{\overset{\displaystyle O}{\|}}{C}\right\}_x OH$$
$$+ (2x-1)\,H_2O$$

Diol   Diacid   Polyester

Here, the resulting product molecule is still difunctional (an alcohol on the left end and an acid on the right; the ester bonds, $\overset{\overset{\displaystyle O}{\|}}{R\text{–}O\text{–}C\text{–}R'}$, are not reactive for polymerization). So, the left end of the molecule can react with another diacid molecule and its right end with another diol molecule. At each step, the growing molecule remains difunctional, so the reaction can continue until all alcohol and acid groups in the reaction mixture are added together (this includes the possibility that two larger molecules can react). Once enough of these molecules (hundreds to even millions of "mers") have condensed together, a polymer is created.

2.2.1.1 Polyesters   In general, the *polycondensation* of x molecules of a diol with x molecules of a diacid to give a *polyester* molecule is written as in the reaction above.

In the polyester (many ester "mers") molecule, the structure in brackets is the repeating unit, and this is what distinguishes one polymer from another. The $\begin{smallmatrix}O\\\|\\\{O-C\}\end{smallmatrix}$ linkage characterizes all polyesters. The generalized organic groups R and R$'$ may vary widely (with a consequent variation in the properties of the polymer), but as long as the repeating unit contains the $\begin{smallmatrix}O\\\|\\\{O-C\}\end{smallmatrix}$ ester linkage, the polymer is a polyester. Of course, there are other functional groups that can react, and thus there are many different types of condensation polymers (polyamides, polyimides, polyethers, etc.).

The quantity x is the *degree of polymerization*, the number of repeating units strung together as identical beads in the polymer chain. It is sometimes also called the *chain length*, but it is a pure number, not a measurable length.

The above nomenclature was introduced by Wallace Hume Carothers, who, with his group at Du Pont, invented two important polymers, neoprene (an addition polymer) and nylon, in the 1930s and was one of the founders of modern polymer science [1].

2.2.1.2 Polyamides (Nylons)   Another functional group that is capable of taking part in a condensation polymerization is the amine ($-NH_2$) group, where one hydrogen from this group reacts with a carboxylic acid in a manner similar to the alcoholic hydrogen to form a polyamide (commonly known as nylon):

$$x\;\underset{}{H\!-\!N\!-\!R\!-\!N\!-\!H} + x\;\underset{}{HO\!-\!C\!-\!R'\!-\!C\!-\!OH} \rightarrow H\{N\!-\!R\!-\!N\!-\!C\!-\!R'\!-\!C\}_x OH$$
$$+ (2x - 1)\,H_2O$$

Diamine   Diacid   Polyamide or nylon

The $\begin{smallmatrix}H\;O\\\|\;\|\\R\!-\!N\!-\!C\!-\!R'\end{smallmatrix}$ linkage is characteristic of all nylons. This bond is also commonly referred to as a peptide bond when it connects two amino acids. Proteins are made from polypeptides, which are one type of natural polymers, and are discussed later.

2.2.1.3 Polyimides   Polyimides are actually a subset of polyamides, but forms when both carboxylic acid groups in diacid monomer react with an amine. As shown, a benzene ring with two acid groups on adjacent carbons (in the ortho position) condenses with an amine to form a single bond. The result is an imide linkage.

Compare the functionality of the diacid in this reaction versus the polyamides or polyesters. Here, the diacid acts as a monofunctional group, while when forming polyamides or polyesters, the diacid reacts separately on both ends. For the formation of a polyimide, the carboxylic acid groups must be close together. Of course, this forms only a single bond. To form a polyimide, there needs to be another diacid as part of the R_1 functional group and another amine in the R_2 group.

2.2.1.4 Polyurethanes   Polyurethanes are made by the reaction of a molecule with two isocyanate groups (a diisocyanate) and two alcohols (a diol). In this reaction, the hydrogen from the alcohol is transferred to the nitrogen atom of the isocyanate group, and a new bond is formed between the isocyanate carbon and the alcohol oxygen.

$$\text{HO–R}_2\text{–OH} \quad + \quad \text{O=C=N–R}_1\text{–N=C=O} \quad \rightarrow \quad \overset{\overset{\displaystyle O \ H}{\displaystyle \| \ \ |}}{\text{HO–(–R}_1\text{–O–C–N–R}_2\text{–)–N=C=O}}$$

Diol   Diisocyanate   Polyurethane

As observed for the polyesters, polyamides, and polyimides, polyurethanes also have a characteristic bond, $-(O-C-N)-$, in their backbone (with the carbon having a $=O$) that makes these polymers easily identifiable. Note that after the reaction to form a urethane bond, the end groups (an alcohol and an isocyanate) are still available to grow the polymer chain further.

2.2.1.5 Polycarbonates   One additional group of condensation polymers highly important in industry is polycarbonates. Polycarbonates are a special subcase of polyesters, in that one monomer has only a single carbon, resulting in the characteristic linkage:

$$\overset{\overset{\displaystyle O}{\displaystyle \|}}{\text{O–C–O}}$$

A common reaction to produce polycarbonates involves a diol (bisphenol-A) and phosgene, which is a diacid chloride (with functionality similar to a dicarboxylic acid) (see part N in Example 2.4).

2.2.1.6 Monomers with Tri- (or Higher) Functionality   The above examples are for polymers formed from difunctional reactants. If either reactant has only one functional group, no polymer can result. However, monomers with functionalities higher than two are also used to make polymers. For example, glycerin is a trifunctional molecule (three alcohol groups) that can be used in polyesterification. Molecules with three or more functional groups tend to make network (or cross-linked) polymers, as the reaction can proceed to form a three-dimensional network instead of just a linear polymer as difunctional monomers do.

$$
\begin{array}{ccc}
H & H & H \\
| & | & | \\
H\!-\!C\!-\!C\!-\!C\!-\!H \\
| & | & | \\
O & O & O \\
| & | & | \\
H & H & H
\end{array}
$$

2.2.1.7 Monomers with Two Different Functional Groups   It is possible that the starting molecule for a condensation polymerization has two (or more) different functional groups and does not require the presence of two different monomers. Two examples of such monomers are amino acids and hydroxy acids:

$$
\underset{\text{Amino acid}}{\overset{H}{\underset{H}{\Large\diagdown}}N-R-\overset{\overset{\textstyle O}{\|}}{C}-OH}
\qquad\qquad
\underset{\text{Hydroxy acid}}{HO-R-\overset{\overset{\textstyle O}{\|}}{C}-OH}
$$

Of course, most of us have heard of amino acids in biology classes, as these are the building blocks for *polypeptides*, which are given the "poly" prefix because they are polymers of amino acids. Multiple polypeptides make up the structure of proteins. An amino acid can react with another amino acid to form an amide bond (or peptide bond if it is an α amino acid) because they have both an amine functional group and a carboxylic acid functional group. Similarly, hydroxy acids condense to form polyesters. However, these reactions may not always proceed in a straightforward fashion. If the R group is large enough, it is possible that the molecule becomes configured so that it can "bite its own tail," condensing to form a cyclic structure rather than a polymer.

$$
\underset{\text{Amino acid}}{H-\overset{\overset{\textstyle H}{|}}{N}-R-\overset{\overset{\textstyle O}{\|}}{C}-OH}
\longrightarrow
\underset{\text{Lactam}}{R\overset{\overset{\textstyle O}{\overset{\|}{C}}}{\underset{N-H}{\diagdown\,|}}}
+ H_2O
$$

This cyclic compound can then undergo a ring-scission polymerization, in which the polymer is formed without splitting out a small molecule (e.g., water), because the small molecule had been eliminated previously in the cyclization step.

$$
x\ R\overset{\overset{\textstyle O}{\overset{\|}{C}}}{\underset{N-H}{\diagdown\,|}}
\longrightarrow
\underset{\text{Polyamide}}{\left[N-R-\overset{\overset{\textstyle O}{\|}}{\underset{}{C}}\right]_x}
\ \ \ \overset{H}{|}
$$

Despite the lack of elimination of a small molecule in the actual polymerization step, the products can be thought of as having been formed by a direct condensation from the monomer and are usually considered condensation polymers. Also of note is that not all condensation polymerizations yield water as the by-product, sometimes carboxylic acids are formed into acid chlorides (with the functional $-OH$ group replaced by a chlorine atom), so instead of an $-OH$ leaving group, a $-Cl$ group leaves, resulting in HCl as the "condensation" by-product.

Note that in this case, the characteristic nylon linkage, $\overset{H\ \ O}{\underset{[N-C]}{|\ \ \|}}$, is split up in the above reaction, and not immediately obvious in the repeating unit as written. If you place a second repeating unit next to the one shown, it becomes evident. This illustrates the somewhat arbitrary location of the brackets, which should not obscure the fact that the polymer is a nylon (though the structure inside the brackets is the repeating "mer"). One important difference between typical synthetic polymers and biological polypeptides is that in most synthetic polymers, the R group is the same throughout the polymer, whereas polypeptides are made up of a sequence of 20 different amino acids (as coded by DNA), each with a unique R group.

Though not formed from a monomer with two different functional groups, another important class of natural polymers is polysaccharides (literally poly sugars). The most common monomer here is glucose ($C_6H_{12}O_6$), and when bonded together by glycosidic linkages (a condensation reaction), it forms starch, cellulose, and other polymers that are important for structural rigidity (particularly in plants) and energy storage in living organisms, as most starches can be broken down (depolymerized) easily.

2.2.2 Addition Polymerization

The second polymer formation reaction is known as addition polymerization and its products as *addition* (or *chain* or *free radical*) *polymers*. Addition polymerizations have two distinct characteristics:

1. No molecule is split out; hence, the repeat unit has the same chemical formula as the monomer.
2. The polymerization reaction involves the opening of a double bond.[1]

Monomers of the general type $\overset{|}{C}=\overset{|}{\underset{|}{C}}$ undergo addition polymerization as

$$x\ \overset{|}{\underset{|}{C}}=\overset{|}{\underset{|}{C}} \rightarrow \left\{\overset{|}{\underset{|}{C}}-\overset{|}{\underset{|}{C}}\right\}_x$$

The double bond "opens up," forming bonds to other monomers at each end, so a carbon–carbon *double bond is difunctional* according to our general definition. (Note that a monomer with two double bonds has a functionality of four; this type of monomer is often referred to as a cross-linking agent as it results in the formation of network structures.) The question of what happens at the ends of addition polymers will be deferred until we reach Chapter 9 on polymerization mechanisms. Also, note that carbon–oxygen double bonds, while common in monomers and polymers, are stable and unreactive for addition polymerization.

An important subclass of the double bond containing monomers is the vinyl monomer:

$$\begin{array}{cc} H & H \\ | & | \\ C & = C \\ | & | \\ H & X \end{array}$$

Addition polymerization is occasionally referred to as vinyl polymerization. Table 2.1 lists some commercially important vinyl monomers, with different substituents replacing X.

TABLE 2.1   Some Commercially Important Vinyl Monomers

Monomer	$-X$
Ethylene	$-H$
Styrene	
Vinyl chloride	$-Cl$
Propylene	$-CH_3$
Acrylonitrile	$-C \equiv N$

[1] Although aromatic rings are often symbolized by ⬡, this is a poor representation of resonance-stabilized structures, which are completely inert to addition polymerization. They are more properly symbolized by ⬡ to avoid confusion with the ordinary double bond. To save space in this book, we will sometimes also use ϕ to represent the aromatic (or phenyl) ring.

Example 2.1 Lactic acid can be dehydrated to form acrylic acid according to the following reaction:

$$
\begin{array}{ccc}
\overset{\text{H}_3\text{C}}{\underset{\text{H}}{\overset{|}{\text{HO-C-C-OH}}}}\overset{\text{O}}{\overset{\|}{}} & \rightarrow & \overset{\text{H}\quad\text{H}\quad\text{O}}{\underset{\text{H}}{\overset{|\quad|\quad\|}{\text{C}=\text{C-C-OH}}}} + \text{H}_2\text{O}
\end{array}
$$

Lactic acid   Acrylic acid

Both these acids can be used as monomers for polymerization. Write the structural formulas for the repeating units for each of the polymers that would be formed if starting from (A) lactic acid and (B) acrylic acid.

Solution. Lactic acid is a hydroxy acid having −OH and −COOH functional groups. It will form a condensation polymer by splitting out water to give

$$
\text{H}\!\!-\!\!\left[\overset{\text{H}_3\text{C}}{\underset{\text{H}}{\overset{|}{\text{O-C-C}}}}\overset{\text{O}}{\overset{\|}{}}\right]_x\!\!\text{OH}
$$

As this is a monomer with two different functional groups, the structure in brackets splits up the characteristic ester linkage.

Acrylic acid is a vinyl monomer with a double bond (since it has only one carboxylic acid, it would not be useful in a condensation polymerization). It undergoes addition polymerization to form

$$
\left[\overset{\text{H}}{\underset{\text{H}}{\overset{|}{\text{C-C}}}}\overset{\text{O}}{\underset{\underset{\text{OH}}{\overset{|}{\text{C}=\text{O}}}}{\overset{\|}{}}}\right]_x
$$

Interestingly, the ester linkage in poly(lactic acid), PLA, is easily hydrolyzed, breaking down the polymer. Thus, this polymer has found applications in biodegradable plastics for products ranging from food packaging to tissue engineering scaffolds for medical applications. Poly(acrylic acid), or PAA, is not degradable, but the pendant carboxylic acid group makes this polymer extremely hydrophilic and thus it is used as a superabsorbent (particularly important in diapers).

Although most molecules with two C=C bonds (dialkenes or simply dienes) have a functionality of 4 for addition polymerization, in *conjugated dienes*, the two double bonds are separated by only one single C−C bond, having a functionality of 2, so they form linear polymers. A conjugated diene is a unique monomer for polymerization because (unlike molecules where the double bonds that act as cross-linking agents are further separated) the double bonds here act in the same way as a single double bond in vinyl polymerizations.

$$
\overset{|\quad|\quad\quad|\quad|}{\underset{|\quad|\quad\quad|\quad|}{\text{C}=\text{C-C}=\text{C}}}
$$

1   2   3   4

The addition polymerization of a conjugated diene results in an *unsaturated* polymer, or a polymer that contains double bonds, either in the backbone of the polymer or in the pendant group. Furthermore, conjugated diene polymerizations can result in several *structural isomers* in the polymer. If the monomer is symmetrical with regard to substituent groups on carbons one through four, it can undergo 1,2-addition and 1,4-addition. If the monomer is asymmetrical, the 3,4 addition will result in yet another structural isomer. (For symmetrical dienes, the 1,2 and 3,4 reactions result in the same structure.) This is illustrated below for the addition polymerization of isoprene (2-methyl-1,3-butadiene), an asymmetrical conjugated diene:

$$
\begin{array}{c}
\text{H} \quad \text{CH}_3 \ \text{H} \ \text{H} \\
| \quad \ | \quad \ | \ \ | \\
\text{+C–C}\!\!=\!\!\text{C–C+}_x \qquad \text{1,4-Polyisoprene} \\
| \qquad\qquad | \\
\text{H} \qquad\qquad \text{H}
\end{array}
$$

$$
\begin{array}{c}
\text{H} \quad \text{CH}_3 \ \text{H} \ \text{H} \\
| \quad \ | \quad \ | \ \ | \\
x \ \text{C=C—C=C} \\
| \qquad\qquad | \\
\text{H} \qquad\qquad \text{H} \\
\text{Isoprene}
\end{array}
\longrightarrow
\begin{array}{c}
\text{H} \ \ \text{CH}_3 \\
| \quad \ | \\
\text{+C–C+}_x \qquad \text{1,2-Polyisoprene} \\
| \quad \ | \\
\text{H} \ \ \text{C–H} \\
\quad \ \ \| \\
\text{H–C–H}
\end{array}
$$

$$
\begin{array}{c}
\text{H} \ \ \text{H} \\
| \ \ | \\
\text{+C–C+}_x \qquad \text{3,4-Polyisoprene} \\
| \ \ | \\
\text{H} \ \ \text{C–CH}_3 \\
\quad \| \\
\text{H–C–H}
\end{array}
$$

Polyisoprene is an important flexible rubbery material, one of the first rubbers to be used as it is found in nature (thus the alternate name "natural rubber"). It is used in belts and hoses in automobiles, in laboratory gloves, and in rubber bands. The 1,2 and 3,4 reactions are sometimes known as vinyl addition, because part of the diene monomer simply acts as an X group in a vinyl monomer.

2.3 STRUCTURE

As an appreciation for the molecular structure of polymers was gained, three major structural categories emerged. These are illustrated schematically in Figure 2.1.

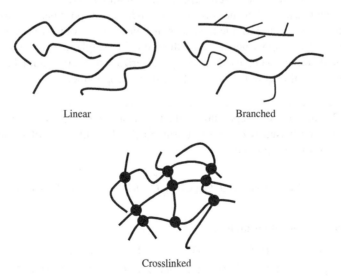

Linear   Branched

Crosslinked

FIGURE 2.1 Schematic representation of polymer structures.

2.3.1 Linear

If a polymer is built from strictly difunctional monomers, the result is a *linear* polymer chain. A scale model of a typical linear polymer molecule made from 0.5 cm diameter clothesline rope would be about 3 m long. This is not a bad analogy: the chains are long, flexible, and essentially one-dimensional structures. The term linear can be somewhat misleading, however, because the molecules do not necessarily assume a geometrically linear conformation as shown in the figure. Some of the better analogies for what these macromolecules look like is a bowl of spaghetti or tangled strands of yarn. The nomenclature for homopolymers usually simply puts "poly" before the name of the repeat unit, which is particularly simple for most addition polymers, such as in polystyrene. If the repeat unit has more than one word, parentheses are used to indicate what is repeated, such as in poly(vinyl chloride). Many condensation polymers would have fairly complex names using this rule, so most are simply referred to by their class (e.g., a polyester) or by their trade names (Dacron®, Nylon®, Lexan®, etc.).

2.3.1.1 Random Copolymers

Polymers consisting of chains that contain a single repeating unit are known as *homopolymers* (this includes many polymers made by addition and condensation polymerization). If, however, the chains contain a random arrangement of two separate and distinct repeating units, the polymer is known as a *random* or *statistical copolymer*, or simply *copolymer*. A random copolymer might be formed by the addition polymerization of a mixture of two different vinyl monomers A and B (the degree of "randomness" depends on the relative amounts and reactivities of A and B, as we shall see later) and can be represented as

AABAAAABBABAAB

and called poly(A-*co*-B), where the first repeating unit listed is the one present in the greater amount. For example, a random synthetic rubber copolymer of 75% butadiene and 25% styrene would be termed poly(butadiene-*co*-styrene). Of course, ter- (3-) and higher multipolymers are possible: nature makes a regular tetrapolymer (RNA and DNA, which are complex polymers with four bases arranged according to genetic codes) and polypeptides that are copolymers with 20 different repeat units (amino acids).

It must be emphasized that the products of condensation polymerizations that require two different monomers to provide the necessary functional groups, for example, a diacid and a diamine, are not copolymers because they contain a single repeating unit. If, however, two different diamines were used in the polymerization, there would be two distinct repeating units and this could be correctly termed a copolymer.

Example 2.2 Illustrate the repeating units that result when 3 mol of hexamethylene diamine (I) are condensed with 2 mol of adipic acid (II) and 1 mol of sebacic acid (III) and name the resulting copolymer

$$H_2N\!\!-\!\!(CH_2)_6\!\!-\!\!NH_2 \quad HO-\overset{\overset{O}{\|}}{C}\!\!-\!\!(CH_2)_4\!\!-\!\!\overset{\overset{O}{\|}}{C}\!\!-\!\!OH \quad HO-\overset{\overset{O}{\|}}{C}\!\!-\!\!(CH_2)_8\!\!-\!\!\overset{\overset{O}{\|}}{C}\!\!-\!\!OH$$

(I)   (II)   (III)

Solution. The two repeating units are

$$\overset{H}{\underset{}{\vert}}\,\overset{H}{\underset{}{\vert}}\,\overset{O}{\underset{}{\|}} \qquad\qquad \overset{H}{\underset{}{\vert}}\,\overset{H}{\underset{}{\vert}}\,\overset{O}{\underset{}{\|}}$$

$$\{N\!\!-\!\!(CH_2)_6\!\!-\!\!N\!\!-\!\!C\!\!-\!\!(CH_2)_4\!\!-\!\!C\} \quad \text{and} \quad \{N\!\!-\!\!(CH_2)_6\!\!-\!\!N\!\!-\!\!C\!\!-\!\!(CH_2)_8\!\!-\!\!C\}$$

From (I) and (II)   From (I) and (III)

The formal (if rather complicated) name for the copolymer containing these two repeating units is poly(hexamethylene adipamide-*co*-hexamethylene sebacamide). Even the linear homopolymer poly(hexamethylene adipamide) is a rather cumbersome name, and thus it is more commonly referred to as nylon 6,6, with the sixes representing the number of carbons in each of the monomers.

2.3.1.2 *Block Copolymers*

Under certain conditions, linear chains can be formed that contain long contiguous blocks of two (or more) repeating units combined in the chains termed a *block copolymer*.

AAAAAAAAAAAAAAAAAAAABBBBBBBBBBBBBBBBBBBBBBBB
AAAAAAAAAAAAAABBBBBBBBBBBBBBBBBBBBAAAAAAAAAAAAAAAA

These structures are called a diblock copolymer, poly(A-*b*-B), and a triblock copolymer, poly(A-*b*-B-*b*-A), respectively. Here, the b (meaning "block") replaces "co" to indicate the organized structure. These structures are important for nanotechnology as several block copolymers can self-assemble into nanospherical micelles, with the A and B blocks arranged to form the core and shell (Figure 2.2). This works especially well if the A and B groups have opposite polarities, so they prefer to segregate rather than intermingle.

2.3.2 Branched

If a few molecules of tri (or higher) functionality are introduced (intentionally or through side reactions) to the reaction, the resulting polymer will have a branched structure. One such example is the grafting of branches made from repeating unit "B" to a linear backbone

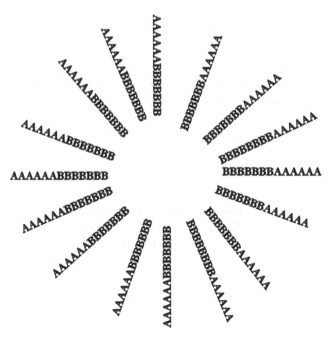

FIGURE 2.2 Block copolymers self-assembling into a micelle with core of B groups and corona of A groups.

of A repeating units. Here, B is said to be grafted onto A:

$$
\begin{array}{c}
_{B}{}^{B\,B}\qquad\qquad\qquad_{B}{}^{B\,B}\\
_{B}{}^{B}\qquad\qquad\qquad\qquad_{B}{}^{B}\\
\text{AAAA--------- A ------A}\\
_{B}{}_{B}{}_{B}
\end{array}
$$

This graft copolymer would be termed poly(A-g-B), as the backbone repeat unit is listed first. Note that "few molecules" is key here, since if even as little as 0.1% of the reaction mixture contains monomers with functionalities of 3 or higher, a network or cross-linked structure is likely to form (particularly at higher reaction conversions).

2.3.3 Cross Linked or Network

As the length and frequency of the branches on polymer chains increase, the probability that the branches will ultimately reach from one backbone chain to another increases. When the backbone chains are connected in this way, the molecular structure becomes a network, with all the chains linked through covalent bonds. This creates a three-dimensional *cross-linked polymer* and the entire mass of the polymer becomes one single tremendously large molecule! One impressive example is a bowling ball, which is a cross-linked polymer, so the entire mass is one molecule; it has a molecular weight on the order of 10^{27} g/mol. Remember this the next time someone suggests that individual molecules are too small to be seen with the naked eye.

Cross-linked or network polymers may be formed in two ways: (1) by starting with reaction masses containing sufficient amounts of tri- or higher functional monomer, or (2) by chemically creating cross-links between previously formed linear or branched molecules ("curing"). The latter is precisely what vulcanization does to natural rubber, and this fact serves to introduce the connection between the phenomenological "reaction to temperature" classification and the more fundamental concept of molecular structure. This important connection will be clarified through a discussion of bonding in polymers.

Example 2.3 Show (a) how a linear, unsaturated polyester is produced from ethylene glycol (I) and maleic anhydride (II), and (b) how the linear, unsaturated polyester is cross-linked with a vinyl monomer such as styrene.

$$
\begin{array}{cc}
\begin{array}{c}
\text{H H}\\
\;|\;\;|\\
\text{HO--C--C--OH}\\
\;|\;\;|\\
\text{H H}
\end{array}
&
\begin{array}{c}
\text{H}\diagdown\;\;\;\;\;\overset{O}{\diagup}\\
\text{C--C}\diagdown\\
\qquad\qquad O\\
\text{C--C}\diagup\\
\text{H}\diagup\;\;\;\;\;\diagdown_{O}
\end{array}
\\[4pt]
\text{(I)} & \text{(II)}
\end{array}
$$

Solution. First, realize that an acid anhydride is simply a diacid with a mole of water split out from the two acid groups. (This is the only common example of acid groups condensing with one another. You cannot ordinarily form polymers this way.) When considering the reaction of an acid anhydride, (conceptually) hydrate it back to the diacid:

$$
\begin{array}{ccc}
\begin{array}{c}
\text{H}\diagdown\;\;\;\;\;\overset{O}{\diagup}\\
\text{C--C}\diagdown\\
\qquad O\\
\text{C--C}\diagup\\
\text{H}\diagup\;\;\;\;\;\diagdown_{O}
\end{array}
+\;H_2O\;\longrightarrow
&
\begin{array}{c}
\;\;\;\;\;\overset{O}{\|}\qquad\quad\overset{O}{\|}\\
\text{HO--C--C=C--C--OH}\\
\qquad\;\;|\;\;\;|\\
\qquad\;\;\text{H H}
\end{array}
\end{array}
$$

Maleic anhydride   Maleic acid

Then condense the diacid with the diol to form a polyester with one double bond per repeating unit:

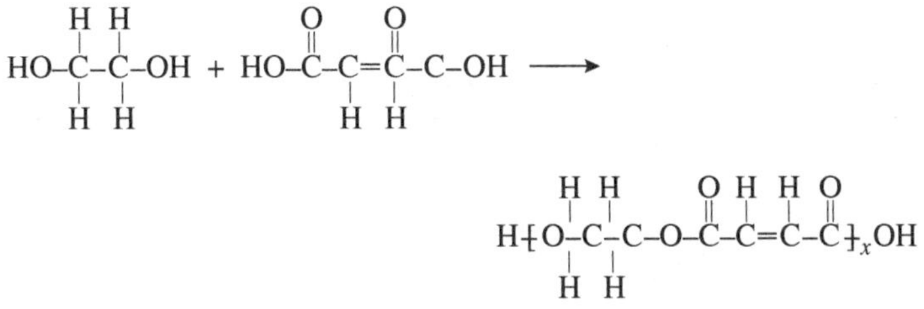

The double bond in the maleic acid is inert toward condensation polymerization. Note that the degree of unsaturation (average number of double bonds per repeating unit) could be varied from zero to one (as it is here) by employing mixtures of a saturated diacid, for example, phthalic anhydride (III), and the maleic anhydride to form copolyesters with saturated and unsaturated repeating units:

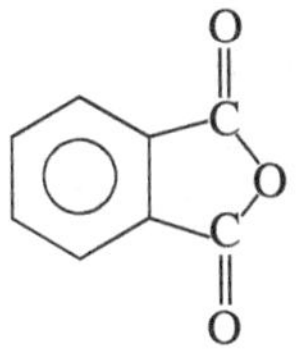

(III)

Commercially, the degree of polymerization x is maintained low (say 8–10 repeat units) so that the product is a viscous liquid. The linear, unsaturated polyester is then diluted with a liquid vinyl monomer, most often styrene. Before use, an initiator chemical that promotes addition polymerization (as discussed in Chapter 10) is added, causing the vinyl monomer to undergo addition copolymerization with the double bonds in the polyester (which has a functionality of 16–20, twice the number of double bonds). This forms a highly cross-linked rigid network.

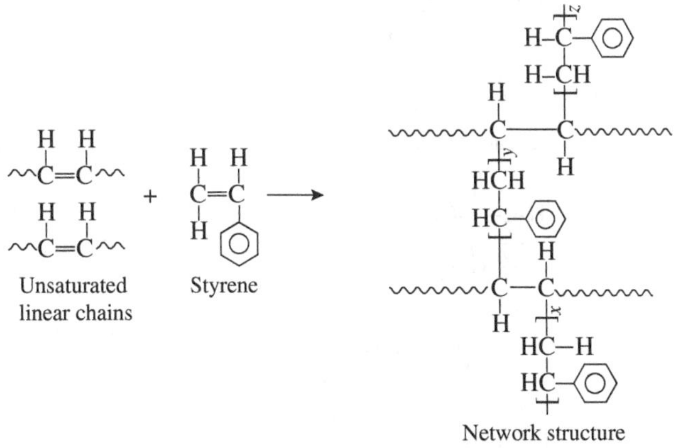

Unsaturated linear chains   Styrene   Network structure

The liquid polyester–styrene mixture is often used to impregnate fiberglass and is cured to form boat hulls, auto (Corvette) bodies, and other so-called fiberglass objects (which are really composite materials of fiberglass reinforced polyester).

Example 2.4 Given the structure of the monomer(s) for the polymers, show the structural formulas of the repeating units for each of the following polymers and classify them according to structure and chemistry of formation. All these polymers are commercially important. Most follow the rules outlined above, but some do not, and have been included here to illustrate their structures, chemistries of formation, and characteristic linkages. The polymer name and starting monomers are shown in the table below:

Part Polymer Name	Monomer Structure(s)	
A Polystyrene	$H–C=CH_2$ Styrene	
B Polyethylene	$H_2C = CH_2$ Ethylene	
C Poly(butylene terephthalate) (PBT)	$HO–(CH_2)_4–OH$ Butylene glycol (1,4-butanediol)	HO–C(=O)–⬡–C(=O)–OH Terephthalic acid
D Poly(ethylene terephthalate) (PET) (Dacron, Mylar®)	$H_3C–O–C(=O)–$⬡$–C(=O)–O–CH_3$ Dimethyl terephthalate	$HO–(CH_2)_2–OH$ Ethylene glycol
E Nylon 6/6 (the numbers designate carbon atoms in the diamine and diacid, respectively)	$H_2N–(CH_2)_6–NH_2$ Hexamethylene diamine	$HO–C(=O)–(CH_2)_4–C(=O)–OH$ Adipic acid
F Nylon 6 (made from a single monomer)	ε-Caprolactam	
G Glyptal (glycerol + phthalic anhydride)	Glycerol	Phthalic anhydride
H Poly(diallyl phthalate)	Diallyl phthalate	
I Melamine-formaldehyde (Melmac®, Formica®)	Melamine	$H–C–H$ with $=O$ Formaldehyde

Part Polymer Name	Monomer Structure(s)
J Polytetra-fluoroethylene (Teflon TFE®)	$F_2C = CF_2$ Tetrafluoroethylene
K Poly(phenylene oxide) (PPO) (*Hint:* polymerized in presence of O_2)	 2,6-Dimethyl phenol
L Polypropylene	 Propylene
M Acetal (polyfor-maldehyde or polyoxy-methylene) (Celcon®, Delrin®)	 Formaldehyde or Trioxane
N Polycarbonate (Calibre®, Lexan®, Makrolon®)	 Bisphenol-A Phosgene
O Epoxy or phenoxy	 Bisphenol-A Epichlorohydrin
P Poly(dimethyl siloxane) (silicone rubber) (*Hint:* polymerized in presence of H_2O)	 Dimethyl dichlorosilane
Q Polyurethane	**HOROH** Diol or glycol $O = C = N-R'-N = C = O$ Diisocyanate
R Polyimide	 Dianhydride $H_2NR'NH_2$ Diamine
S Polysulfone	

Solution (see Table 2.2).

TABLE 2.2 Solution to Example 2.4

Part	Polymer Name	Polymer Structure	Polymerization Type	Structure Type	Leaving Group	Comments
A	Polystyrene	$\left[\begin{array}{c} CH_2{-}CH(C_6H_5) \end{array}\right]_x$	Addition	Linear	None	
B	Polyethylene	$\left[\begin{array}{c} CH_2{-}CH_2 \end{array}\right]_x$	Addition	Linear or branched from side reactions	None	
C	Poly(butylene terephthalate) (PBT)	$H{-}[O{-}(CH_2)_4{-}O{-}C(=O){-}C_6H_4{-}C(=O){-}]_x{-}OH$	Condensation	Linear	H_2O	
D	Poly(ethylene terephthalate) (PET) (Dacron, Mylar)	$H_3C{-}[O{-}C(=O){-}C_6H_4{-}C(=O){-}O{-}(CH_2)_2{-}]_x{-}OH$	Condensation	Linear	CH_3OH	Can also be made from the diacid above, splitting out H_2O
E	Nylon 6/6	$H{-}[N(H){-}(CH_2)_6{-}N(H){-}C(=O){-}(CH_2)_4{-}C(=O){-}]_x{-}OH$	Condensation	Linear	H_2O	Characteristic nylon linkage circled
F	Nylon 6	$[C(=O){-}N(H){-}(CH_2)_5{-}]_x$	Condensation	Linear	None	Amide in caprolactam splits

G	Glyptal	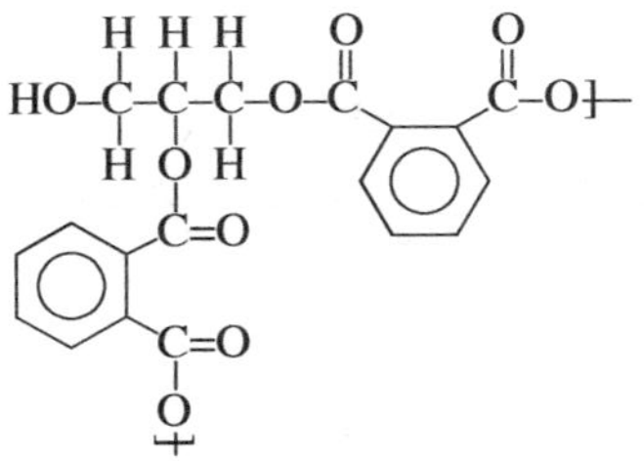 Condensation	Branched or cross-linked	None	Structure depends on ratio of reactants, usually highly cross-linked see Chapter 8
H	Poly(diallyl phthalate)	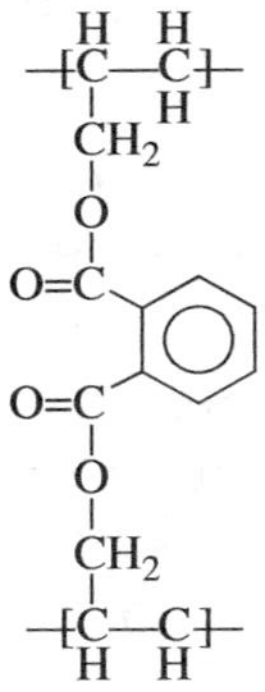 Addition	Cross-linked	None	Monomer has functionality of 4 with two double bonds
I	Melamine formaldehyde (Melmac, Formica)	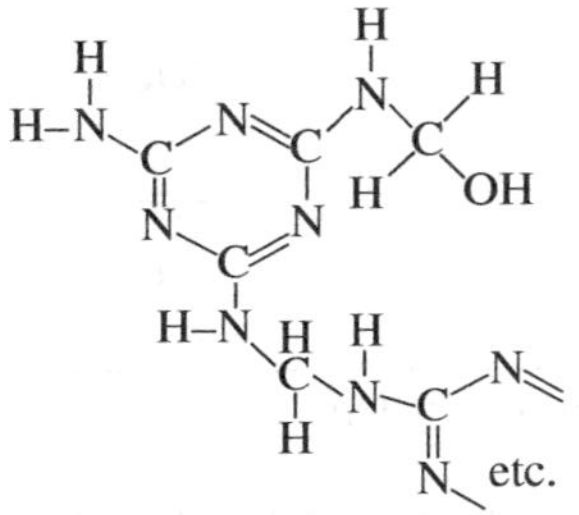 Condensation	Cross-linked	H₂O	Similar structures result when formaldehyde is condensed with urea, (NH₂)₂CO, or phenol, φ-OH

(Continued)

27

TABLE 2.2 (*Continued*)

Part	Polymer Name	Polymer Structure	Polymerization Type	Structure Type	Leaving Group	Comments
J	Polytetrafluoro-ethylene (Teflon TFE)	$\left[\begin{array}{cc} F & F \\ C & C \\ F & F \end{array}\right]_x$	Addition	Linear	None	
K	Poly(phenylene oxide)	(aromatic ring with two CH_3 groups and O) $_x$	Condensation	Linear	H_2O	Oxidative coupling mechanism
L	Polypropylene	$\left[\begin{array}{cc} H & H \\ C & C \\ H & CH_3 \end{array}\right]_x$	Addition	Linear	None	
M	Acetal (Celcon®, Delrin®)	$\left[\begin{array}{cc} H \\ C & O \\ H \end{array}\right]_x$	Addition	Linear	None	Only other double bond that forms addition polymers
N	Polycarbonate (Calibre, Lexan, Makrolon)	$H-O-\text{(aromatic)}-C(CH_3)_2-\text{(aromatic)}-O-\left[C=O\right]_x-Cl$	Condensation	Linear	HCl (if alcohol and acid, will split out H_2O)	Characteristic carbonate linkage: $\left(O-\overset{\overset{\displaystyle O}{\|}}{C}-O\right)$
O	Epoxy or phenoxy	$H-O-\text{(aromatic)}-C(CH_3)_2-\text{(aromatic)}-O-\left[\begin{array}{ccc} H & H & H \\ C & C & C \\ H & OH & H \end{array}\right]_x$	Condensation and epoxide ring scission	Linear	HCl	For $x<8$, a liquid epoxy, if $8<x<20$, a solid epoxy, and for $x\approx100$, a "phen-oxy" plastic

	Polymer	Structure / Reaction	Type	Chain	By-product	Comments
		$-OH + H_2C-C-C-Cl + HO- \rightarrow -O-C-C-C-O- + HCl$ (with H, H, O / H, H)	Condensation	Linear	HCl	Will get side reaction with no bisphenol A; can be cross-linked with diamines or anhydrides through —OH termini
P	Poly(dimethyl siloxane) (silicone rubber)	$\begin{array}{c} CH_3 \\ -[\,Si-O\,]_x- \\ CH_3 \end{array}$	Condensation	Linear	HCl	Commercial materials contain some three-functional cross-linking agent; also acetate groups can be substituted for Cl
Q	Polyurethane	$-[\,O-R-(O-C-N-R'-N-C\,]_x-$ (urethane link circled; O, H / H, O)	Condensation	Linear	None (the C=N group opens)	Characteristic urethane link is circled. R and R′ can vary widely
R	Polyimide	polyimide structure (imide linkage circled), $-[R'-N\cdots C=O\;\;O=C\cdots]_x-$ with R	Condensation	Linear! (not network!) Both amine H's react with anhydride, so it has only a functionality of 2	H$_2$O	The characteristic imide linkage is circled
S	Polysulfone	$-[\,O-\!\!\bigcirc\!\!-\overset{CH_3}{\underset{CH_3}{C}}-\!\!\bigcirc\!\!-O-\!\!\bigcirc\!\!-\overset{O}{\underset{O}{S}}-\!\!\bigcirc\!\!-]_x-$	Condensation	Linear	NaCl	

TABLE 2.3 Examples of the Versatile Applications of Polymers

Bulletproof vests
Food containers and bottles
Furniture cushions and upholstery
Carpeting and vinyl flooring
Pharmaceutical coatings (e.g., gelcaps)
Paints (latex and oil-based)
Clothing (nearly all types)
Chewing gum
Sports equipment
Wetsuits
Shatterproof windshields
Casing for electronics

2.4 CONCLUSIONS

The division of polymers along the three categories discussed in this chapter included response to temperature (thermosets versus thermoplastics), polymerization mechanism (condensation versus addition), and structural organization (linear, branched, or network). Although these are very handy distinctions used to describe a polymer, we will see in further chapters that in some cases many polymers behave in a similar way despite having a variety of compositions and structures, there are many variables (molecular weight, additives, thermal processing history, etc.) that yield a wide range of properties for any one polymer. This wide range of properties allows polymers to be used in numerous applications (Table 2.3), as we will discover. The applications for the specific polymers in Example 2.4 are left as a homework exercise.

PROBLEMS

1   A 40-ft "fiberglass" boat hull weighs 2000 kg. Sixty percent of that is a fully cured cross-linked polyester resin (the remainder is the glass reinforcing fibers, pigment, fillers, etc.). What is the molecular weight of the polymer (in g/gmol)?

2   Pyromellitic dianhydride, PMDA, is useful in a variety of polymerization reactions.

(a)   What is its functionality in polyesterification reactions? Structurally, what type of polymer would you expect to get by reacting PMDA with a glycol (diol)?

(b) Show the repeating unit of the polymer formed by reacting PMDA with hexamethylene diamine (Part E in Example 2.4).

3 The thermotropic liquid crystal polymer Xydar® is reported to be made by condensing terephthalic acid (HOOC−ϕ−COOH), p,p'-dihydroxybiphenyl (HO−ϕ−ϕ−OH), and p-hydroxybenzoic acid (HO−ϕ−COOH). What repeating unit(s) would you expect to find in this polymer?

4 Complete the following additions to Example 2.4 by drawing the repeat unit for each of these polymers.

T. Polyacrylamide (PAAm)—starting monomer: acrylamide

U. Polyurea—starting monomers: diamine (see part E) and diisocyanate (part Q)

V. Polyarylate—starting monomers: terephthalic acid (part C) and bisphenol-A (part N)

W. Polyphenylene sulfide (PPS) (Ryton®, Supec®)—starting monomers: *para*-dichlorobenzene and Na_2S

X. Poly(pivalolactone)—starting monomer

Y. Poly(methyl methacrylate) (PMMA)—starting monomer: methyl methacrylate

5 By consulting the Internet (search for the Macrogalleria) or using a library resource, list one common product that is made using each of the polymers in Example 2.4 and in Problem 4 (A–Y).

6 Styrene ($H_2C{=}CH{-}\phi$) can be polymerized along with divinylbenzene (DVB) ($H_2C{=}CH{-}\phi{-}HC{=}CH_2$) to form a cross-linked polymer. If a mixture containing 99 mol% styrene and 1 mol% DVB goes to completion,

(a) What is the average number of repeat units between cross-links, $\bar{x}_c$?

(b)   What is the average molecular weight between cross-links, $\bar{M}_c$?

(c)   Draw a portion of the chemical structure of the cross-linked polymer. *Note*: ϕ = benzene ring (C_6H_6 or in the case of styrene: C_6H_5 or DVB: C_6H_4)

7   A polymer is made by reacting (assume completely) 2 mol of glycerin (Part G in Example 2.4) with 3 mol of terephthalic acid (Part C in Example 2.4). Will the polymer be linear, branched, or cross-linked? What is the molecular weight of the resulting polymer?

8   Assume that you have supplies of the five reagents below. Consider only the reactions discussed in the chapter and assume that they can be carried out completely and irreversibly.

(I)   (II)   (III)   (IV)   (V)

(a)   Describe (show the reagents and reaction steps) how you would make a graft copolymer with the branches distributed randomly along the main chain and with no possibility of cross-linking. Show the main chain and branched repeating units.

(b)   Describe how you would make a block copolymer ($A_xB_yA_z$, with x, y, and z not necessarily being the same). Show the A and B repeating units.

9   Given supplies of isocyanatoethyl methacrylate (IEM), and a glycol, $HO-R-OH$, and using only reactions discussed in this chapter, do the following.

(a)   Show the repeating unit(s) of all the linear homopolymers that could be produced.

(b)   Illustrate, showing the reactions involved, two different methods of producing a cross-linked polymer, again using only the IEM and the glycol.

10   Given supplies of acrylic acid (I), adipic acid (II), and propylene glycol (III)

$H_2C{=}CHCOOH$ $HOOC{\left(CH_2\right)}_4COOH$ $HO{\left(CH_2\right)}_3OH$

(a)   Show the repeating unit(s) of all the linear homopolymers that could be produced.

(b) Describe three different procedures by which cross-linked polymers could be produced (again using only I, II, and III).

11 Find the chemical structures of sebacic acid and propylene glycol (use your textbook, a chemistry handbook, or the Internet). Will these react by free radical (addition) polymerization or condensation (step growth) polymerization? Write the order of the atoms along the repeat unit on the polymer backbone (do not include pendant groups or groups found only at the end of the polymer chain).

12 Nylon 6,6 has carbons (CH_2's) between successive amide linkages in the repeating structure. Nylon-11 is used by NASA for electronic coatings and has 11 carbons between each amide linkage. Unlike Nylon 6,6, which is produced using two different monomers, Nylon-11 is made using a single monomer structure. Draw the monomer used to produce Nylon-11. What is the sequence of atoms that make up the polymer backbone? (Do not include pendant groups or atoms that are present only at the end of the polymer chain.)

13 The sidewalls of a car tire are made from polyisoprene rubber and is fully cross-linked through the process of vulcanization.

(a) Draw the structures of the monomer (isoprene) and the repeat unit once polymerized.

(b) Briefly describe the process of vulcanization.

14 If the polyisoprene tire sidewalls in Problem 13 weigh 5 kg, determine the molecular weight of the cross-linked polyisoprene in g/gmol. (The answer may be surprisingly high.)

15 As a class assignment, each student should select one common polymer from the list below and prepare a one page infosheet that includes monomer and polymer structures, physical properties of the polymer, and applications or end products that the polymer is used in. The infosheets can be presented in class and copies made for students to become part of the class notes.

Polyethylene (HDPE/LDPE)

Polypropylene, PP

Poly(ethylene terephthalate), PET

Poly(tetrafluoroethylene), PTFE, Teflon

Poly(vinyl chloride), PVC

Polystyrene, PS

Poly(methyl methacrylate), PMMA

Polycarbonates

Polyamides (Nylon 6,6)

Polyurethanes, PU

Poly(acrylic acid), PAA

Poly(*N*-vinyl pyrrolidone), PVP or PNVP

Polyisoprene

Polybutadiene (*cis*, *trans*, and vinyl)

Poly(acrylonitrile–butadiene–styrene), ABS

Aramids
Poly(cyanoacrylate)s
Poly(dimethyl siloxane), Silicone
Poly(ethylene oxide)

A visit to the website for the Macrogalleria may be helpful for most (if not all) of these polymers.

REFERENCE

[1]   Hounshell, D.A. and J.K. Smith, Jr., *Invention and Technology*, Fall 1988, p. 40.

CHAPTER 3

MOLECULAR STRUCTURE OF POLYMERS

This chapter explores the types of bonds formed in polymers, and focuses on how weak bonds can impact material properties. The arrangement of these bonds and side groups along the polymer backbone also lead to a wide range of possible isomers (same chemical formula, different arrangement of atoms) that are also important in determining polymer properties.

3.1 TYPES OF BONDS

Various types of bonds hold together the atoms in polymeric materials, unlike in metals, for example, where only one type of bond (metallic) exists. These types are: (1) primary covalent, (2) hydrogen bond, (3) dipole interaction, (4) van der Waals, and (5) ionic. Examples of each are shown in Figure 3.1. Hydrogen bonds, dipole interactions, van der Waals bonds, and ionic bonds are known collectively as secondary (or weak) bonds. The distinctions are not always clear-cut, that is, hydrogen bonds may be considered as the extreme of dipole interactions. The secondary bonds are generally weaker bonds and are responsible for many of the bonds between different polymer chains (intermolecular bonds).

3.2 BOND DISTANCES AND STRENGTHS

Regardless of the type of bond, the potential energy of the interacting atoms as a function of the separation between them is represented qualitatively by the potential function sketched in Figure 3.2. As the interacting centers are brought together from large separation, an increasingly great attraction tends to draw them together (negative potential energy). Beyond the separation r_m, as the atoms are brought closer together, their electronic "atmospheres" begin to interact and a powerful repulsion is set up. At r_m, the system is at a

Fundamental Principles of Polymeric Materials, Third Edition. Christopher S. Brazel and Stephen L. Rosen.
© 2012 John Wiley & Sons, Inc. Published 2012 by John Wiley & Sons, Inc.

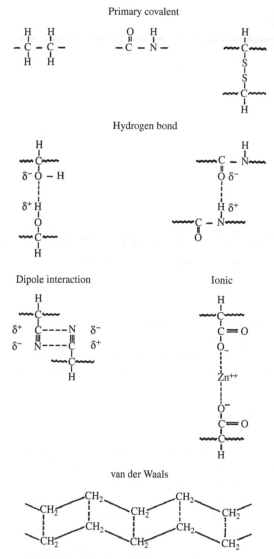

FIGURE 3.1 Bonding in polymer systems.

minimum potential energy, its most probable or equilibrium separation, r_m being the equilibrium bond distance. The "depth" of the potential well ε is the energy required to break the bond, separating the atoms completely.

Table 3.1 lists the approximate bond strengths and interatomic distances of the bonds encountered in polymeric materials. The important fact to notice here is how much stronger the primary covalent bonds are than the others. As the material's temperature is raised and its thermal energy (kT)[1] is thereby increased, the primary covalent bonds will be the last to dissociate when the available thermal energy exceeds their dissociation energy.

---

[1] *Note*: kT is a representation of molecular energy from thermodynamics. k is the Boltzmann constant and T is the absolute temperature.

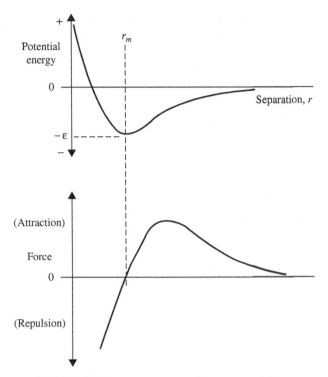

FIGURE 3.2   Interatomic potential energy and force.

3.3   BONDING AND RESPONSE TO TEMPERATURE

In linear and branched polymers, only the secondary bonds hold the individual polymer chains together (neglecting temporary mechanical entanglements). Thus, as the temperature is raised, a point will be reached where the forces (the weak bonds of types 2–5) holding the chains together become insignificant, and the chains are then free to slide past one another, that is, to *flow* upon the application of stress. Therefore, *linear* and *branched* polymers are generally *thermoplastic*. The crosslinks in a *network* polymer, on the other hand, are held together by the same primary covalent bonds as are the main chains. When the thermal energy exceeds the dissociation energy of the primary covalent bonds, both main-chain and crosslink bonds randomly fail, and the polymer degrades. Hence, *crosslinked* polymers are *thermosetting*.

TABLE 3.1   Bond Parameters [1,2]

Bond Type	Interatomic Distance, r_m, nm	Dissociation Energy, ε, kcal/mol
Primary covalent	0.1–0.2	50–200
Hydrogen bond	0.2–0.3	3–7
Dipole interaction	0.2–0.3	1.5–3
van der Waals bond	0.3–0.5	0.5–2
Ionic bond	0.2–0.3	10–20

There are some exceptions to these generalizations. It is occasionally possible for secondary bonds to make up for in quantity what they lack in the quality (or strength) of a single bond. For example, polyacrylonitrile $\left(\begin{smallmatrix} H & H \\ | & | \\ C & -C \\ | & | \\ H & C\equiv N \end{smallmatrix}\right)_x$ is capable of strong dipole interactions between the pendent nitrile groups on every other carbon atom along the polymer backbone. If these secondary bonds could be broken one by one (i.e., "unzippped"), polyacrylonitrile would behave as a typical thermoplastic. This is impossible, of course, due to the random nature of polymer configurations. By the time enough of the secondary bonds have been dissociated to free the chains and allow flow, the dissociation energy of some carbon–carbon backbone covalent bonds will have been exceeded and the materials will degrade. Extreme stiffness of the polymer chain also contributes to this sort of behavior. Cellulose has a bulky, complex repeat unit that contains three hydroxyl groups. Though linear, its chains are therefore stiff and strongly hydrogen bonded, and, thus, it is not thermoplastic. If the hydroxyls are reacted with acids, such as nitric, acetic, or butyric, the resulting derivative of cellulose (a cellulose ester) behaves as a typical thermoplastic largely because of the reduced hydrogen bonding:

$$\underset{\text{Cellulose}}{\{R(OH_3)\}_x} + \underset{\text{Acetic acid}}{3x\, HO-\overset{\displaystyle O}{\overset{\|}{C}}-CH_3} \rightarrow \underset{\text{Cellulose acetate}}{\{R(O-\overset{\displaystyle O}{\overset{\|}{C}}-CH_3)_3\}_x} + 3x\, H_2O$$

Here, three −OH groups in the basic repeat unit for cellulose, glucose or $C_6H_{12}O_6$, are converted to acetate esters. Since cellulose is already a polymer, the acetate esters simply modify side groups, making the structure of these cellulose derivatives more "workable." They are commonly used in pharmaceutical coatings and a number of foodstuffs, even being a component in some soft-serve ice creams.

Polytetrafluoroethylene (Teflon, TFE) with the repeat unit $-(CF_2-CF_2)-$ is another example of a thermosetting material, as the close packing and extensive secondary bonding of the main chains prevents flow when the polymer is heated.

3.4   ACTION OF SOLVENTS

The action of solvents on polymers is in many ways similar to that of heat. Appropriate solvents, that is, those that can form strong secondary bonds with the polymer chains, can penetrate, replace the interchain secondary bonds, and thereby pull apart and dissolve linear and branched polymers. The polymer–solvent secondary bonds cannot overcome primary covalent crosslinks, however, so crosslinked polymers are not soluble, although they may swell considerably. (Try soaking a rubber band in toluene overnight or take apart a diaper and add your favorite beverage.) The amount of swelling is, in fact, a convenient measure of the extent of crosslinking. A lightly crosslinked polymer, such as the rubber band or superabsorbent polymers in diapers, will swell tremendously, while one with extensive crosslinking, for example, an ebonite ("hard rubber") bowling ball, will not swell noticeably at all.

3.5   BONDING AND MOLECULAR STRUCTURE

It is obvious that the chemical nature of a polymer is of considerable importance in determining the polymer's properties. Of comparable significance is the way the atoms are arranged geometrically within the individual polymer chains.

A look into protein structure illustrates the different types of bonds and their effects on the three-dimensional organization of a polymer. The bonds that are found in proteins (polypeptides) are the same as those listed here for polymers, yet weak bonds play a very important role in determining the three-dimensional organization and biological activity of proteins. These bonds are divided into interactions at four levels (and demonstrated for a protein in Figure 3.3).

1. *Primary Bonds:* These are covalent bonds that link monomers and make up the backbone of the polymer. This is also termed the one-dimensional structure, as it can be represented by laying out repeat units in a straight line (which is handy for demonstrations in a book, but hardly representative of the complex three-dimensional structure of polymers, as depicted as a bowl of spaghetti).

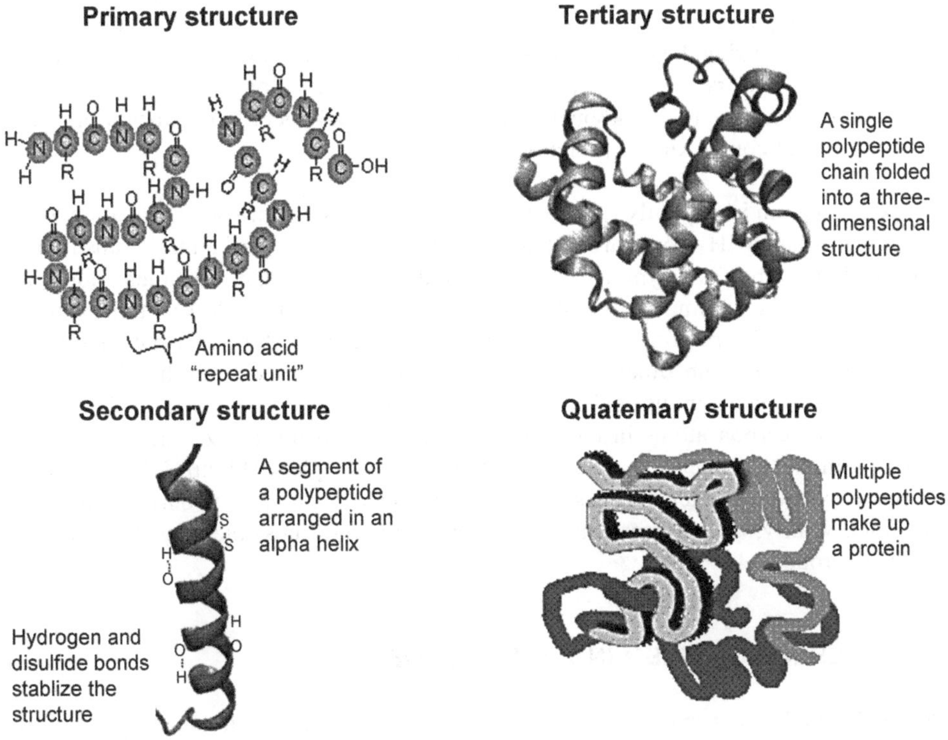

FIGURE 3.3 Demonstration of the different levels of bonding and structural arrangement in polymers (here, for a polypeptide). The primary structure links amino acids (monomers) and starts to form shapes at the secondary level through weak bonds (such as the hydrogen bonds and cysteine disulfide bonds visible in the alpha helix). Higher levels of structure are quite regular in proteins, allowing for complex shapes that are unique for each protein. For synthetic polymers, the shapes are much more random.

FIGURE 3.4 The geometry of a polyethylene chain. The carbon backbone is shown in black, with C−C−C bond angle of 109.5°, and the gray hydrogen atoms alternating positions along the backbone to minimize repulsive forces between consecutive side groups.

2. *Secondary Bonds:* These are weak bonds between near-neighbor repeat units, covering a range of perhaps 10–20 repeat units. Hydrogen bonds are common here and cause an impact on the formation of coils or zig-zag structures.

3. *Tertiary Bonds:* These are weak intramolecular bonds (those between a polymer and itself) between repeat units that are widely separated in the one-dimensional structure (i.e., perhaps an alcohol in the side group of the third repeat unit hydrogen bonds to a carboxylic acid in the 8000th repeat unit). These bonds hold the three-dimensional structure of an individual polymer in place.

4. *Quaternary Bonds:* These are weak intermolecular bonds (those between two different polymers). They stabilize the three-dimensional structure of a polymeric material, linking the different chains without forming covalent bonds (which would cause a network to form). This breakdown of bonding in proteins is essentially the same for all polymers, although commercial homopolymers do not display the same structural complexity of proteins.

The carbon atom is normally (exclusively, for our purposes) tetravalent. In compounds such as methane (CH_4) and carbon tetrachloride (CCl_4), the four identical substituents surround the carbon in a symmetrical tetrahedral geometry. If the substituent atoms are not identical, the symmetry is destroyed, but the general tetrahedral pattern is maintained. This is still true for each carbon atom in the interior of a linear polymer chain, where two of the substituents are the extensions of the polymer backbone. If a polyethylene chain (in normal spaghetti-type coil) were to be stretched out, for example, the carbon atoms in the chain backbone would lie in a zigzag fashion in a plane, with the hydrogen substituents on either side of the plane (Figure 3.4). In the case of polyethylene, in which all the substituents are the same, this is the only arrangement possible. With vinyl polymers, however, there are several possible ways of arranging the side groups.

3.6 STEREOISOMERISM IN VINYL POLYMERS

Before beginning a discussion of the isomerism in vinyl polymers, it must be pointed out that the monomers ($H_2C=CHX$, where X can be Cl, −OH, an organic group, etc.) polymerize almost exclusively in a head-to-tail fashion, placing the X groups on every other carbon atom along the chain. Although head-to-head (or tail-to-tail) connections are possible, steric hindrances between successive X groups (particularly if the X group is bulky) and electrostatic repulsion between groups with similar polarities generally keeps most of the linkages as head-to-tail. Ignoring head-to-head connections, there are

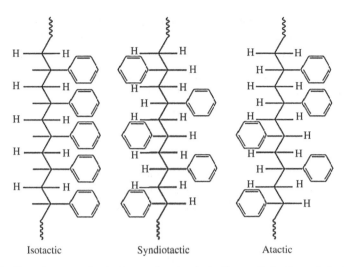

Isotactic   Syndiotactic   Atactic

FIGURE 3.5 Stereoisomers of polystyrene showing isotactic, syndiotactic, and atactic structures.

then three possible ways in which the X side group may be arranged with respect to the carbon backbone plane (see Figure 3.5). These arrangements represent three types of *stereoisomers*:

1. *Atactic:* A random arrangement of the X groups. Its lack of regularity has important consequences.
2. *Isotactic:* The structure in which all the X groups are lined up on the same side of the backbone plane.
3. *Syndiotactic:* Alternating placement of the X group on either side of the plane.

These three terms were coined by Dr. Giulio Natta, who shared the 1964 Nobel Chemistry Prize for his work in this area. Although atactic polymers are certainly most common, the arrangement and packing (and thus the properties) of *stereoregular* (syndiotactic and isotactic) polymers makes them important for certain applications. Methods to synthesize the stereoregular isomers will be discussed in Part II of this book.

Although useful for descriptive purposes, the planar zigzag arrangement of the main-chain carbon atoms is not always the one preferred by nature, that is, it is not necessarily the minimum free-energy configuration. In the case of polyethylene, it is, but for isotactic and syndiotactic polypropylene (which has a pendent methyl, $-CH_3$, group), the preferred (minimum-energy) configurations are quite regular, with the backbone forming helical twists to maximize the distance between consecutive $-CH_3$ groups. The atactic polymer, however, lacks regular twists and has an irregular shape.

Atactic polypropylene has a consistency somewhat like used chewing gum, whereas the stereoregular forms are hard, rigid plastics. The reason why this regularity (or lack of it) has a profound effect on mechanical properties is discussed in the next chapter.

The type of stereoregularity described above is a direct result of the dissymmetry of vinyl monomers. It is established in the polymerization reaction and no amount of twisting and turning the chain about its bonds can convert the three-dimensional geometry of one stereoisomer into another (molecular models are a real help, here).

The situation is even more complex for monomers of the form $HXC=CHX'$, where X and X' are different substituent groups. This is discussed by Natta [3], but is currently of no commercial importance.

Example 3.1 Both isotactic and atactic polymers of propylene oxide

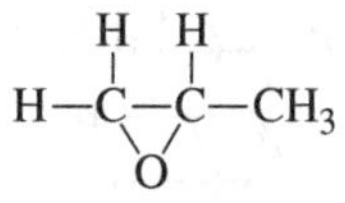

$$H-\underset{\underset{\displaystyle O}{|}}{\overset{\overset{\displaystyle H}{|}}{C}}-\underset{}{\overset{\overset{\displaystyle H}{|}}{C}}-CH_3$$

have been prepared by ring scission polymerization.
(a) Write the general structural formula of the polymer.
(b) Indicate how the atactic, isotactic, and syndiotactic structures differ.

Solution.
(a)

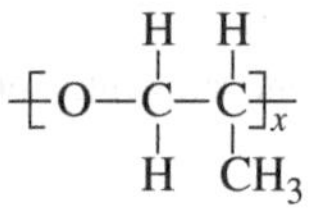

$$\begin{bmatrix} O-\overset{\overset{\displaystyle H}{|}}{\underset{\underset{\displaystyle H}{|}}{C}}-\overset{\overset{\displaystyle H}{|}}{\underset{\underset{\displaystyle CH_3}{|}}{C}} \end{bmatrix}_x$$

(b)

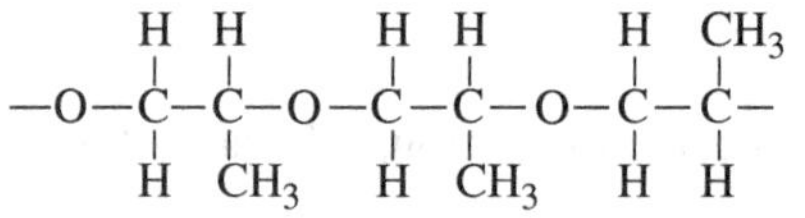

$$-O-\overset{H}{\underset{H}{C}}-\overset{H}{\underset{CH_3}{C}}-O-\overset{H}{\underset{H}{C}}-\overset{H}{\underset{CH_3}{C}}-O-\overset{H}{\underset{H}{C}}-\overset{CH_3}{\underset{H}{C}}- \qquad \text{Atactic}$$

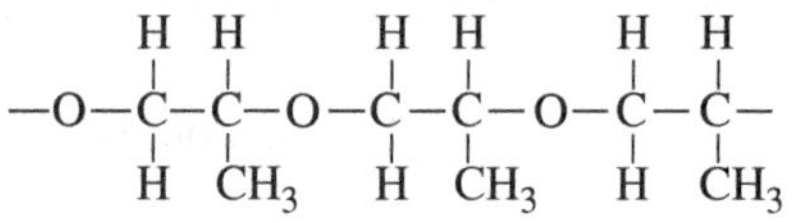

$$-O-\overset{H}{\underset{H}{C}}-\overset{H}{\underset{CH_3}{C}}-O-\overset{H}{\underset{H}{C}}-\overset{H}{\underset{CH_3}{C}}-O-\overset{H}{\underset{H}{C}}-\overset{H}{\underset{CH_3}{C}}- \qquad \text{Isotactic}$$

$$-O-\overset{H}{\underset{H}{C}}-\overset{H}{\underset{CH_3}{C}}-O-\overset{H}{\underset{CH_3}{C}}-\overset{CH_3}{\underset{H}{C}}-O-\overset{H}{\underset{H}{C}}-\overset{H}{\underset{CH_3}{C}}- \qquad \text{Syndiotactic}$$

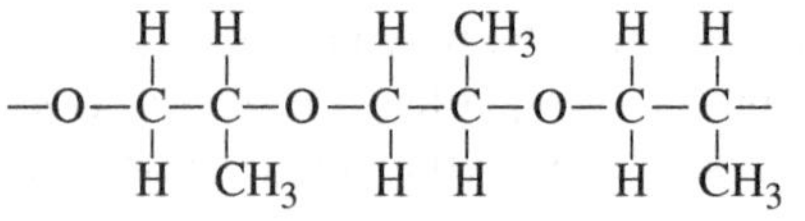

3.7   STEREOISOMERISM IN DIENE POLYMERS

Another type of stereoisomerism arises in the case of poly-1,4-dienes because carbon–carbon double bonds are rigid and do not allow rotation. The substituent groups on the double-bonded carbons may be either on the same side of the chain (*cis*) or on the opposite sides (*trans*), as shown in Figure 3.6 for 1,4 polyisoprene.

The chains of *cis*-1,4-polyisoprene assume a tortured, irregular configuration because of the steric interference of the substituents adjacent to the double bonds. This stereoisomer is familiar as natural rubber (made by the rubber tree) and is used in rubber bands. *trans*-1,4-polyisoprene chains assume a regular structure. This polymer is known as gutta-percha, a tough but not elastic material, long used as a golf-ball cover.

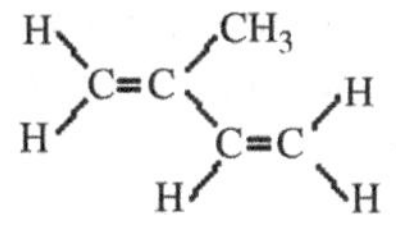

Isoprene monomer

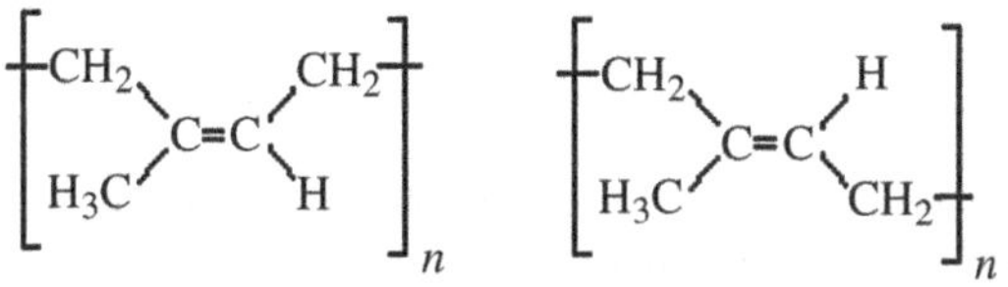

(a) *cis*-polyisoprene
(natural rubber)

(b) *trans*-polyisoprene

FIGURE 3.6   *cis* and *trans* isomers result from polymerization of isoprene.

Note that stereoisomerism in poly-1,4-dienes does not depend on the dissymmetry of the repeating unit. Atactic, syndiotactic, and isotactic isomers are possible with butadiene, even though all of the carbon substituents are hydrogen, as shown in Example 3.2.

Example 3.2 Identify all possible structural and stereoisomers that can result from the addition polymerization of butadiene:

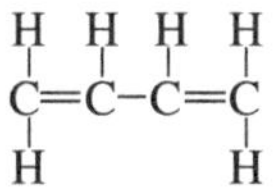

Solution. Recall from Chapter 2 that butadiene can undergo 1,4-addition or 1,2-addition (3,4-addition results in the same structure as 1,2-addition). The 1,4-polymer can have *cis* and *trans* isomers like the 1,4-polyisoprene in Figure 3.6. The 1,2-polymer can have atactic, isotactic, and syndiotactic stereoisomers like any vinyl polymer. The 1,4-polymer results in a different structure, but cannot have stereoisomers because all of the carbons are part of the backbone of the polymer and both the substituents remaining on the double bond in the backbone are hydrogens.

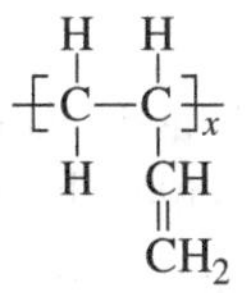

Example 3.3 Identify all the possible structural and stereoisomers that can result from the addition polymerization of chloroprene (this rubbery polymer is commercially known as Neoprene and is used in wetsuits).

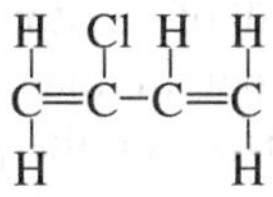

Solution. As in Example 3.2, the 1,4-polymer can have *cis* and *trans* stereoisomers (two different possibilities). This unsymmetrical diene monomer can also undergo both 1,2-and 3,4-addition:

$$
\begin{array}{cc}
\underset{\substack{\\ \text{1, 2}}}{
\begin{array}{c}
\text{H} \quad \text{Cl} \\
\fleft C - C \fright_x \\
\text{H} \quad \text{CH} \\
\quad \parallel \\
\quad \text{HCH}
\end{array}}
&
\underset{\substack{\\ \text{3, 4}}}{
\begin{array}{c}
\text{H} \quad \text{H} \\
\fleft C - C \fright_x \\
\text{Cl}-\text{C} \quad \text{H} \\
\quad \parallel \\
\quad \text{HCH}
\end{array}}
\end{array}
$$

Each of these structural isomers may have atactic, isotactic, or syndiotactic stereoisomers (making for six more possible isomers). Thus, in principle at least, there are *eight* different isomers of polychloroprene possible.

3.8 SUMMARY

Bonding in polymers is significantly more complex than in metals. The variety of bonds and the way side groups are organized in isomers make for the wide range of properties polymers can have, as will be demonstrated in later chapters.

PROBLEMS

1 Given the monomers below:

$$
\text{H}-\underset{\text{H}}{\text{N}}\fleft CH_2 \fright_6 \underset{\text{H}}{\text{N}}-\text{H}
\qquad
\text{HO}-\overset{O}{\underset{}{C}}-\overset{H}{\underset{H}{C}}-\overset{H}{\underset{}{C}}{=}\overset{H}{\underset{}{C}}-\overset{H}{\underset{H}{C}}-\overset{O}{\underset{}{C}}-\text{OH}
\qquad
\overset{H}{\underset{H}{C}}{=}\overset{H}{\underset{}{C}}-\overset{O}{\underset{}{C}}-\text{OH}
$$

(a) Show the repeating units of all the *linear homopolymers* that could be produced.

(b) Which of the polymers, if any, could have *cis* and *trans* isomers? Illustrate. How would this differ from the isomerism that arises in 1,4-polydienes?

(c) Which of the polymers, if any, could have isotactic and syndiotactic isomers? Illustrate.

2 Draw chemical structures of isotactic and syndiotactic oligomers of vinyl alcohol with five repeat units; include hydrogen atoms for the end groups.

3 (a) Draw a *structure* to represent the triblock polymer, poly(ethylene oxide-b-propylene oxide-b-ethylene oxide), PEO_m-PPO_n-PEO_m. This polymer is used as a surfactant and has the tradename Pluronic®. For your structure, use Pluronic F127, where $m = 99$ and $n = 65$.

(b) If you were to stretch out a Pluronic F127 in planar zigzag form, without violating required bond angles, so that all of the atoms in the polymer backbone were in the same plane, estimate the order of magnitude for the polymer chain length, including the end groups ($-$H's), on the polymer backbone.

Bond	Bond Length, Å	Bond Type	Bond Angle, °
C–C	1.54	C–C–C	109.5
C=C	1.34		
C–H	1.20	C–C=C	122
C–O	1.43		
C–N	1.47	C–O–C	108
C=O	1.23	O–C–C	110

REFERENCES

[1] Platzer, N., *Ind. Eng. Chem.* **61**(5) 10, 1969.

[2] Miller, M.L., *The Structure of Polymers.* Reinhold, New York, 1966.

[3] Natta, G., *Sci. Am.* **205**(2), 33 (1961).

CHAPTER 4

POLYMER MORPHOLOGY

It was pointed out in the previous chapter that the geometric arrangement of the atoms in polymer chains can exert a significant influence on the properties of the bulk polymer. To appreciate why this is so, the subject of *polymer morphology*, the structural arrangement of the chains in the polymer, is introduced here.

Covalent bonds hold the polymer together but with macromolecules (another term for polymers), the long-range structure is more like a bowl of spaghetti (Figure 4.1), where the backbones of linear polymers twist and turn in a random fashion. This means that the "length" of a polymer is never the actual distance between the end groups in a stretched-out polymer. The dimensional space occupied by a polymer is best described by two parameters, the end-to-end distance, $<r>$, and the radius of gyration $<r_g>$. Here, the $<>$ brackets indicate that the terms are statistical averages over many polymer chains. $<r_g>$ is the average distance from the centroid (mass-weighted center) of a polymer chain to each of the monomer units. It would be equal to the radius of a sphere if a polymer were ever balled up perfectly. Even though that is unlikely, $<r_g>$ is a useful term to represent the three-dimensional space taken up by a polymer molecule.

The end-to-end distance is a physical length separating the first and the last repeat units on a single polymer chain. This can have two extremes: on one hand, the polymer can be stretched out entirely along the backbone, limited by bond lengths and bond angles (highly unlikely), while on the other, the polymer can twist and turn and end up with the first and the last repeat unit in the same location ($r = 0$). When statistically averaged, $<r>$, is somewhat more useful, as it can indicate whether a polymeric material consists of entwined stretched-out fibers with many weak intermolecular bonds, or if the material is made up of distinct blobs of polymers that prefer weak intramolecular bonds.

The arrangement of individual polymers in a material is broken down to general regions: amorphous (unstructured) and crystalline. This chapter covers the types of crystallinity in

Fundamental Principles of Polymeric Materials, Third Edition. Christopher S. Brazel and Stephen L. Rosen.
© 2012 John Wiley & Sons, Inc. Published 2012 by John Wiley & Sons, Inc.

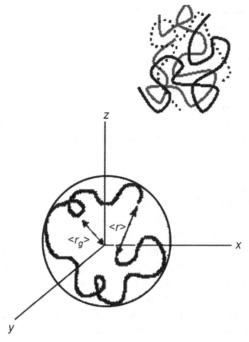

FIGURE 4.1 Depiction of (a) Three polymer chains randomly oriented and (b) a depiction of the end-to-end distance, $<r>$, for a single polymer. Also shown in figure (b) is the radius of gyration, $<r_g>$, which represents the average distance from the center of gravity to each monomer unit.

polymers and describes how crystals (and the degree of crystallinity) impact the optical and the mechanical properties of the polymers.

4.1 AMORPHOUS AND CRYSTALLINE POLYMERS

What is a crystal? For salts and metals, crystals are fairly straightforward because these materials exist primarily as crystals when solid. The arrangement of atoms can affect the type of crystallinity (e.g., face-centered cubic), but the main effect of crystals is to bring atoms together in a regular arrangement. Thinking back to the bowl of cooked spaghetti representation of polymers, regular arrangement may seem nearly impossible (and it is for some polymers, particularly those with bulky side groups). For polymers with small side groups (such as polyethylene), large segments of polymers can arrange into crystals, but there are no polymers that are 100% crystalline.

Although the precise nature of crystallinity in polymers is still under investigation, a number of facts have long been known about the requirements for polymer crystallinity. First, an ordered, regular chain structure (such as an isotactic or a syndiotactic isomer) is necessary to allow the chains to pack into an ordered, regular, three-dimensional crystal lattice. Thus, stereoregular polymers are more likely to have sizable crystalline portions than those that have irregular chain structures. Similarly, large pendent groups, particularly if they are spaced irregularly, hinder crystallinity. Second, no matter how regular the chains, the secondary forces holding the chains together in the crystal lattice must be strong enough to overcome the disordering effect of thermal energy; so hydrogen bonding or strong dipole interactions promote crystallinity and, other things being equal, raise the crystalline melting temperature.

X-ray diffraction studies show that there are numerous polymers that do not meet the above criteria and show no traces of crystallinity, that is, they are completely *amorphous*. In contrast to the regular, ordered arrangement in a crystal lattice, the chains in an amorphous polymer mass assume a more-or-less random, twisted, entangled, "balled-up" configuration, as in the bowl of cooked spaghetti. A better analogy, in view of the constant thermal motion of the chain segments, is a bowl of wriggling worms. The latter analogy forms the basis of a currently popular quantitative model of polymer behavior known as reptation theory (named for the motion of reptiles).

At the other extreme, despite intensive efforts, no one has succeeded in producing a completely crystalline polymer. The crystalline content may in certain cases (in very regular polymers without bulky side groups) reach 98%, but at least a few percent noncrystalline material always remains. In the case of metals, crystal defect concentrations are normally on the order of parts per million, so metals are generally considered perfectly crystalline in comparison with even the most highly crystalline polymers. Interestingly, recently developed amorphous metals have expanded the range of material properties that can be achieved by many metals. While crystal defects in metals are often the sites of fracture, polymers have covalent bonds linking the crystalline and amorphous regions, making the interfaces between these regions more continuous. The ability to tailor and adjust crystal size and degree of crystallinity (percent crystalline versus percent amorphous) allows for a wide range of mechanical properties in polymers, from rubbery and flexible (largely amorphous with small crystals) to rigid and strong (largely crystalline).

References 1–5 contain extensive discussions of the techniques used to study crystallinity in polymers as well as reviews of research results.

4.2 THE EFFECT OF POLYMER STRUCTURE, TEMPERATURE, AND SOLVENT ON CRYSTALLINITY

Polymer crystals are highly ordered arrangements of polymer segments. Because they are stabilized by weak bonds, anything that can alter or disrupt these bonds will cause a drop in the degree of crystallinity. Perhaps most straightforward is the effect of temperature. As with all crystalline solids, as the temperature rises, bonds holding a crystal together begin to come apart, resulting in a melting transition (T_m). For polymers, it should be noted that only the crystalline portion melts, so a 100% amorphous polymer will not exhibit a T_m (but it will degrade as the temperature continues to rise). The effect of temperature on polymer structure is of great significance, therefore Chapter 6 is devoted entirely to this subject.

Solvents can also disrupt weak polymer bonds, thus they have an effect on crystals similar to melting. Solvents are small molecules (at least when compared to polymers) that can diffuse into a polymer and form new weak bonds, thus replacing polymer–polymer interactions with polymer–solvent interactions. Because polymers are so large, the solvation process can be lengthy (hours to days to dissolve in some cases). In a good solvent, especially in dilute solutions, polymer crystallinity disappears. Upon drying (or cooling a molten polymer), the chains can reform crystals with the crystal size and the degree of crystallinity dependent on the rate of solvent evaporation (or rate of cooling). The behavior of polymers in solution is covered in Chapter 7.

Because polymer crystals have a regular arrangement of the polymer backbone, the side groups (and their tacticity) are crucial in determining whether a polymer will crystallize. Bulky side groups (large groups that extend from the backbone) prevent this alignment, so polymers such as poly(methyl methacrylate), shown below, are entirely amorphous. It should be noted that most polymers do not display appreciable crystallinity, particularly those with bulky side groups that prevent chain alignment. The irregular arrangement of these groups (in atactic polymers) also hinders the formation of crystals. On the other hand, side groups with organic functionalities that can form weak bonds (e.g., −OH) can increase the tendency for a polymer to crystallize.

4.3   THE EFFECT OF CRYSTALLINITY ON POLYMER DENSITY

For simplicity, let us assume that only crystalline and amorphous phases are present. With a given polymer, the properties of each phase remain the same but the relative amounts of the phases can vary, and this can strongly influence the bulk properties, particularly the mechanical properties. This is illustrated for polyethylene below, the most crystallizable polymer due to its simple repeat unit structure $(-CH_2-CH_2-)$.

Since polymer chains are packed together more efficiently and tightly in crystalline areas than in amorphous areas, crystallites will have a higher density. Thus, low-density $(0.92 \, g/cm^3)$ polyethylene is estimated to be about 43% crystalline, while high density $(0.97 \, g/cm^3)$ polyethylene is about 76% crystalline. Density is, in fact, a convenient measure of the degree of crystallinity. Because the volumes of the crystalline and amorphous phases are additive, density and degree of crystallinity are related by

$$\frac{1}{\rho} = \frac{w_c}{\rho_c} + \frac{w_a}{\rho_a} = \frac{w_c}{\rho_c} + \frac{(1 - w_c)}{\rho_a} \tag{4.1}$$

where the w's are weight fractions and the subscripts c and a refer to the crystalline and amorphous phases, respectively.

In the case of polyethylene, the differences in degree of crystallinity arise largely from branching that occurs during polymerization (although it is also influenced by the molecular weight and the rate of cooling, as will be discussed later in Chapter 6). The branch points sterically hinder packing into a crystal lattice in their immediate vicinity, and thus lower the degree of crystallinity.

TABLE 4.1   The Influence of Crystallinity on Some of the Properties of Polyethylene[a]

Commercial Product	Low Density	Medium Density	High Density
Density range, g/cm^3	0.910–0.925	0.926–0.940	0.941–0.965
Approximate % crystallinity	42–53	54–63	64–80
Branching, equivalent CH_3 groups/1000 carbon atoms	15–30	5–15	1–5
Crystalline melting point, °C	110–120	120–130	130–136
Hardness, Shore D	41–46	50–60	60–70
Tensile modulus, psi, N/m^2	0.14–0.38×10^5 $(0.97$–$2.6 \times 10^8)$	0.25–0.55×10^5 $(1.7$–$3.8 \times 10^8)$	0.6–1.8×10^5 $(4.1$–$12.4 \times 10^8)$
Tensile strength, psi, N/m^2	600–2300 $(0.41$–$1.6 \times 10^7)$	1200–3500 $(0.83$–$2.4 \times 10^7)$	3100–5500 $(2.1$–$3.8 \times 10^7)$
Flexural modulus, psi, N/m^2	0.08–0.6×10^5 $(0.34$–$4.1 \times 10^8)$	0.6–1.15×10^5 $(4.1$–$7.9 \times 10^8)$	1.0–2.6×10^5 $(6.9$–$18 \times 10^8)$

[a]It must be kept in mind that mechanical properties are influenced by factors other than the degree of crystallinity (molecular weight, in particular).

Low-density polyethylene (LDPE) has traditionally been made by a high-pressure process (25,000–50,000 psi) and high-density polyethylene (HDPE) by a low-pressure process (~100 psi). Thus, LDPE is sometimes referred to as high-pressure polyethylene and HDPE as low-pressure polyethylene. If they were not confusing enough, LDPE are now also made by low-pressure processes similar to those used for HDPE. There are also polyethylenes with small grafted organic groups along the backbone. These materials have most unfortunately (and inaccurately) been termed linear, low-density polyethylene (LLDPE). If they were truly linear, they would not be of low density.

Traditional (high pressure) LDPE is a homopolymer. Its long, "branched branches" arise from a side reaction during polymerization. LLDPE, however, has short branches that are introduced by random copolymerization with minor amounts (say 8–10%) of one or more α-olefins (vinyl monomers with hydrocarbon X groups, such as 1-butene, 1-hexene, 1-octene, and 4-methyl-1-pentene). This same approach is extended (by adding more comonomer) to make very-low density polyethylene (VLDPE) or ultralow density polyethylene (ULDPE). $(\rho < 0.915 \, g/cm^3)$. The nature of the branching affects some properties to a certain extent. For example, LLDPEs form stronger, tougher films than LDPEs of equivalent density. The various polyethylenes are summarized in Table 4.1 and their molecular architectures are sketched in Figure 4.2. Equation 4.1 must be used with caution where the density is varied with a comonomer, as in LLDPE, because ρ_a will, in general, vary with copolymer composition.

4.4   THE EFFECT OF CRYSTALLINITY ON MECHANICAL PROPERTIES

Since the polymer chains are more closely packed in the crystalline areas than in the amorphous, there are more of them available per unit area to support a stress. Also, since they are in close and regular contact over relatively long distances in the crystallites, the secondary forces holding them together are cumulatively greater than in the amorphous regions. Thus, crystallinity can significantly increase the strength and rigidity of a polymer. For this reason, stereoregular polypropylenes, which can and do crystallize, are fairly hard,

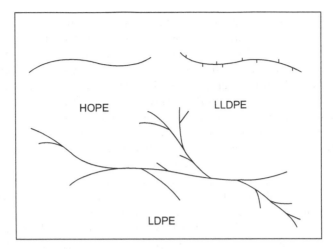

FIGURE 4.2 Molecular architecture of various polyethylenes.

rigid plastics, while irregular, atactic polypropylenes are amorphous, soft, and sticky. And, other things being equal, the greater the ratio of crystalline to amorphous phase, the stronger, harder, more rigid, and less easily deformable the polymer will be, as illustrated in Table 4.1 for the hardness, tensile modulus, tensile strength, and flexural modulus. These parameters (and several others) will be discussed in detail in Chapter 16, but each of these parameters tends higher with an increase in the crystallinity of polyethylene.

4.5 THE EFFECT OF CRYSTALLINITY ON OPTICAL PROPERTIES

The optical properties of polymers are largely influenced by crystallinity. A polymer with no additives is normally either clear or white and opaque (or transluscent, somewhere in-between the two extremes), as illustrated in Figure 4.3. When light passes between two

FIGURE 4.3 Typical plastic objects that are transparent, translucent, and opaque.

phases with different refractive indices, some of it is scattered at the interface if the dimensions of the discontinuities are comparable to or greater than the wavelength of visible light (400–700 nm). Thus, a block of ice is transparent, but snow appears white because light passes alternately from air to ice crystals many times. Similarly, in a crystalline polymer, the usually denser crystalline areas have a higher refractive index than the amorphous areas, so *crystalline polymers are either opaque or translucent* because light is scattered as it passes from one phase to the other. So, in general, *transparent polymer's are completely amorphous*, and a simple visual inspection can give clues about a polymer's crystallinity (although additives are often used that can change the optical properties). An interesting exception arises in the case of isotactic poly (4-methyl-1-pentene). The refractive indices (and densities) of the amorphous and crystalline phases are almost identical. Thus, as far as light is concerned, it is a homogeneous material and is the single transparent polymer that is known to be highly crystalline. Also, as the dimensions of the dispersed-phase particles become smaller than the wavelength of visible light (consider nanoparticle additives), scattering decreases, so a polymer with very small crystals and a low degree of crystallinity might appear nearly transparent, particularly in thin sections.

The converse is not necessarily true; lack of transparency in a polymer may be due to crystallinity, but it can be caused by an added second phase, such as a filler. If the polymer is known to be a pure homopolymer, however, translucency is a sure sign of crystallinity. Thus, commercial homopolymers of styrene, which is atactic and, therefore, completely amorphous because of the irregular arrangement of the bulky phenyl side groups, is perfectly transparent. Ironically, it is often called "crystal" polystyrene because of its crystal clarity. Isotactic polystyrene has been synthesized in the laboratory and does crystallize. It has the white, translucent appearance typical of polyethylene. Foamed polystyrene (cups, packing peanuts) is white because light passes between atactic polystyrene and gas bubbles. High-impact polystyrene consists of a dispersion of 1–10 μm polybutadiene rubber particles in a continuous phase of atactic polystyrene. Thus, while each phase is completely amorphous and individually transparent, they have different refractive indices, so the composite material scatters light and appears white and translucent.

Example 4.1 Explain the following facts.

(a) Polyethylene and polypropylene produced with stereospecific catalysts each are fairly rigid, translucent plastics, while a 65–35 copolymer of the two, produced in exactly the same manner, is a soft, transparent rubber.

(b) A plastic is similar in appearance and mechanical properties to the polyethylene and polypropylene described in (a), but it consists of 65% ethylene and 35% propylene units. The two components of this plastic cannot be separated by any physical or chemical means without degrading the polymer.

Solution.

(a) The polyethylene produced with these catalysts is linear and thus highly crystalline. The polypropylene is isotactic and also highly crystalline. The crystallinity confers mechanical strength and translucency to each polymer individually. The 65–35 copolymer, ethylene-propylene rubber (EPR), is a random copolymer, so the CH_3

groups from the propylene monomer are arranged at irregular intervals along the chain, preventing packing in a regular crystal lattice, giving an amorphous, rubbery polymer.

(b) Since the components cannot be separated, they must be covalently bound within the chains. The properties indicate a crystalline polymer, so the CH_3 groups from the propylene cannot be spaced irregularly, as in the random copolymer in (a). Thus, these materials must be block copolymers of ethylene and stereoregular polypropylene, poly(ethylene-b-propylene). The long blocks of ethylene pack into a polyethylene crystal lattice and the propylene blocks into a polypropylene lattice.

4.6 MODELS FOR THE CRYSTALLINE STRUCTURE OF POLYMERS

4.6.1 Fringed Micelle Model

The first attempt to explain the observed properties of crystalline (the word should be prefaced by "semi-" but rarely is) polymers was the *fringed micelle* model (Figure 4.4). This model pictures crystalline regions known as fringed micelles or *crystallites* interspersed in an amorphous matrix. The crystallites, whose dimensions are in the order of tens of nanometers, are small volumes in which portions of the chains are regularly aligned parallel to one another, tightly packed into a crystal lattice. The individual chains, however, are many times longer than the dimensions of a crystallite, so they pass from one crystallite to another through amorphous areas.

This model nicely explains the coexistence of crystalline and amorphous material in polymers, and also explains the increase in crystallinity that is observed when fibers are drawn (stretched). Stretching the polymer orients the chains in the direction of the stress, increasing the alignment in the amorphous areas and producing greater degrees of crystallinity (Figure 4.4b). Since the chains pass randomly from one crystallite to another, it is easy to know why perfect crystallinity can never be achieved. This also explains why the effects of crystallinity on properties are in many ways similar to those of crosslinking, because, like crosslinks, the crystallites tie the individual chains together. Unlike crosslinks, though, the crystallites do not form covalent bonds between polymer chains and will generally melt before the polymer degrades and solvents that form strong secondary bonds with the chains can dissolve them (rather than just swell a crosslinked polymer).

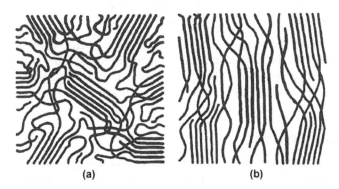

(a)   (b)

FIGURE 4.4 The fringed micelle model: (a) unoriented; (b) chains oriented by applied stress.

The fringed micelle model has been superseded as new information has led to the development of more advanced models. Nevertheless, it still does a good job of qualitatively predicting the effects of crystallinity on the mechanical properties of the polymers.

Example 4.2 Because the monomer vinyl alcohol is unstable, poly vinyl alcohol, PVA, is made by polymerization of vinyl acetate followed by the hydrolysis of poly vinyl acetate to form −OH side groups:

Polyvinyl acetate
Polyvinyl alcohol
Acetic acid

The extent of reaction may be controlled to yield polymers with anywhere from 0% to 100% of the original acetate groups hydrolyzed. Pure poly(vinyl acetate), 0% hydrolyzed, is insoluble in water. It has been observed, however, that as the extent of hydrolysis is increased, the polymers become more water soluble, up to about 87% hydrolysis, after which further hydrolysis decreases water solubility at room temperature. Explain briefly.

Solution. The normal poly vinyl acetate is atactic, and the irregular arrangement of the acetate side groups renders it completely amorphous. Water cannot form strong enough secondary bonds with the chains to dissolve it with the generally hydrophobic pendent group. As the acetate groups are replaced by hydroxyls in PVA (much more hydrophilic), sites are introduced that can form hydrogen bonds with water, thereby increasing the solubility. At high degrees of substitution, the more compact hydroxyl pendent groups allow the chains to pack into a crystal lattice. The hydroxyl groups provide hydrogen-bonding sites between the chains that help hold them in the lattice, and thus solubility in water is reduced.

Example 4.3 Explain the following experiment. A weight is tied to the end of a poly(vinyl alcohol) fiber. The weight and part of the fiber are dunked in a beaker of boiling water. As long as the weight remains suspended, the situation is stable, but when the weight is rested on the bottom of the beaker, the fiber dissolves.

Solution. As long as the weight is suspended from the fiber, the stress maintains the alignment of the chains in a crystal lattice against the disordering effects of thermal energy and solvent (water) penetration. When the weight is rested on the bottom, the stress is removed, allowing water to penetrate between and dissolving the polymer chains.

4.6.2 Lamellar Crystals [1–7]

The first direct observation of the nature of polymer crystallinity resulted from the growth of single crystals from dilute solution. Either by cooling or by evaporation of solvent, thin pyramidal or plate-like polymer crystals (lamellae) were precipitated from dilute solutions

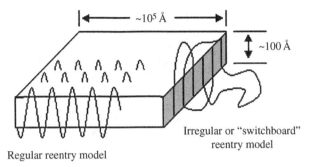

FIGURE 4.5   Polymer single crystals in flat lamellae. Two concepts of chain reentry are illustrated.

(Figure 4.5). These crystals were on the order of 10 μm along a side and only about 0.01 μm thick. This was fine, except that X-ray measurements showed that the polymer chains were aligned perpendicular to the large flat faces of the crystals, and it was known that the extended length of the individual chains was on the order of 0.1 μm. How could a chain fit into a crystal one-tenth of its length? The only answer is that the chain must fold back on itself, as shown in Figure 4.5. Two competing models of chain folding are illustrated: either the polymer folds back neatly on itself or several chains align for short distances with amorphous regions outside the crystal.

This folded-chain model has been well substantiated for single polymer crystals. The lamellae are about 50–60 carbon atoms thick, with about five carbon atoms in a direct reentry fold. The atoms in a fold, whether direct or indirect reentry, can never be part of a crystal lattice.

It is now well established that similar lamellar crystallites exist in bulk polymer samples crystallized from polymer melts, although the lamellae may be up to 1 μm thick. Recent results support the presence of a third, interfacial region between the crystalline lamellae and the amorphous phase. This interfacial phase can make up about 10–20% of the material. Furthermore, there does not seem to be much, if any, direct-reentry folding of chains in bulk (or melt)-crystallized lamellae. This is illustrated in Figure 4.6. Orientation of the lamellae along with additional orientation and crystallization in the interlamellar amorphous regions, as in the fringed micelle model, is usually invoked to explain the increase in the degree of crystallinity with drawing (stretching).

4.6.3 Spherulites

Not only are polymer chains often arranged to form crystallites but also these crystallites often aggregate into supermolecular structures known as *spherulites*. Spherulites are in some ways similar to the grain structure found in metals. They grow radially from a point of nucleation until other spherulites are encountered. Thus, the size of the spherulites can be controlled by the number of nucleation sites present, with more nuclei resulting in more but smaller spherulites. They are typically about 0.01 mm in diameter and have a Maltese cross appearance when viewed between polarizing filters (Figure 4.7a). Figure 4.7 shows how the polymer chains are thought to be arranged in the spherulites.

Large spherulites, which can grow as a polymer cools, contribute to brittleness in polymers and also scatters a lot of light. To minimize brittleness and enhance transparency, nucleating agents are often added or the polymer is shock-cooled (which increases the nucleation rate) to promote smaller spherulites.

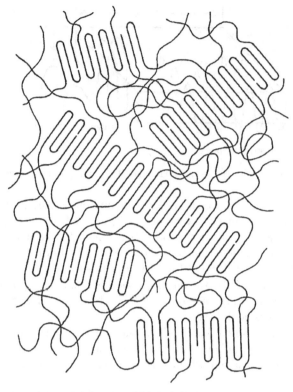

FIGURE 4.6 Compromise model showing folded-chain lamellae tied together by interlamellar amorphous chains.

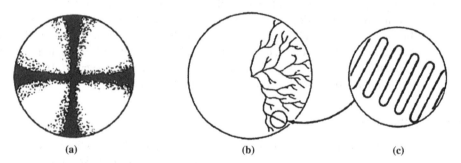

(a)   (b)   (c)

FIGURE 4.7 Spherulite crystals: (a) appearance between crossed polarizing filters; (b) branching of lamellae; (c) orientation of chains in lamellae.

4.7 EXTENDED CHAIN CRYSTALS

Polymers crystallized from a quiescent melt (such as those in Table 4.1) will have, in general, a random orientation of the chains at the macroscopic level. While the chains in any crystallite are oriented in a particular direction, the multiple crystallites are randomly oriented. It has long been known that if the chain axes could share a common orientation at the macroscopic level, the material would have superior mechanical properties in the orientation direction, since the chains would be arranged most efficiently to support a stress. In fact, fiber drawing does just this to a certain extent, which is why it is practiced.

Polymers crystallized while being subjected to an extensional flow (drawing), which tends to disentangle the chains and align them in the direction of flow, form fibrillar structures. These are believed to be extended-chain crystals, in which the chains are aligned parallel to one another over great distances, with little or no chain folding [8]. When fibers of such materials are further drawn, a super-strong fiber results. Very high molecular weight polymers are favored for this purpose. The longer chains align more readily, and there are fewer crystalline defects due to chain ends. Also, formation of the fibers from relatively dilute solution allows easier disentanglement and better alignment of the chains.

For linear polyethylene fibers made in this fashion, the tensile modulus (a measure of the strength of a material, which is defined more rigorously in Chapter 16) of 44 GPa (6.4×10^6 psi) compares well with steel (which is about 206 GPa or 30×10^6 psi) but has a density 7.6 times as great! PE fibers with a tensile strength of 1.8 GPa (2.6×10^5 psi) have also been claimed [9]. Compare these figures with those for ordinary, quiescently crystallized HDPE in Table 4.1.

Extended-chain crystals have also been identified as tying together lamellae in bulk-crystallized polymers and forming the core ("skewer") of the interesting "shish kebab" structures grown from dilute solutions subjected to shearing [10].

4.8 LIQUID CRYSTAL POLYMERS [11]

As noted above, if polymer chains can be aligned in the molten state (the polymer melt) prior to crystallization, high degrees of extended-chain crystallinity may be formed, imparting remarkable mechanical properties, at least in the chain direction. Another way of doing it is by starting with molecules that show a degree of order in the liquid phase; these are generally made from rigid monomers that have phenyl groups that become part of the polymer backbone. Such liquid-crystalline materials have been known for years, but until fairly recently were limited to relatively small molecules. Now, however, several types of polymers are known that exhibit liquid-crystalline order, either *lyotropic* (that become ordered in solution) or *thermotropic* (that become ordered in the melt) [12].

Kevlar® is an aromatic polyamid ("aramid") with the repeat unit:

It is spun into fibers from a lyotropic liquid-crystalline solution in concentrated H_2SO_4. The solution is extruded through small holes into a bath that leaches the acid out of the polymer, forming the fibers (through a "wet spinning" process, as discussed in Chapter 17). Because the molecules are oriented prior to crystallization, the fibers maintain a high degree of extended-chain crystalline order in the fiber direction, imparting remarkable strength. Table 4.2 compares some properties of these fibers with those of ordinary nylon 6/6 fibers. With a density of only 1.44 g/cm^3, they compare very favorably with other reinforcing fibers (glass, graphite, etc.) on a strength per weight basis. As a result, Kevlar

TABLE 4.2   Comparison of Liquid-Crystalline Kevar 49 Fibers with Nylon 6/6 Fibers [13]

	Kevlar 49	Nylon 6/6
Tensile strength, GPa, psi	2.95 (4.3×10^5)	1.28 (1.8×10^5)
Tensile modulus, GPa, psi	130 (19×10^6)	6.2 (3.0×10^5)
Elongation to break, %	2.3	19

fibers are used in bullet-proof vests and other ballistic armor, as a tire cord (replacing steel belts to improve strength and offering sidewall puncture resistance as well), and for reinforcement in high-strength composites.

Several thermotropic aromatic copolyesters have also been commercialized. Vectra A® is reported to have the repeating units

and Xydar have the repeat units

These thermotropic liquid-crystalline polymers have high melting points but can be melt-processed like other thermoplastics. The macroscopic orientation of the extended-chain crystals depends on the orientation imparted by flow during processing (molding, extrusion, etc.). Because of the fibrous nature of the extended-chain crystals, these plastics behave as "self-reinforced composites," with excellent mechanical properties, at least in the chain direction. This is illustrated in Table 4.3 for molded specimens of a liquid-crystalline copolyester of ethylene glycol, terephthalic acid, and p-hydroxybenzoic acid [14]. In the direction parallel to the flow, the properties listed in Table 4.3 favorably compare with ordinary crystalline thermoplastics (nylons, polyesters) reinforced with up to 30% glass fibers.

TABLE 4.3   Properties of a Molded Liquid-Crystalline Copolyester [14]

	‖ to Flow	⊥ to Flow
Tensile strength, MPa	107	28.4
Elongation, %	8	10
Flexural modulus, GPa	11.8	1.67
Linear coefficient of expansion, °C^{-1}	0	4.5×10^{-5}

It is obvious from the repeat units above that the liquid-crystalline behavior is promoted by highly aromatic chains. The phenyl rings inhibit rotation, resulting in stiff, rigid, extended chains, and because they are nearly planar, they stack next to one another, promoting long-range order in both the liquid and the solid states.

Thermotropic and lyotropic liquid crystals behave in a reversible manner as the temperature and the solvent are changed, respectively. The crystals change their alignment from highly ordered to amorphous as the crystals reversibly come apart and reform. These kinds of polymer crystals have found wide usage in electronics, in the form of liquid crystal displays (LCDs). Here, the crystallinity can be switched on and off thermally, electrically, or by a change in solution properties. Taking advantage of the opaque/transparent transition as polymers change from crystalline to amorphous, a display either allows or prevents viewing a backlit display (or alternately change the pattern of reflected light, as in many calculators).

PROBLEMS

1 Derive Equation 4.1.

2 X-ray diffraction measurements show that polyethylene crystallizes in a body-centered orthorhombic unit cell with lattice parameters $a = 0.7417$ nm, $b = 0.4945$ nm, and $c = 0.2547$ nm at 25 °C. The a, b, and c axes are orthogonal. The chain axes run in the c direction and there are two repeating units per unit cell (one running up the center and one-quarter in each of the four corners). Calculate the crystal density of polyethylene.

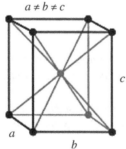

3 The density of amorphous polyethylene is estimated to be 0.855 g/cm^3 at 25 °C by extrapolating values from above the melting point. Use this value and your answer to Problem 4.2 to estimate

    (a)   the degree of crystallinity of a 0.93 g/cm^3 polyethylene sample, and

    (b)   the density of a 72% crystalline polyethylene sample.

4 Suppose that you know the theoretical enthalpy of melting for a 100% crystalline polymer. Show how you could obtain the degree of crystallinity for an actual sample by measuring its enthalpy of melting.

5 The entry in Table 4.1 "Branching, equivalent CH$_3$ groups/1000 C atoms" is determined by calculating that number for a sample of poly(ethylene-co-propylene) with the same density. How many CH$_3$ groups are there per 1000 main-chain C atoms in a 65 (weight)% ethylene-35% propylene copolymer? This copolymer, ethylene–propylene rubber (EPR) is completely amorphous. (*Note*: You may ignore the ends of the chains.)

6 Traditional introductory materials science texts, concerned mostly with metals and ceramics, devote considerable space to crystal structures and crystal defects. These topics have been studied for polymers, but they are hardly mentioned here. Why are they much less important in polymers?

7 Equation 4.1 relates the density of a (semi)crystalline polymer to its composition and the densities of the individual phases. Develop two simple models that relate the tensile modulus of a crystalline polymer, E, to the moduli of the individual phases, E_c and E_a. The densities of each phase are known, as are the % crystallinity. Base each model on a layered structure of the two phases (OK, so this is pretty naive). In one, the layers are assumed to be perpendicular to the applied stress and in the other they are assumed to be parallel.

8 Both Kevlar (Section 4.7) and extended-chain linear polyethylene (Section 4.2) fibers contain a high degree of extended-chain crystallinity, yet the former is somewhat stronger than the latter. Why?

9 Nylon 6/6 (Chapter 2, Example 4E) is opaque when pure. However, if isophthalic acid (a diacid in which the two acid groups are arranged in meta positions on a phenyl ring) is substituted for the adipic acid, the resulting polymer is transparent. Explain.

10 Two perfectly transparent polymers are physically mixed. The resulting mixture is also perfectly transparent. What does this mean? *Hint*: There is more than one possibility.

REFERENCES

[1] Schultz, J.M., *Polymer Material Science*, Prentice-Hall, Englewood Cliffs, NJ, 1974.

[2] Wunderlich, B., *Macromolecular Physics*, Vol. 1, *Crystal Structure, Morphology, Defects*, Academic, New York, 1973.

[3] Tadukoru, H. *Structure of Crystalline Polymers*, Wiley-Interscience, New York, 1979.

[4] Mandlekern, L. Chapter 4 in *Physical Properties of Polymers*, American Chemical Society, Washington, DC, 1984.

[5] Bovey, F.A., Chapter 5 in *Macromolecules: An Introduction to Polymer Science*, F.A. Boveyand F.H. Winslow (eds), Academic, New York, 1979.

[6] Geil, P.H., *Polymer Single Crystals*, Interscience, New York, 1963.

[7] Oppenlander, G.C., *Science* 159, 1311 (1968).

[8] Southern, J.H. and R.S. Porter, *J. Appl. Polym. Sci.* **14**, 2305 (1970); *J. Macromol. Sci. Phys.* **B4**, 541 (1970).

[9] Kavesh, S. and D.C. Prevorsek, U.S. Patent 4,413,110.

[10] Pennings, A.J., *J. Polym. Sci. (Symposia)* **59**, 55 (1977).

[11] Weiss, R.A. and C.K. Ober (eds), *Liquid-Crystalline Polymers*, American Chemical Society, Washington, DC, 1990.

[12] Blumstein, A. (ed.) *Liquid Crystalline Order in Polymers*, Academic, New York, 1978.

[13] DuPont product literature.

[14] Menges, G. and G. Hahn, *Modern Plastics*, October 1981, p. 56.

CHAPTER 5

CHARACTERIZATION OF MOLECULAR WEIGHT

5.1 INTRODUCTION

With the exception of a few naturally occurring polymers (such as proteins and DNA), all polymers consist of molecules with a distribution of chain lengths. Compared to molecules such as water that have a definite structure (H_2O) and molecular weight (18), the polymer chains making up a sample consist of molecules with a range of molecular weights. For example, poly(ethylene glycol) or PEG is often used at a range of molecular weights, around 400 (which is actually just an oligomer of approximately nine repeat units), PEG is a viscous liquid, while with increasing molecular weight, PEG becomes waxy (around 600), and can be formed into a solid powder (around 3000). Because the molecular weight is strongly related to the mechanical and chemical properties of polymers, it is necessary to characterize the entire distribution quantitatively, or at least to define and measure average chain lengths or molecular weights for these materials. Extensive reviews are available [1,2] concerning the effects of molecular weight and its distribution on the mechanical properties of polymers.

First, the molecular weight of a single polymer chain, M, is easily calculated, if the chain length, x, and chemical composition of the repeat unit (and thus the molecular weight of the repeat unit, M_r) are known:

$$M = xM_r \tag{5.1}$$

Note that the terms chain length and degree of polymerization can be used interchangeably, and both are represented by x.

When considering a collection of polymer chains, an average molecular weight must be considered. However, the method for determining this average depends on how you

Fundamental Principles of Polymeric Materials, Third Edition. Christopher S. Brazel and Stephen L. Rosen.
© 2012 John Wiley & Sons, Inc. Published 2012 by John Wiley & Sons, Inc.

count. In normal averaging (e.g., your grades in a class), you sum up each of the individual chain molecular weights (or scores) and divide by the number of chains (or assignments):

$$M = W/N = \text{Total weight} \,/\, \text{Total number of chains} \qquad (5.2)$$

This is a perfectly valid way to calculate molecular weight, but we must consider statistics and distribution patterns to get a more accurate picture of how polymers behave, so this chapter will discuss different ways of calculating molecular weights, introduce distribution profiles, and discuss laboratory methods used to measure or estimate the average molecular weight of polymer samples.

5.2   AVERAGE MOLECULAR WEIGHTS

Since nearly every collection of polymer molecules contains a distribution of molecular weights, a *number-average* molecular weight $\bar{M}_n$ may be defined in an analogous fashion to Equation 5.3:

$$\bar{M}_n = \frac{W}{N} = \frac{\displaystyle\sum_{x=1}^{\infty} n_x M_x}{\displaystyle\sum_{x=1}^{\infty} n_x} = \frac{n_1 M_1}{\sum n_x} + \frac{n_2 M_2}{\sum n_x} + \cdots = \sum_{x=1}^{\infty} \left(\frac{n_x}{N}\right) M_x \qquad (5.3)$$

where $\quad W = $ total sample weight $= \displaystyle\sum_{x=1}^{\infty} w_x = \sum_{x=1}^{\infty} n_x M_x$

$\qquad\quad w_x = $ total *weight* of x-mer

$\qquad\quad N = $ total *number of moles* in the sample (of all sizes) $= \displaystyle\sum_{x=1}^{\infty} n_x$

$\qquad\quad n_x = $ *number of moles* of x-mer

$\qquad\quad M_x = $ molecular weight of x-mer

$\qquad (n_x/N) = $ mole fraction of x-mer

Although this equation appears complicated, if you work with it, you will see that it reduces to the common equation used to average grades, except that it gives the average length of all the chains in a polymer sample. Any analytical technique that determines the *number* of moles present in a sample of known weight, regardless of their size, will give the number-average molecular weight.

Rather than count the number of molecules of each size present in a sample, it is possible to define an average in terms of the *weights* of molecules present at each size level. This is the weight-average molecular weight, $\bar{M}_w$. $\bar{M}_w$ places an emphasis on the larger molecular weight chains, which is particularly useful when considering the mechanical properties of polymers, as the larger chains contribute more to a polymer's strength than smaller-sized chains.

Consider a bowl of fruit that contains three apples, five oranges, and a bunch of grapes. If you consider each grape as a separate fruit, the average size ("number average molecular weight") will be much closer to the size of a grape, but the majority of the mass in the bowl is due to the larger apples and oranges, so a weight-average size will represent that the larger fruits make up more of the mass. For a polymer, the weight-average molecular weight, $\bar{M}_w$, is

$$\bar{M}_w = \frac{\sum w_x M_x}{\sum w_x} = \frac{w_1 M_1}{\sum w_x} + \frac{w_2 M_2}{\sum w_x} + \cdots = \sum \left(\frac{w_x}{W}\right) M_x = \frac{\sum n_x M_x^2}{\sum n_x M_x} \qquad (5.4)$$

Analytical procedures that, in effect, determine the weight of molecules at a given size level result in the weight-average molecular weight. Thus, $\bar{M}_w$ is greater than $\bar{M}_n$ and would be closer to the size of the apples and oranges.

The number-average molecular weight is the *first moment* of the molecular weight distribution, analogous to the center of gravity (the first moment of the mass distribution) in mechanics. The weight-average molecular weight, the *second moment* of the distribution, corresponds to the radius of gyration in mechanics. Higher moments, for example, $\bar{M}_z$, the third moment, may also be defined and are used occasionally.

It is sometimes more convenient to represent the size of polymer molecules in terms of the degree of polymerization or chain length x, rather than molecular weight, as related by Equation 5.1. Thus, $\bar{x}_n$ and $\bar{x}_w$ can also be used to represent average polymer sizes, and

$$M_x = mx \qquad (5.5a)$$

$$\bar{M}_n = m\bar{x}_n \qquad (5.5b)$$

$$\bar{M}_w = m\bar{x}_w \qquad (5.5c)$$

where m = molecular weight of a repeating unit (also sometimes called M_r)
$\bar{x}_n$ = number-average degree of polymerization or chain length
$\bar{x}_w$ = weight-average degree of polymerization or chain length

(These relations neglect the difference between the end groups on the molecule and the repeating units. This is perfectly justifiable in most cases since the end groups are an insignificant portion of a typical large polymer molecule.) It is also possible to use these relations for copolymers, but m should be modified to reflect the average composition of a repeat unit in the chain. In terms of chain lengths, Equations 5.3 and 5.4 become

$$\bar{x}_n = \frac{\sum n_x x}{\sum n_x} = \sum \left(\frac{n_x}{N}\right) x \qquad (5.3a)$$

$$\bar{x}_w = \frac{\sum n_x x^2}{\sum n_x x} = \sum \left(\frac{w_x}{W}\right) x \qquad (5.4a)$$

It may be shown that $\bar{M}_w \geq \bar{M}_n$ ($\bar{x}_w \geq \bar{x}_n$). The averages are equal only for a *monodisperse* polymer, where all of the chains are exactly the same length. The ratio $\bar{M}_w/\bar{M}_n = \bar{x}_w/\bar{x}_n$ is known as the *polydispersity index*, PI. PI is a measure of the breadth of the molecular weight distribution. A monodisperse sample would have a PI of 1.0, but typical values range from about 1.02 for carefully fractionated polymers synthesized by anionic addition reactions to over 50 for some commercial polymers.

Example 5.1 Measurements on two essentially monodisperse fractions of a linear poly(methyl methacrylate) (PMMA), A and B, yield molecular weights of 100,000 and 400 000, respectively. Mixture 1 is prepared from one part by weight of A and two parts by weight of B. Mixture 2 contains two parts by weight of A and one of B. Determine the weight- and number-average molecular weights of mixtures 1 and 2, as well as their weight- and number-average degrees of polymerization. The repeat unit for PMMA is $CH_2C(CH_3)COOCH_3$, which has a molecular weight, m, of 100.

Solution. For mixture 1:

$$n_A = \frac{1}{100\,000} = 1 \times 10^{-5}$$

$$n_B = \frac{2}{400\,000} = 0.5 \times 10^{-5}$$

$$\bar{M}_n = \frac{\sum n_i M_i}{\sum n_i} = \frac{(1 \times 10^{-5})(10^5) + (0.5 \times 10^{-5})(4 \times 10^5)}{1 \times 10^{-5} + 0.5 \times 10^{-5}} = 2.0 \times 10^5$$

$$\bar{M}_w = \sum \left(\frac{w_i}{W}\right) M_i = \frac{1}{3}(1 \times 10^5) + \frac{2}{3}(4 \times 10^5) = 3 \times 10^5$$

For PMMA, $m = 100$, thus
from Equations 5.5*b* and 5.5*c*:

$$\bar{x}_w = \bar{M}_w/m = 3.0 \times 10^5/100 = 3.0 \times 10^3$$

$$\bar{x}_n = M_n/m = 2.0 \times 10^5/100 = 2.0 \times 10^3$$

For mixture 2:

$$n_A = \frac{2}{100\,000} = 2 \times 10^{-5}$$

$$n_B = \frac{1}{400\,000} = 0.25 \times 10^{-5}$$

$$\bar{M}_n = \frac{(2 \times 10^{-5})(10^5) + (0.25 \times 10^{-5})(4 \times 10^5)}{2 \times 10^{-5} + 0.25 \times 10^{-5}} = 1.33 \times 10^5$$

$$\bar{M}_w = \frac{2}{3}(1 \times 10^5) + \frac{1}{3}(4 \times 10^5) = 2 \times 10^5$$

Again, for PMMA, $m = 100$, thus

$$\bar{x}_w = \bar{M}_w/m = 2 \times 10^5/100 = 2.0 \times 10^3$$

$$\bar{x}_n = \bar{M}_n/m = 1.33 \times 10^5/100 = 1.33 \times 10^3$$

Example 5.2 Two *polydisperse* samples are mixed in equal weights. Sample A has $\bar{M}_n = 100,000$ and $\bar{M}_w = 200,000$. Sample B has $\bar{M}_n = 200,000$ and $\bar{M}_w = 400,000$. What are $\bar{M}_n$ and $\bar{M}_w$ of the mixture?

Solution. First, let us derive general expressions for calculating the averages of mixtures

$$\bar{M}_n \equiv \frac{W}{N} = \frac{\displaystyle\sum_i W_i}{\displaystyle\sum_i N_i}$$

where subscript i refers to the various polydisperse components of the mixture. Now, for a given component,

$$N_i = \frac{W_i}{\bar{M}_{ni}}$$

$$\bar{M}_n(\text{mixture}) = \frac{\displaystyle\sum_i W_i}{\displaystyle\sum_i (W_i/\bar{M}_{ni})} \tag{5.6}$$

$$\bar{M}_w = \frac{\sum w_x M_x}{W} = \frac{\displaystyle\sum_i \left(\displaystyle\sum_i w_x M_x\right)_i}{\displaystyle\sum_i W_i}$$

$$\bar{M}_{wi} = \frac{\left(\displaystyle\sum_x w_x M_x\right)_i}{W_i} \tag{5.7}$$

$$\bar{M}_w(\text{mixture}) = \frac{\displaystyle\sum_i (\bar{M}_{wi} W_i)}{\displaystyle\sum_i W_i} = \sum_i \left(\frac{w_i}{\displaystyle\sum_i W_i}\right) \bar{M}_{wi}$$

where $(w_i/\Sigma W_i)$ is the weight fraction of component i in this mixture. In this case, let $W_A = 1\text{g}$ and $W_B = 1\text{ g}$. Then

$$\bar{M}_n = \frac{W_A + W_B}{(W_A/\bar{M}_{n_A}) + (W_B/\bar{M}_{n_B})} = \frac{1+1}{(1/10^5) + (1/2 \times 10^5)} = 133,000$$

$$\bar{M}_w = \left(\frac{W_A}{W_A + W_B}\right)\bar{M}_{W_A} + \left(\frac{W_B}{W_A + W_B}\right)\bar{M}_{W_B}$$

$$= \left(\frac{1}{2}\right)2 \times 10^5 + \left(\frac{1}{2}\right)4 \times 10^5 = 300,000$$

Note that even though the polydispersity index of each component of the mixture is 2.0, the PI of the mixture is greater, that is, 2.25.

TABLE 5.1 Methods of Determining Average Polymer Molecular Weights [5]

Method	Type of Molecular Weight Determined
End group analysis	$\bar{M}_n$
Boiling-point elevation	$\bar{M}_n$
Freezing-point depression	$\bar{M}_n$
Osmotic pressure	$\bar{M}_n$
Light scattering	$\bar{M}_w$
Ultracentrifugation (sedimentation)	$\bar{M}_w$
Viscometry	$\bar{M}_v$
Gel permeation chromatography	$\bar{M}_n, \bar{M}_w, MWD$, and PI

5.3 DETERMINATION OF AVERAGE MOLECULAR WEIGHTS

Everything would be quite easy if we knew the molecular weight of each chain in a polymer sample, but this is quite difficult to determine. This challenge has led to the development of several different techniques that attempt to measure or at least estimate the molecular weight of polymers [3]. This section covers some of the more common ways to determine molecular weight, including an overview of the operation of the instruments and (hopefully) giving an appreciation for the associated advantages and limitations.

In general, techniques for determination of average molecular weights fall into two categories: absolute and relative. In the former, measured quantities are theoretically related to the average molecular weight; in the latter, a quantity is measured that is in some way related to the molecular weight, but the exact relation must be established by calibration with one of the absolute methods.

Another important way to categorize these techniques is the type by which the average molecular weight is determined: $\bar{M}_n$ or $\bar{M}_w$ (or a viscosity-average molecular weight, $\bar{M}_v$). Another type of molecular weight, $\bar{M}_c$, molecular weight between crosslinks, is only applicable for polymer networks and is discussed in later chapters. The techniques covered in this chapter are listed in Table 5.1, along with the type(s) of molecular weight determined by each.

5.3.1 Absolute Methods

These methods allow the determination of the total number of moles of chains in a sample, N. If the total sample weight, W, is known,

$$\bar{M}_n = \frac{W}{N} \tag{5.8}$$

5.3.1.1 End-Group Analysis   If the chemical nature of the end groups on the polymer chains is known (as in many condensation polymers), standard analytical techniques may sometimes be employed to determine the concentration of the end groups and thereby of the polymer molecules, giving directly the number-average molecular weight. For example, in linear polyesters formed from a stoichiometrically equivalent batch of monomers, there are, on average, one unreacted acid group and one unreacted −OH group per molecule. These groups may sometimes be analyzed by appropriate titration. If an addition polymer is known to terminate by disproportionation (see Chapter 9), there will be one double bond for every two polymer molecules, which may be detectable quantitatively by halogenation or by infrared measurements. Other possibilities include the use of a radioactively tagged initiator that remains attached to the end of each chain.

These methods have one drawback in addition to the necessity of knowing the nature of the end groups. As the molecular weight increases, the concentration of end groups (number per unit volume) decreases, and the measurement sensitivity drops off rapidly. For this reason, these methods are generally limited to smaller polymers in the range of $\bar{M}_n < 10,000$.

5.3.1.2 Colligative Property Measurements

You may recall from an introductory chemistry course that addition of salts to a liquid can lead to boiling-point elevation or freezing-point depression. When a solute is added to a solvent, it causes a change in the activity and chemical potential (partial molal Gibbs free energy) of the solvent. The magnitude of the change is directly related to the solute concentration. For example, when pure water is boiled, the chemical potentials of the liquid and the vapor in equilibrium with it are the same. If, now, some salt is added to the water, it lowers the chemical potential of the liquid water. To reestablish equilibrium with the pure water vapor above the salt solution, the temperature of the system must be raised, causing a boiling-point elevation. In a similar fashion, the addition of ethylene glycol antifreeze depresses the freezing point of water.

Freezing-point depression, boiling-point elevation, and, a third technique, osmotic pressure, may be used to determine the number of moles of polymer per unit volume of solution and thereby establish the number-average molecular weight. The following are the relevant thermodynamic equations for the three techniques:

$$\lim_{c \to 0} \frac{\Delta T_b}{c} = \frac{RT^2}{\rho \Delta H_v \bar{M}_n} \qquad \text{(boiling-point elevation)} \tag{5.9}$$

$$\lim_{c \to 0} \frac{\Delta T_f}{c} = \frac{-RT^2}{\rho \Delta H_f \bar{M}_n} \qquad \text{(freezing-point depression)} \tag{5.10}$$

$$\lim_{c \to 0} \frac{\pi}{c} = \frac{RT}{\bar{M}_n} \qquad \text{(osmotic pressure)} \tag{5.11}$$

where
c = solute (polymer concentration), mass/volume
T = absolute temperature
R = gas constant
ΔH_v = solvent enthalpy of vaporization
ΔH_v = solvent enthalpy of fusion
ΔT_b = boiling-point elevation
ΔT_f = freezing-point depression
π = osmotic pressure
ρ = density

It is important to note that the thermodynamic equations apply only for ideal solutions, a condition that can be reached only in the limit of infinite dilution of the solute. For real situations, a dilute solution can be assumed in most cases if the concentration, c, is less than 1 wt%. It should also be noted that for these techniques to work, the polymer must be fully dissolved in a good solvent, whereas polymers precipitate in a bad solvent. If not dilute enough,

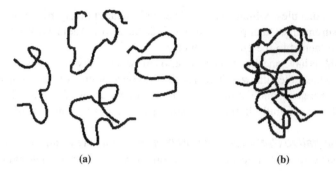

FIGURE 5.1 (a) Dilute polymer chains that have no interchain interactions (though they may have intrachain interactions), and (b) concentrated polymer solution showing significant interchain interactions.

polymer chains may interact in the solution (Figure 5.1), causing serious error in molecular weight estimation. In each of these equations, data collected for different concentrations can be plotted versus concentration, with the value at $c = 0$ used to determine $\bar{M}_n$.

Freezing-point depression and boiling-point elevation require precise measurements of very small temperature differences. Although they are used occasionally, the difficulties involved have prevented their widespread application. For example, a ΔT_f of 0.0002 °C might be the only observed change for a polymer with molecular weight 25,000. Osmotic pressure, on the other hand, is the most common absolute method of determining $\bar{M}_n$. A schematic diagram of an osmometer is shown in Figure 5.2. The solution and solvent chambers are separated by a "semipermeable" membrane, one that ideally allows passage

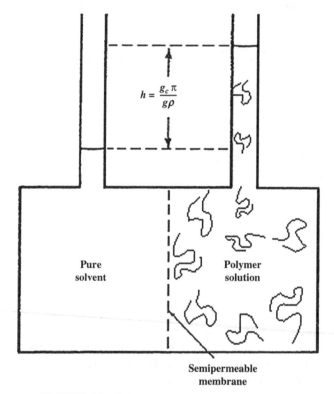

$$h = \frac{g_c\,\pi}{g\rho}$$

Pure
solvent

Polymer
solution

Semipermeable
membrane

FIGURE 5.2 Schematic diagram of an osmometer.

of solvent molecules but not solute molecules. The solvent flows through the membrane to dilute the solution. This is a natural consequence of the tendency of the system to increase its entropy, which is accomplished by the dilution of the solution. This dilution continues until the tendency toward further dilution is counterbalanced by the increased pressure in the solution chamber. At this point, the chemical potential of the solvent is the same in both chambers and the pressure difference between the chambers is the osmotic pressure, π. By making measurements at several concentrations, plotting (π/c) versus c and extrapolating to zero concentration, the number-average molecular weight is established through Equation 5.11.

One of the major difficulties with membrane osmometry is finding suitable semiper-meable membranes. Ordinary cellophane or modifications of it are commonly used. Unfortunately, if the membrane is sufficiently tight to prevent the passage of low molecular weight chains, the rate of solvent passage is slow and it takes longer to reach equilibrium. In practice, all membranes allow some of the low molecular weight polymer in a distribution to sneak through. Also, as the average molecular weight increases, π increases, so the measurement precision also decreases. These factors usually limit the applicability of osmometry to $50,000 < \bar{M}_n < 1,000,000$.

High-speed, automated membrane osmometers monitor the fluid volume in one of the chambers and externally *apply* the osmotic pressure to the solution chamber to prevent *flow*. Since no flow through the membrane is necessary, they can reduce measurement time from days to hours or even minutes.

A related colligative technique is vapor-pressure osmometry. Two thermistors are placed in a carefully thermostatted chamber that contains a pure-solvent reservoir so that the atmosphere is saturated with solvent vapor. A drop of solvent is placed on one thermistor and a drop of polymer solution is placed on the other. Because of the solvent's lower chemical potential in the solution, solvent vapor condenses on the solution drop, giving up its heat of condensation, warming the solution drop relative to the pure solvent drop. In principle, the equilibrium ΔT is thermodynamically related to the molar solution concentration, thereby allowing calculation of $\bar{M}_n$. In practice, heat losses (mainly along the thermistor leads) require that the instrument be calibrated for precise results, really making it a relative technique (as opposed to absolute). On a routine basis, commercial instruments are probably limited to maximum $\bar{M}_n$ values of 40,000–50,000 [4].

5.3.1.3 Weight-Average Techniques

The absolute techniques discussed to this point establish the number of molecules present per unit volume of solution, regardless of their size or shape. Other methods measure quantities that are related to the average *mass* of the molecules in solution, thereby giving the weight-average molecular weight. One of the more common of these is *light scattering*, which is based on the fact that the intensity of light scattered by a polymer molecule is proportional, among other things, to the square of its mass. A light-scattering photometer measures the intensity of scattered light as a function of the scattering angle (Figure 5.3). Measurements are made at several concentrations. By a double extrapolation to zero angle and zero concentration (known as *Zimm plot*) and with a knowledge of the dependence of the solution refractive index versus concentration, $\bar{M}_w$ is established. This technique also provides information on the solvent–polymer interaction and on the configuration of the polymer molecules in solution, since the quantitative nature of the scattering depends also on the size of the particles. Light scattering is generally applicable over a wider range $10,000 < \bar{M}_w < 10,000,000$. The weight-average molecular weight can also be obtained with an ultracentrifuge, which distributes the molecules according to their mass in a centrifugal force field.

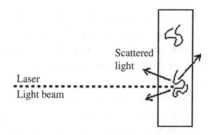

FIGURE 5.3   Light-scattering experiment to determine $\bar{M}_w$ of polymers in dilute solution.

It must be noted that for light-scattering and sedimentation (by ultracentrifuge) techniques (as well as most relative methods, below) to give an accurate measurement of molecular weight, dilute solutions are required. If two polymer molecules are entangled, they will appear as a single molecule and skew the results to higher molecular weights (that is why many techniques extrapolate to zero concentration).

5.3.2 Relative Methods

The molecular weight determination techniques discussed to this point allow direct calculation of the average molecular weight from experimentally measured quantities through known theoretical relations. Sometimes, however, these relations are not known, and although something is measured that is known to be related to molecular weight, one of the absolute methods above must be used to calibrate the technique.

A case in point is solution viscosity. It has long been known that relatively small amounts of dissolved polymer could cause tremendous increases in viscosity (ever made Jell-O®?). It is logical to assume that, other things being equal, larger molecules will impede flow more than smaller ones and give a higher solution viscosity. The solution viscosity will also depend on the solvent viscosity, temperature, solute concentration, as well as the particular polymer and solvent, because the interactions between the polymer and the solvent influence the conformation of the polymer molecules in solution and entanglements between the polymer molecules. In functional form, we may write

$$\eta = \eta \ (\eta_s, \ T, \ \text{polymer, solvent, } c, \ \text{entanglements, } M)$$
where $\eta =$ solution viscosity
$\eta_s =$ pure solvent viscosity

We can get rid of the effect of solvent viscosity by calculating the fractional increase in viscosity caused by the added solute (polymer) through the *specific viscosity* η_{sp}:

$$\eta_{sp} = \frac{\eta - \eta_s}{\eta_s} = \frac{\eta}{\eta_s} - 1 = \eta_r - 1 \tag{5.12}$$

where $\eta_r = \eta/\eta_s$ is known as the *relative viscosity*. Similarly, we can normalize for concentration by dividing the specific viscosity by concentration to get the reduced viscosity, η_{red}:

$$\eta_{red} = \frac{\eta_{sp}}{c} = \frac{(\eta/\eta_s) - 1}{c} \tag{5.13}$$

To get rid of the influence of entanglements on viscosity, we extrapolate the reduced viscosity to zero concentration to get the intrinsic viscosity $[\eta]$:

$$[\eta] = \lim_{c \to 0} \frac{(\eta/\eta_s) - 1}{c} \tag{5.14}$$

The intrinsic viscosity, then, should be a function of the molecular weight of the polymer in solution, the polymer–solvent system, and the temperature. If measurements are made at constant temperature using a specified solvent for a particular polymer, it should be quantitatively related to the polymer's molecular weight.

Huggins proposed a relation between reduced viscosity and concentration for dilute polymer solutions ($\eta_r < 2$):

$$\frac{\eta_{sp}}{c} = [\eta] + k'[\eta]^2 c \quad \text{(Huggins equation)} \tag{5.15}$$

Interesting enough, k' turns out to be approximately equal to 0.4 for a variety of polymer–solvent systems, providing a convenient means of estimating dilute solution viscosity versus concentration if the intrinsic viscosity is known. By expanding the natural logarithm in the definition of inherent viscosity, η_{inh}, into a power series it may be shown that an equivalent form of the Huggins equation is

$$\eta_{inh} = \frac{\ln \eta_r}{c} = [\eta] + k''[\eta]^2 c \tag{5.16}$$

where $k'' = k' - 0.5$

Equation From 5.14 it is seen that an alternative definition of intrinsic viscosity is

$$[\eta] = \lim_{c \to 0} \eta_{inh} = \lim_{c \to 0} \left[\frac{\ln(\eta/\eta_s)}{c} \right] \tag{5.17}$$

Plots of the reduced and inherent viscosities are linear with concentration, at least at low concentrations, in accord with Equations 5.16 and 5.17, and have a common intercept, the intrinsic viscosity (Figure 5.4). Exceptions occur with polyelectrolytes, where the degree of ionization and, therefore, the chemical nature of the polymer changes with concentration.

Note that the intrinsic viscosity has dimensions of reciprocal concentration. For some strange reason, concentrations were traditionally given in g/dL ($=100$ mL), although g/mL is now becoming more common. In fact, the relative, specific, reduced, intrinsic, and inherent viscosities are not true viscosities, and do not have dimensions of viscosity. More appropriate terminology has been proposed, but not widely adopted. Table 5.2 summarizes the various quantities defined and typical units.

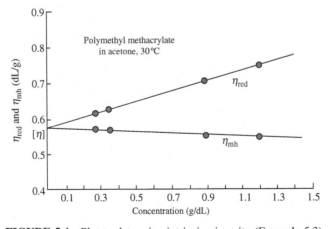

FIGURE 5.4 Plot to determine intrinsic viscosity (Example 5.3).

TABLE 5.2 Solution Viscosity Terminology [5]

Quantity	Common Units	Common Name	Recommended Name
η	Centipoise	Solution viscosity	Solution viscosity
η_s	Centipoise	Solvent viscosity	Solvent viscosity
$\eta_r = \eta/\eta_s$	Dimensionless	Relative viscosity	Viscosity ratio
$\eta_{sp} = (\eta - \eta_s)/\eta_s = \eta_r - 1$	Dimensionless	Specific viscosity	–
$\eta_{red} = \eta_{sp}/c = \eta_r - 1/c$	dL/g	Reduced viscosity	Viscosity number
$\eta_{inh} = \ln(\eta_r)/c$	dL/g	Inherent viscosity	Logarithmic viscosity number
$[\eta] = \lim\limits_{c \to 0} \eta_{red} = \lim\limits_{c \to 0} \eta_{inh}$	dL/g	Intrinsic viscosity	Limiting viscosity number

Now that the intrinsic viscosity has been established, how is it related to molecular weight? Studies of the intrinsic viscosity of essentially monodisperse polymer fractions whose molecular weights have been established by one of the absolute methods indicate a rather simple relation (Figure 5.5):

$$[\eta]_x = K(M_x)^a \quad (0.5 < a < 1) \tag{5.18}$$

where the subscript x refers to a monodisperse sample of a particular molecular weight. Equation 5.18 is known as the *Mark–Houwink–Sakurada* (MHS) relation.

What about an unfractionated, polydisperse sample? Experimentally, the measured intrinsic viscosity of a mixture of monodisperse fractions is a weight average

$$[\eta] = \frac{\sum [\eta]_x w_x}{\sum w_x} = \sum \left(\frac{w_x}{W}\right)[\eta]_x \tag{5.19}$$

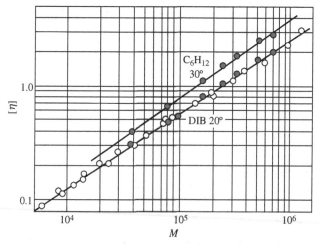

FIGURE 5.5 Intrinsic viscosity-molecular weight relations for polyisobutylene in cyclohexane at 30 °C and diisobutylene at 20 °C. Reprinted from P.J. Flsory, *Principles of Polymer Chemistry*. Copyright 1953 by Cornell University. Used by permission of Cornell University Press.

A *viscosity-average* molecular weight, $\bar{M}_v$, is defined in terms of this measured intrinsic viscosity as

$$\bar{M}_v = \left(\frac{[\eta]}{K}\right)^{1/a} = \left\{\frac{\sum M_x^a w_x}{W}\right\}^{1/a} = \left\{\frac{\sum M_x^a n_x M_x}{\sum n_x M_x}\right\}^{1/a} = \left\{\frac{\sum n_x M_x^{(1+a)}}{\sum n_x M_x}\right\} \quad (5.20)$$

With $0.5 < a < 1.0$ for polymers, $\bar{M}_n < \bar{M}_v < \bar{M}_w$, but $\bar{M}_v$ is closer to $\bar{M}_w$ than $\bar{M}_n$. If the molecular weight distribution of a series of samples does not differ too much, that is, if the ratios of the various averages remain nearly the same, approximate equations of the type $[\eta] = K'(\bar{M}_w)^{a'}$ may be applicable.

Why bother with calculating $\bar{M}_v$ if it is different from either of the molecular weight averages that we defined (Eqs 5.3 and 5.4 for $\bar{M}_n$ and $\bar{M}_w$)? Viscosity is an easily measured property of polymer solutions, and $\bar{M}_v$ can be used for quality control in polymerization processes, so it is still quite useful.

Example 5.3 The following data were obtained for a sample of poly(methyl methacrylate), PMMA, in acetone at 30 °C:

η_r	c (g/100 mL)
1.170	0.275
1.215	0.344
1.629	0.896
1.892	1.199

For PMMA in acetone at 30 °C, $[\eta] = 5.83 \times 10^{-5} (\bar{M}_v)^{0.72}$. Determine $[\eta]$ and $\bar{M}_v$ for the sample and K', the constant in the Huggins equation.

Solution. For the data above, calculations give

η_{sp}	η_{red} (dL/g)	$\ln \eta_r$	η_{inh} (dL/g)
0.170	0.618	0.157	0.571
0.215	0.625	0.195	0.567
0.629	0.702	0.488	0.545
0.892	0.744	0.638	0.532

In Figure 5.4, η_{red} and η_{inh} are plotted against concentration and extrapolated to a common intercept at zero concentration, the intrinsic viscosity, $[\eta] = 0.580$ dL/g.

$$\bar{M}_v = \left(\frac{[\eta]}{5.83 \times 10^{-5}}\right)^{1/0.72} = \left(\frac{0.580}{5.83 \times 10^{-5}}\right)^{1.39} = 357,000$$

The regression slope of the upper line is 0.137. From Equation 5.17, $k' = \text{slope}/[\eta]^2$ $= 0.408$ (dimensionless).

Example 5.4 For the PMMA fractions in Example 5.1, calculate $\bar{M}_v$ for mixtures 1 and 2 in acetone at 30 °C and compare with $\bar{M}_n$ and $\bar{M}_w$ (a is given in Example 5.3).

Solution. From Equation 5.20, for mixture 1:

$$\bar{M}_v = \left\{ \sum \left(\frac{w_x}{W}\right) M_x^a \right\}^{1/a} = \left\{ \frac{1}{3}(1 \times 10^5)^{0.72} \right.$$

$$\left. + \frac{2}{3}(4 \times 10^5)^{0.72} \right\}^{1/0.72} = 288,000 \quad \begin{array}{l} \bar{M}_n = 200,000 \\ \bar{M}_w = 300,000 \end{array}$$

In mixture 2:

$$\bar{M}_v = \left\{ \frac{2}{3}(1 \times 10^5)^{0.72} \right.$$

$$\left. + \frac{1}{3}(4 \times 10^5)^{0.72} \right\}^{1/0.72} = 187,000 \quad \begin{array}{l} \bar{M}_n = 133,000 \\ \bar{M}w = 200,000 \end{array}$$

Viscosities for molecular weight determination are usually measured in glass capillary viscometers, in which the solution flows through a capillary under its own head. Two common types, the Ostwald and Ubbelohde, are sketched in Figure 5.6. (Since polymer solutions are non-Newtonian, intrinsic viscosity must be defined, strictly speaking, in terms of the zero-shear or lower-Newtonian viscosity (see Chapter 14). This is rarely a problem, because the low shear rates in the usual glassware viscometers give just that. Occasionally, however, extrapolation to zero-shear conditions is required.)

In these types of viscometers, fluid is drawn up through a capillary into bulbs, with timing lines marked on the glass. As the fluid flows back down through the capillary (the

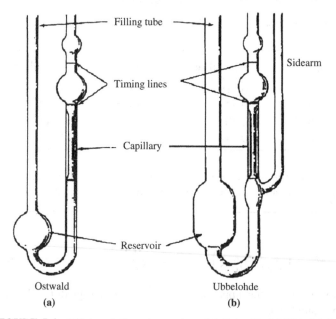

Filling tube

Timing lines

Sidearm

Capillary

Reservoir

Ostwald
(a)

Ubbelohde
(b)

FIGURE 5.6 Dilute-solution viscometers: (a) Ostwald; (b) Ubbelohde.

TABLE 5.3   K and *a* Values of the MHS Equation for Selected Polymer–Solvent Systems [6]

Polymer	Solvent	Temperature	K (10^3 mL/g)	a
Polystyrene, PS	Tetrahydrofuran, THF	22 °C	11.0	0.725
Poly(methyl methacrylate)	Acetone	30 °C	5.83	0.720
Atactic polypropylene	Benzene	25 °C	27.0	0.71
Poly(vinyl alcohol)	Water	25 °C	20.0	0.76
Poly(vinyl chloride)	THF	25 °C	16.3	0.766
Poly(vinyl chloride)	THF	30 °C	63.8	0.65

right side of both viscometers in Figure 5.6), the flow time is recorded. These flow times are related to the viscosity by an equation of the form

$$\frac{\eta}{\rho} = v = at + \frac{b}{t} \qquad (5.21)$$

where a and b are instrument constants (found through calibration), ρ is the solution density, and v is the kinematic viscosity (with typical units of cm^2/s). The last term (b/t), the kinetic energy correction, is generally negligible for flow times of over a minute, and since the densities of the dilute polymer solutions differ little from that of the solvent, the ratio of the viscosity of the polymer solution to the solvent viscosity can be approximated by the flow times recorded for these solutions, respectively:

$$\frac{\eta}{\eta_s} \approx \frac{t}{t_s} \qquad (5.22)$$

The Ubbelohde viscometer has the distinct advantage that the driving fluid head is independent of the amount of solution in it; hence, dilution can be carried out right in the instrument.

The equipment necessary for intrinsic viscosity determination is inexpensive, and the measurements straightforward and rapid. The MHS constants K and a in Equations 5.18 and 5.20 are extensively tabulated for a wide variety of polymer–solvent systems and temperatures (Table 5.3) [6]. Even more simple devices (such as cup viscometers–which are metal cups with fixed holes in the bottom) are commonly used in industry for rapid viscosity estimation, this is particularly true in paint formulations. Automated viscometers, such as rotational viscometers, are also commonly used in analytical laboratories and offer rapid and accurate results.

5.4   MOLECULAR WEIGHT DISTRIBUTIONS

Before continuing to one additional molecular weight characterization technique (gel permeation chromatography), let us delve a bit deeper into the concept of molecular weight distributions. A typical synthetic polymer might consist of a mixture of molecules with degrees of polymerization x ranging from one to perhaps millions. The complete molecular weight distribution specifies the mole (number) or mass (weight) fraction of molecules at each size level in a sample. (Actually, moles or masses could be specified, but they vary linearly with sample size, that is, are extensive quantities, while the mole or mass fractions are intensive, independent of sample size, and are therefore preferable.)

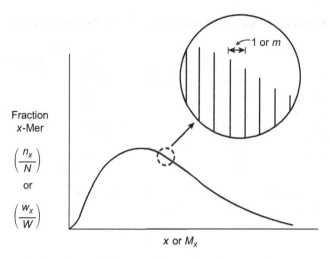

FIGURE 5.7 Molecular weight distributions illustrating the actual discrete distribution and the usual continuous approximation to it.

Distributions are often presented in the form of a plot of mole (n_x/N) or mass (w_x/W) fraction of x-mer versus either chain length, x, or chain mass, M_x. Since x and M_x differ by a constant factor, the molecular weight of the repeating unit M_r (Eq. 5.1), it makes little difference which is used. Because x can assume only integral values, a true distribution must consist of a series of spikes, one at each integral value of x, or separated by M_r molecular weight units if plotted against M_x. The height of the spike represents the mole or mass fraction of that particular x-mer. No analytical technique is capable of resolving the individual x-mers, so distributions are drawn (and represented mathematically) as continuous curves, drawn through the tops of the spikes as sketched in Figure 5.7.

The averages may be calculated from either the mole- or the mass-fraction distributions. For continuous distributions, the summations in Equations 5.3 and 5.4 must be replaced by the corresponding integrals (the subscript x has been dropped for clarity and because of the continuous nature of the distributions).

From the mole (number)-fraction distribution (n/N),

$$\bar{M}_n = \text{First moment} = \frac{\int_0^\infty nM \, dM}{\int_0^\infty n \, dM} = \frac{\int_0^\infty (n/N)M \, dM}{\int_0^\infty (n/N) \, dM} \tag{5.23}$$

Since $\bar{M}_n = m\bar{x}_n$ and $M = mx$:

$$\bar{x}_n = \int_0^\infty \left(\frac{n}{N}\right) x \, dx \tag{5.24}$$

Note that the integrals replace the summation symbols used in the definition of $\bar{M}_n$ (Eq. 5.3). Also,

$$\int_0^\infty \left(\frac{n}{N}\right) dx = 1 \tag{5.25}$$

that is, the mole fractions must sum to 1. Also,

$$\bar{M}_w = \text{Second moment} = \frac{\int_0^\infty nM^2 \, dM}{\int_0^\infty nM \, dM} = \frac{\int_0^\infty (n/N)M^2 \, dM}{\int_0^\infty (n/N)M \, dM} \qquad (5.26)$$

And, since $\bar{M}_w = m\bar{x}_w$,

$$\bar{x}_w = \frac{\int_0^\infty (n/N)x^2 \, dx}{\int_0^\infty (n/N)x \, dx} = \frac{1}{\bar{x}_n} \int_0^\infty \left(\frac{n}{N}\right)x^2 dx \qquad (5.27)$$

If the mass (weight)-fraction distribution (w/W) is known, the average molecular weights and chain lengths can be determined:

$$\bar{M}_n = \frac{\int_0^\infty nM \, dM}{\int_0^\infty n \, dM} = \frac{\int_0^\infty w \, dM}{\int_0^\infty (w/M) \, dM} = \frac{\int_0^\infty (w/W) \, dM}{\int_0^\infty (1/M)(w/W) \, dM} \qquad (5.28)$$

$$\bar{x}_n = \frac{\int_0^\infty (w/W) \, dx}{\int_0^\infty (1/x)(w/W) \, dx} \qquad (5.29)$$

$$\bar{M}_w = \frac{\int_0^\infty nM^2 \, dM}{\int_0^\infty nM \, dM} = \frac{\int_0^\infty (nM)M \, dM}{\int_0^\infty (nM) \, dM} = \frac{\int_0^\infty wM \, dM}{\int_0^\infty w \, dM} = \frac{\int_0^\infty (w/W)M \, dM}{\int_0^\infty (w/W) \, dM} \qquad (5.30)$$

$$\bar{x}_w = \int_0^\infty \left(\frac{w}{W}\right)x \, dx \qquad (5.31)$$

Note that

$$\int_0^\infty \left(\frac{w}{W}\right) dx = 1 \qquad (5.32)$$

that is, the mass fractions must sum to 1.

Given the mole (number)-fraction distribution, the mass (weight)-fraction distribution may be calculated, and vice versa. Since $w = nM$ and, by definition of $\bar{M}_n$ (Eq. 5.3), $W = N\bar{M}_n$,

$$\frac{w}{W} = \frac{M}{\bar{M}_n}\left(\frac{n}{N}\right) = \frac{x}{\bar{x}_n}\left(\frac{n}{N}\right) \tag{5.33}$$

Example 5.5 One analytic representation of a distribution of chain lengths that finds some practical application (as we will see later) is

$$\frac{n}{N} = \frac{1}{C}e^{-(x/c)}$$

where C is a constant.

(a) Determine $\bar{x}_n$ for this distribution.
(b) Determine x_w and the polydispersity index, $PI = \bar{x}_w/\bar{x}_n$, for this distribution.
(c) Obtain the expression for the weight-fraction distribution (w/W).
(d) Sketch the number- and weight-fraction distributions in the form of $(n/N)\,\bar{x}_n$ and $(w/W)\,\bar{x}_n$ versus $x/\bar{x}_n$.

Solution.
(a)

$$\bar{x}_n = \int_0^\infty \left(\frac{n}{N}\right) x\, dx = \frac{1}{C}\int_0^\infty xe^{-x/c}dx$$

Fortunately, for most of us, the above definite integral is tabulated in standard references. The result is

$$\bar{x}_n = C$$

(b)

$$\bar{x}_w = \frac{1}{\bar{x}_n}\int_0^\infty \left(\frac{n}{N}\right) x^2 dx = \frac{1}{\bar{x}_n^2}\int_0^\infty x^2 e^{-x/\bar{x}_n}dx$$

This integral is also tabulated, and gives the simple result

$$\bar{x}_w = 2\,\bar{x}_n$$

that is, $PI = \bar{x}_w/\bar{x}_n = 2$ for this distribution.
(c)

$$\frac{w}{W} = \left(\frac{n}{N}\right)\frac{x}{\bar{x}_n} = \frac{x}{\bar{x}_n^2}e^{-x/\bar{x}_n}$$

(d) The distributions are shown in Figure 5.8. Note that while the number of molecules decreases exponentially with x, the weight of a molecule increases linearly with x. The weight fraction therefore goes through a maximum in this case.

This distribution is derived for a particular kind of polymerization reaction in Chapter 9.

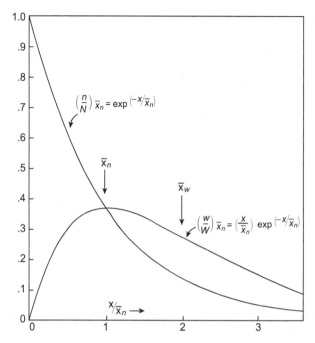

FIGURE 5.8   Molecular weight distribution of Example 5.5 plotted in reduced form.

5.5 GEL PERMEATION (OR SIZE-EXCLUSION) CHROMATOGRAPHY (GPC, SEC) [7]

Until fairly recently, the approximate molecular weight distribution of a polymer could be determined only by laborious fractionation of the sample according to molecular weight followed by determination of the molecular weights of the individual fractions with one of the techniques previously discussed. The bases for these fractionation techniques are discussed in Chapter 7. Suffice it to say that they tend to be difficult and time-consuming and so, in general, are avoided whenever possible.

Now, however, gel permeation chromatography (GPC, also known as size-exclusion chromatography, SEC) has been firmly established as a means of rapidly determining molecular weight averages ($\bar{M}_n$, $\bar{M}_w$, $\bar{M}_z$) and the corresponding distributions. GPC makes use of a column, or series of columns, packed with particles of a porous substrate (known as the stationary phase). The term gel in gel-permeation chromatography refers to a cross-linked polymer that is swollen by the solvent used. This is perhaps the most common type of substrate, but others, for example, porous glass beads, are also used. The column is maintained at a constant temperature, and the solvent (known as the mobile phase) is passed through it at a constant rate. At the start of a run, a small amount of polymer solution is injected into the flowing mobile phase just before it enters the column. The solvent flow carries the polymer through the column. The smaller molecules in the sample have easy access to the substrate pores and diffuse in and out of the pores, creating a highly tortuous, lengthy pathway from the entrance to the exit from the column. The larger polymer chains simply cannot fit into the pores in the stationary phase and pass around the outside, and because they have a shorter pathway from entrance to exit, they elute from the column first. Thus, polymer chains of different molecular weights are effectively separated, with the molecular weight of the chains exiting the column decreasing with time.

A concentration-sensitive detector is placed at the outlet of the column. The most common detector is a differential refractometer (which measures the difference in refractive index between the pure solvent and that of the eluting fluid, containing the polymer), but ultraviolet (UV) or infrared (IR) detectors are also useful as long as the polymer has a functional group that can be detected in these light ranges. Regardless of the type of detector, it is essential that it measure some quantity Q that is proportional to the *mass* concentration of polymer at the column outlet (Q must be independent of M):

$$Q = kc \tag{5.34}$$

where Q = detector readout
k = proportionality constant
c = mass concentration of polymer (g/cm^3)

Thus, a GPC curve consists of a plot of Q (usually in arbitrary scale divisions, but related to the quantity of chains eluting at a particular time) versus v, the volume of solvent that has passed through the detector since sample injection, called *elution* or *retention volume*. The x-axis can easily be converted to time by dividing the elution volume by the volumetric flow rate of the solvent.

An example of a GPC curve is shown in Figure 5.9. Such a curve is often immediately useful for quality-control purposes, as the width of the distribution indicates sample polydispersity. The graph can also be compared directly to the graph for a known standard material, revealing a quick qualitative comparison. This may be enough to take corrective action in processing, but the usefulness of GPC extends well beyond qualitative comparisons, because, with appropriate calibration, it can determine all types of average molecular weights ($\bar{M}_n$, $\bar{M}_w$, $\bar{M}_z$), and the PI of polymer samples. An additional benefit is that GPC can also be useful in detecting small molecular weight impurities or additives (such as plasticizers or stabilizers—see Chapter 18). These materials show up as peaks at the low molecular weight (high v or long t) end of the spectrum.

GPC chromatograms might look like one of the curves plotted in Figure 5.9. Nearly monodisperse polystyrene standards are used to calibrate a GPC. If run through the column as a mixture, the elution curve will show sharp peaks for each molecular weight standard, with the highest molecular weight eluting first. After the column has been calibrated, polymers with unknown molecular weight distributions can be tested. In Figure 5.9, one sample with two narrowly dispersed samples are shown in curve B (note that the curves do not overlap and that the area under the curves is proportional to the amount of polymer of each molecular weight), while a mixture of highly polydisperse polymer samples will have very broad peaks that overlap (curve C).

The total area under a chromatographic peak is proportional to the amount (mass) of polymer eluted. Thus, for polymer mixtures, it is possible to use GPC to estimate the weight fractions of different components, as long as their peaks do not overlap on the elution (x) axis.

It should be emphasized that GPC requires dilute solutions, so that each polymer molecule passes through the porous stationary phase with no entanglements. Because the hydrodynamic volume occupied by grafted or branched polymers is somewhat less than that occupied by a linear polymer with the same molecular weight, GPC typically underpredicts $\bar{M}_n$ and $\bar{M}_w$ for nonlinear polymers.

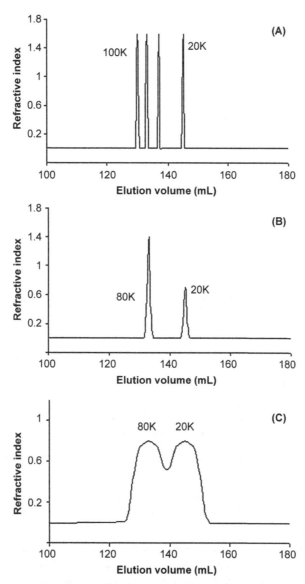

FIGURE 5.9 A depiction of typical results from GPC. Three GPC chromatograms showing the refractive index (which is proportional to the quantity of sample eluted) against elution volume for (a) four monodisperse polystyrene standards used to calibrate the GPC, (b) a sample that contains a 2:1 polymer blend of narrowly disperse polymers with $\bar{M}_w$ 80,000 and 20,000, and (c) a sample that contains a 1:1 polymer blend of widely polydisperse polymer sample $\bar{M}_w$ 80,000 and 20,000. The elution volumes match the calibration curve shown in Figure 5.10.

Conventional GPC is a relative method. Thus, to provide quantitative results, the relation between M and v (or elution time, t) must be established by calibration with monodisperse polymer standards. A calibration curve is shown in Figure 5.10. Typically, when the data are plotted in the form $\log M$ versus v, the curve is linear over much of the range, but sometimes turns up sharply at low v (high M).

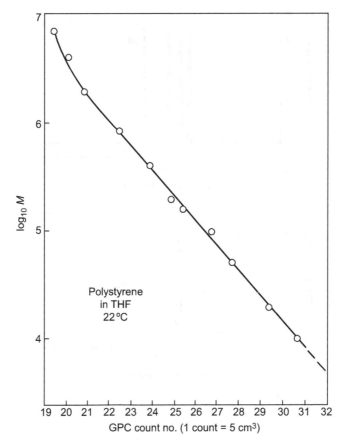

FIGURE 5.10   GPC calibration curve. Narrow polydispersity ($\bar{M}_w/\bar{M}_n$) < 1.1 polystyrene samples in tetrahydrofuran at 22 °C.

Example 5.6 Discuss the physical significance of an upturn in the calibration curve at low *v*.

Solution. The point at which the curve begins to shoot up is set by the largest pore size in the column. All of the chains that are too large to fit in those pores will pass through the column at the same rate, giving the infinite slope; that is, the packing material in a particular column cannot discriminate between molecules above that size. If the polymer sample to be analyzed contains significant material at high molecular weight, a column with larger pores should either replace the original column or be added in series to extend the calibration and obtain a good separation.

The calibration curve, strictly speaking, applies only to the particular polymer, solvent, temperature, flow rate, and column for which it was established. Change any one, and the calibration is no longer valid. Most SEC calibrations are obtained with polystyrene, because the necessary monodisperse standards are readily available at reasonable cost. What do you do when you want to analyze another polymer? If absolute values are not needed, the molecular weights can be reported with respect to the polystryene (or other polymer) standards used to calibrate. A more sophisticated approach was suggested by Grubisic *et al.* [8]. When they plotted $\log([\eta]M)$ (the product $[\eta]M$ is proportional to the hydrodynamic volume of the molecules) versus v, they obtained a single, universal calibration curve for a variety of different polymers. Thus, if you measure intrinsic viscosity along with SEC elution volume for your polystyrene standards (or calculate $[\eta]$ for them with Eq. 5.18 and literature values for K and a), you can plot the universal calibration curve for your column.

To back out a calibration curve for a different polymer, you would need to know K and a under the new conditions:

$$[\eta]_0 M_0 = [\eta]M = KM^{(a+1)} \tag{5.35}$$

$$M = \left(\frac{[\eta]_0 M_0}{K}\right)^{1/(a+1)} \tag{5.36}$$

where the subscript 0 above refers to the calibration conditions, that is, values from the universal calibration.

Balke et al. [9] have proposed a calibration method that requires only a single polydisperse sample of known $\bar{M}_n$ and $\bar{M}_w$. The method assumes that the relation between $\log M$ and v is linear, and therefore it can be characterized by two parameters, a slope and an intercept. Given the GPC curve and the calibration, $\bar{M}_n$ and $\bar{M}_w$ can be calculated (using techniques outlined below). With the GPC data on the standard, a computer program adjusts the two calibration parameters until the known $\bar{M}_n$ and $\bar{M}_w$ are obtained, in effect using the two known average molecular weights to solve for the two unknown calibration parameters. If the calibration curve is not linear (and you would have no way of knowing), serious errors can result.

Most GPC packages include the software necessary to calculate molecular weight averages and the complete distribution from the data and the calibration, so one needs only to push the appropriate buttons. Nevertheless, the equations are outlined below to provide an understanding of what the software is doing and what its limitations might be.

Molecular weight averages may be approximated directly from the GPC curve and calibration by breaking the curve into arbitrary volume increments Δv. Usually, Δv is taken as 5 cm^3, but smaller Δv's improve accuracy. The number of moles of polymer in a volume increment Δv is n_i:

$$n_i = \frac{c_i \Delta v}{M_i} \tag{5.37}$$

where c_i = polymer mass concentration in the ith volume increment
$\quad M_i$ = molecular weight of polymer in the ith volume increment (assumed essentially constant over small Δv)Because $c_i = Q_i/k$ (Eq. 5.34),

$$n_i = \frac{Q_i \Delta v}{k M_i} \tag{5.38}$$

If the volume increment Δv is constant, inserting Equation 5.38 into Equations 5.3 and 5.4 gives

$$\bar{M}_n = \frac{\sum_i n_i M_i}{\sum_i n_i} = \frac{\sum_i Q_i}{\sum_i (Q_i/M_i)} \tag{5.39}$$

$$\bar{M}_w = \frac{\sum_i n_i M_i^2}{\sum_i n_i M_i} = \frac{\sum_i Q_i M_i}{\sum_i Q_i} \tag{5.40}$$

Note that the proportionality constant k between concentration and GPC readout cancels out, and therefore need not be known. Q_i is normally read from the GPC curve as the *distance above the baseline* (in any convenient units) at a given v. The baseline presumably represents the detector output with the pure solvent. Establishing a good baseline is not always a trivial procedure, and the results can be quite sensitive to its location. M_i is read from the calibration curve at the same v. For greater accuracy, if necessary, the integral analogs of Equations 5.39 and 5.40 may be used, with the integrals evaluated numerically:

$$\bar{M}_n = \frac{\int_0^\infty Q \, dv}{\int_0^\infty (Q/M) dv} \tag{5.41}$$

$$\bar{M}_w = \frac{\int_0^\infty QM \, dv}{\int_0^\infty Q \, dv} \tag{5.42}$$

In most cases, the averages as calculated above and the qualitative information about the distribution provided by the GPC curve are all that are necessary. However, the technique is capable of providing the true distributions, if desired. The necessary calculations are outlined below:

$$\text{Moles of polymer in volume increment } dv = \frac{c \, dv}{M} = \frac{Q \, dv}{k \, M} \frac{(\text{g/cm}^3)(\text{cm}^3)}{(\text{g/g mol})} \tag{5.43}$$

$$\text{Moles of polymer/unit of } M = \frac{c}{M} \frac{dv}{dM} = \frac{Q \, dv}{kM \, dM} \tag{5.44}$$

$$n = \text{moles of polymer in a molecular weight interval } m = \frac{mc \, dv}{M \, dM} = \frac{Qm \, dv}{kM \, dM}$$

$$= \frac{mc}{2.303 \, M^2} \left(\frac{dv}{d\log M} \right) = \frac{Qm}{2.303 \, kM^2} \left(\frac{dv}{d\log M} \right) \tag{5.45}$$

Note:

$$\frac{dv}{d\log M} = \frac{1}{\text{slope of calibration}}$$

The preceding calculations assign all the moles of polymer in a range m of the continuous GPC curve to a single spike (see Section 5.4) to calculate the correct height of the distribution:

$$N = \int_0^\infty \frac{c}{M} \, dv = \frac{1}{k} \int_0^\infty \frac{Q}{M} \, dv \tag{5.46}$$

$$\frac{n}{N} = \frac{(Qm/2.303M^2)(dv/d\log M)}{\int_0^\infty (Q/M) dv} \tag{5.47}$$

$$w = \text{Weight of polymer in a mol wt interval } m = nM$$

$$= \frac{mc}{2.303M}\left(\frac{dv}{d\log M}\right) = \frac{Qm}{2.303kM}\left(\frac{dv}{d\log M}\right) \tag{5.48}$$

$$W = \text{Total sample wt} = \int_0^\infty c\,dv = \frac{1}{k}\int_0^\infty Q\,dv$$
$$= \frac{\text{Area under GPC curve}}{k} \tag{5.49}$$

$$\frac{w}{W} = \frac{(Qm/2.303M)(dv/d\log M)}{\displaystyle\int_0^\infty Q\,dv} \tag{5.50}$$

Again, k cancels out in these equations. Once the distributions are known, the exact averages may be calculated by the techniques in Section 5.4. For example, insertion of Equation 5.47 into Equations 5.23 and 5.26 gives Equations 5.41 and 5.42.

GPC columns can be selected that have a wide spectrum of pore sizes (and thus separate a wide range of molecular weights–from thousands to millions). Alternatively, columns with more narrow pore size distributions can be used to get a better separation of polymers that have similar molecular weights—say from 1000 to 20,000. These columns are designed to be used with organic solvents (required for most polymers) or water that flow through as the mobile phase. The selection of which columns to use depends on the polymer structure and the expected molecular weight distribution.

A multiangle laser light-scattering photometer is often added to the usual concentration detector. This combination performs an online light-scattering determination of the M of the polymer in the eluting stream. This makes GPC an absolute technique, eliminating the need for calibration. Similarly, a differential viscometer is commercially available which performs an online intrinsic viscosity measurement on the column effluent [10]. Recall that the ordinate of the universal calibration is $[\eta]M$. This device requires simple calibration (in effect locating the universal calibration curve for the particular column, temperature, solvent, etc.) to provide absolute molecular weight distributions and also provides information on the hydrodynamic volume of the molecules in solution and therefore on branching, because a branched molecule in solution has a smaller hydrodynamic volume than a linear molecule of the same molecular weight. As mentioned earlier, an important caution regarding GPC/SEC is as follows: because the molecules are separated based on their size, linear polymers will flow differently through the stationary phase (the packed column) than branched or grafted polymers, so molecular weight estimation by this technique is normally less accurate for branched, grafted, or dendritic (very highly branched) polymers.

5.6 SUMMARY

Molecular weight has an enormous impact on polymer properties, as we will discover in later chapters. Because polymers are made of many copies of the same repeat unit, traditional methods to determine absolute molecular weights are not applicable. Thus, several techniques are discussed in this chapter to measure or estimate polymer molecular weights. Depending on the techniques used, the number-average or weight-average molecular weights are determined, with most sensitive techniques requiring that the polymer be in dilute solution so that physically entangled polymers are not mistakenly recorded as one large polymer. Of the techniques discussed so far, gel permeation chromatography offers the greatest amount of information about a polymer sample, giving the size distribution of a polymer sample, which allows calculation of $\bar{M}_n$ and $\bar{M}_w$ and the polydispersity index.

PROBLEMS

1 Given the following chain lengths (x's) for poly(ethylene terephthalate), determine the number average molecular weight, weight average molecular weight, and polydispersity index. The system consists of 10 polymer chains, with repeat unit lengths of 12, 25, 175, 186, 192, 194, 199, 200, 202, and 212.

2 The weights of the (extremely) offensive linemen of the MIT (Monongahela Institute of Technology) football team are:

Cussler	Split end	180 lb
Miller	Left tackle	270 lb
Westerberg	Left guard	256 lb
Brenner	Center	285 lb
Anderson	Right guard	260 lb
Jain	Right tackle	305 lb
Prieve	Tight end	250 lb

Calculate the number- and weight-average line weights.

3 Using end group analysis to find molecular weight, 6.50 moles of ethanediamine ($H_2N-CH_2CH_2-NH_2$) were reacted with 6.50 moles of adipic acid ($HOOC-CH_2-CH_2-CH_2-CH_2-COOH$). After the reaction, a titration was performed using NaOH (which reacts with COOH) and an indicating dye to determine the number of acid groups remaining free after the polymerization. If 73.0 mL of 3.0 M NaOH were required to neutralize the polymer, what is the average molecular weight (M_n) of the polymer?

4 An osmometry experiment was done to estimate the molecular weight of a sample of polystyrene. The concentration of polymer in toluene was varied and the height difference between the solution and the solvent was measured. The temperature was 27 °C, where the specific gravity of toluene is 0.8667. Based on the following data, determine the molecular weight of PS. (A graph is required.) Is this $\bar{M}_w$ or $\bar{M}_n$?

Concentration, C_A g/mL	Osmotic Pressure, π_A, cm of Solvent
0.0320	0.700
0.0660	1.82
0.100	3.10
0.140	5.44
0.190	9.30

5 Six students have taken viscometry data for a monodisperse sample of atactic polystyrene at six different concentrations in toluene. Unfortunately, the techniques were not all identical, so they reported different versions of viscosity. The density of toluene at 30 °C is 0.8665 g/mL, 0.8656 g/mL at 40 °C. The viscosity of toluene is 0.5260 cP at 30 °C, 0.4710 cP at 40 °C.

Given the data below, resolve the experiments so that you can determine the viscosity-average molecular weight, $\bar{M}_v$, of the polymer.

Experiment 1: Polymer concentration in toluene C = 0.1000 g/dL; inherent viscosity = 0.060254 dL/g, 30 °C.

Experiment 2: C = 0.0500 g/dL; used an Ostwald viscometer at 30 °C to determine a flow time of 4 min, 36.50 s; in the viscometer calibration equation $\eta/\rho = at + b/t$, $a = 0.001340$, and $b = 65.760$ (where time is in seconds, density in g/mL, and viscosity in cP)

Experiment 3: C = 0.1100 g/mL; solution viscosity = 1.07360 cP at 40 °C.

Experiment 4: C = 0.1650 g/dL; determined the relative viscosity solution to be 1.0130 at 30 °C.

Experiment 5: C = 8.0 g/mL; determined specific viscosity was 10.896 at 30 °C

Experiment 6: C = 0.0290 g/dL; used a cone-and-plate viscometer at 30 °C to find a solution viscosity of 0.52655 cP

MHS constants for atactic polystyrene at 30 °C are: $K = 0.2 \times 10^{-3}$ mL/g; $a = 0.72$. Verify the validity of each experiment for estimating $\bar{M}_v$.

Note: Discount experiments if they were not properly conducted.

6 Consider a copolymer with repeating units A and B in which the mole fraction of A is y_A. The molecular weights of the repeating units are m_A and m_B. Obtain the analog of Equation 5.5a for the copolymer, in which x represents the total number of repeating units of both kinds.

7 A *discrete* distribution of chain lengths has the same number of moles of each species over the range $1 \le x \le a$ and nothing outside of that range. Obtain expressions for $\bar{x}_n$, $\bar{x}_w$, and the polydispersity index for this distribution. *Hints:* Start with Equations 5.3a and 5.4a. You may have to look up some series sums in a math table.

8 Consider the *continuous* distribution of chain lengths in which the number of moles of each species is constant over the range $0 \le x \le a$ and zero outside that range.

(a) Obtain expressions for $\bar{x}_n$, $\bar{x}_w$, and the polydispersity index.

(b) What is the constant value of (n/N) between 0 and a?

(c) Obtain an expression for the weight-fraction distribution (w/W) as a function of x and sketch it.

(d) Show that your answers to Problem 5.7 above agree with these when $a \gg 1$.

9 Crud Chemicals is using GPC to analyze polystyrene samples plasticized with low molecular weight mixed hydrocarbon oil, $\bar{M}_n = 400$. Their calibration curve is identical to the one in Figure 5.10 and from past experience they know that the unplasticized polymer gives a GPC curve just like Figure 5.9c.

(a) Sketch what you think a GPC curve for a plasticized material would look like.

(b) If you know that $k_{oil} = 0.70\,k_{PS}$ (in Equation 5.34), could you use the GPC curve to determine the weight fraction plasticizer in the sample? If so, clearly describe how. If not, what additional information would you need?

10 Given the hypothetical distribution of chain lengths:

$$\frac{n}{N} = C(1 - 10^{-3}x) \quad 0 \le x \le 10^3$$

(a) Determine the value of the constant C.

(b) Calculate $\bar{x}_n$, $\bar{x}_w$, and the polydispersity index for this distribution.

(c) Find the expression for the weight fraction of chains as a function of molecular length x, w/W, and roughly sketch this distribution as w/W versus x. Sketch n/N versus x on the same plot.

11 Some authorities prefer to represent molecular weight distributions in cumulative (integral) form rather than the differential form used here. Cumulative distributions are the number or weight fraction of material in the sample with chain length $\le x$ (or molecular weight $\le M$), so that the maximum value of the y-axis is 1.
For the distribution of chains in Example 5.5, obtain expressions for the cumulative mole and weight distributions in terms of $x/\bar{x}_n$ and sketch them.

12 A polystyrene sample has the same distribution of chain lengths as given in Example 5.5. A light-scattering measurement gives a molecular weight of 208,000. Using the result from Example 5.5(c), calculate the mass of material with $x = 100$ in a 100 g sample of this polystyrene.

13 Many commercially used polymers are actually blends of two or more different molecular weights, which gives a net composition with a bimodal molecular weight distribution.

(a) If an equal weight of polymers A, B, and C (below) are blended, what are the new $\bar{M}_n$, $\bar{M}_w$, and polydispersity of the blend?

	$\bar{M}_n$	$\bar{M}_w$	PDI
Polymer A:	200,000	600,000	3.00
Polymer B:	350,000	400,000	1.14
Polymer C:	150,000	150,000	1.00

(b) Draw (sketch) a graph to represent the results of running this blend through a gel permeation chromatograph. The y-axis is an indicator of intensity (refractive index or absorbance) and is proportional to the mass of polymers that are leaving the GPC column. Label the x- and y-axes and label the parts of the chromatograph to show the different polymers in the blend (A, B, and C). Consider that the A, B, and C peaks may overlap.

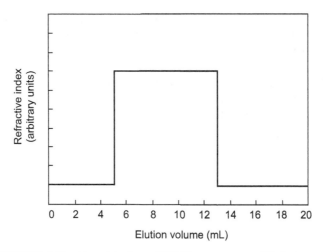

FIGURE 5.11 Raw data from GPC Chromatogram of Polystyrene.

14 The people at Crud Chemicals eagerly await the first GPC analysis of the polystyrene from their new polymerization process. There is a hushed silence as the first sample is injected into the GPC column. Suddenly, the chart pen begins to move, and to everyone's amazement, the GPC curve in Figure 5.11 emerges:
Not knowing what to make of this, they call you in as a consultant. Their GPC operates at 22 °C with THF solvent. Its calibration curve is identical to Figure 5.10.

(a) Sketch the number-fraction molecular weight distribution for their sample, including quantitative molecular weight extremes.

(b) Calculate $\bar{M}_n$ and $\bar{M}_w$ for their sample. You can do this approximately using the discrete equations but really impress them and do it exactly by treating the continuous distribution.

15 Consider the polymer mixture formed in Example 5.2. Each component of the mixture follows the distribution of chain lengths given in Example 5.5. For simplicity, the molecular weight of a repeat unit may be taken as $m = 100$. What are the mole fractions of material with $x = 10$ (n_{10}/N) and (w_{10}/W) in the mixture?

REFERENCES

[1] Martin, J.R., J.F. Johnson, and A.R. Cooper, *J. Macromol. Sci. Rev.* **C8**(1), 57 (1972).

[2] Nunes, R.W., J.R. Martin, and J.F. Johnson, *Polym. Eng. Sci.* **22**(4), 193 (1982).

[3] Collins, E.A., *et al.*, *Experiments in Polymer Science*, Wiley, New York, 1973.

[4] Burge, D.E., *American Laboratory Raw data from GPC Chromatogram of Polystyrene* **9**(7), 41 (1977).

[5] Billmeyer, F., *Textbook of Polymer Science*, 2nd ed., Wiley, New York, 1971, Chapter 3.

[6] Kurata, M. and Y. Tsunashima, Viscosity-molecular weight relationships and unperturbed dimensions of linear chain molecules, Chapter VII/1 in *Polymer Handbook*, 3rd ed., J Brandrupand E.H. Immergut (eds), Wiley, New York, 1975.

[7] Provder, T. (ed.), *Size Exclusion Chromatography: Methodology and Characterization of Polymers and Related Materials*, ACS Symposium Series No. 245, American Chemical Society, Washington, DC, 1984.

[8] Grubisic, Z. *et al.*, *J. Polym. Sci.* **B-5**, 753 (1967).

[9] Balke, S.T., *et al.*, *Ind. Eng. Chem., Prod. Res. Dev.* **8**, 54 (1969).

[10] Yau, W.W. and S.W. Rementer, *J. Liquid Chromatogr.* **13**(4), 626; Wang, P.J. and B.S Glassbrenner, *ibid,* **11**(16) 3321 (1988).

CHAPTER 6

THERMAL TRANSITIONS IN POLYMERS

6.1 INTRODUCTION

The reader should be familiar with thermal transitions, particularly the melting (or freezing) and the boiling points of pure substances. Polymers are a bit more complex; their size leads to the definition of a new term, the *glass transition temperature,* which divides glassy from rubbery behavior. The polydispersity of polymeric samples also leads to melting point ranges. Alternately, the boiling point is largely unimportant for polymers, since they degrade well before macromolecules vaporize. This chapter focuses on some of the unique thermal behavior observed in polymers and explains why polymers have a thermal history that is crucial in determining physical properties.

6.2 THE GLASS TRANSITION

It has long been known that amorphous polymers can exhibit two distinctly different types of mechanical behavior. Some, such as poly(methyl methacrylate) (PMMA), trademarked as Lucite® or Plexiglas®, and polystyrene (PS), are hard, rigid, *glassy* plastics at room temperature. Other polymers, for example, polybutadiene, poly(ethyl acrylate), and polyisoprene, are soft, flexible rubbery materials. If, however, PS and PMMA are heated to around 125 °C, they exhibit typical rubbery properties; when a rubber ball is cooled in liquid nitrogen, it becomes rigid and glassy and shatters when an attempt is made to bounce it. So, there is some temperature, or narrow range of temperatures, below which an amorphous polymer is in a glassy state and above which it is rubbery. This temperature is known as the *glass transition temperature,* T_g. The glass transition temperature is a property of the polymer, and whether the polymer has glassy or rubbery properties depends on whether its application temperature is above or below its glass transition temperature.

Fundamental Principles of Polymeric Materials, Third Edition. Christopher S. Brazel and Stephen L. Rosen.
© 2012 John Wiley & Sons, Inc. Published 2012 by John Wiley & Sons, Inc.

Note that the T_g is a property of the amorphous regions of polymers; however, since no polymer is 100% crystalline, the T_g is important for all polymeric materials.

6.3 MOLECULAR MOTIONS IN AN AMORPHOUS POLYMER

To understand the molecular basis for the glass transition, the various molecular motions occurring in an amorphous polymer mass may be broken into four categories.

1. Translational motion of entire molecules that permits flow.
2. Cooperative wriggling and jumping of segments of molecules approximately 40–50 carbon atoms in length, permitting flexing and uncoiling, that lead to elasticity.
3. Motions of a few atoms along the main chain (five or six, or so) or of side groups on the main chains.
4. Vibrations of atoms about equilibrium positions, as occurs in crystal lattices, except that the atomic centers are not in a regular arrangement in an amorphous polymer.

Motions 1–4 above are arranged in order of decreasing activation energy (as well as decreasing number of atoms involved in the motion), that is, smaller amounts of thermal energy (kT^1) are required to produce them. The glass transition temperature is thought to be that temperature at which motions 1 and 2 are pretty much "frozen out," and there is only sufficient energy available for motions 3 and 4. Of course, not all molecules possess the same energies at a given temperature. The molecular energies follow a Boltzmann distribution, and even below T_g, there will be occasional type 2 and even type 1 motions, which can manifest themselves over extremely long periods of time.

6.4 DETERMINATION OF T_g

How is the glass transition temperature studied? A common method is to observe the variation of some thermodynamic property with T, for example, the specific volume, as shown in Figure 6.1. As temperature rises, the polymer expands, with a change to a higher slope in the v versus T plot above the glass transition temperature.

The value of T_g determined in this fashion will vary somewhat with the rate of cooling or heating. This reflects the fact that long, entangled polymer chains cannot respond instantaneously to changes in temperature and illustrates the difficulty in making thermodynamic measurements on polymers. It often takes an extremely long time to reach equilibrium, if indeed it is ever reached, and it is difficult to be sure if and when it is reached. Strictly speaking, the glass transition temperature should be defined in terms of equilibrium properties or at least those measured with very low rates of temperature changes. Also, a sharp "break" in the property is never observed, but T_g can always be established within a couple of degrees by extrapolation of the linear regions (as shown in Figure 6.1). Other properties such as refractive index may also be used to establish T_g.

---

[1] kT is the product of the Boltzmann constant and absolute temperature. It is related to energy on a molecular level and is discussed in more detail in physical chemistry.

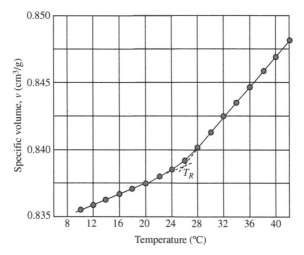

FIGURE 6.1 Specific volume (v) versus temperature for poly(vinyl acetate) [1].

In contrast to a change in *slope* at the glass transition, a thermodynamic property such as specific volume exhibits a *discontinuity* with temperature at the crystalline melting point in polymers as in other materials (Figure 6.2). The glass transition is therefore known as a *second-order* thermodynamic transition (where v versus T is continuous and dv/dT versus T is discontinuous) in contrast to a first-order transition such as the melting point (where v versus T is discontinuous).

T_g characterizes the amorphous phase. Since all polymers have at least some amorphous material (they cannot be 100% crystalline), they all have a T_g, but not all polymers have a crystalline melting point, they cannot have if they do not crystallize (and many polymers will degrade before they melt).

Transitions in polymers are rapidly and conveniently studied using differential scanning calorimetry (DSC) [3]. Small samples ($\sim 10\,mg$) of the polymer and an inert reference

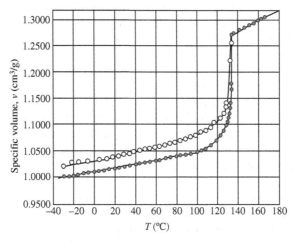

FIGURE 6.2 Specific volume–temperature relations for linear polyethylene (Marlex-50). ○: Specimen slowly cooled from melt to room temperature prior to fusion. ●: Specimen crystallized at 130 °C for 40 days, then cooled to room temperature prior to fusion [2].

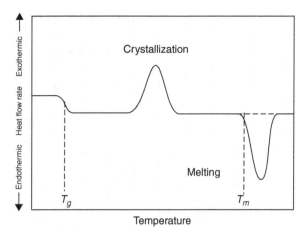

FIGURE 6.3 Schematic DSC curve. This is what would be observed on heating a material from state 2 (Example 6.8, Figure 6.5) to state 1.

substance (one that undergoes no transitions in the temperature range of interest) are mounted in a block with a heater for each and thermocouples to monitor temperatures. The thermodynamic property monitored here is the enthalpy. The power supplied to each heater is monitored and adjusted to keep the sample and the reference at the same temperature as both are heated at a programmed rate (typically 5–20 °C/min). At T_g, the heat capacity of the sample suddenly increases, requiring more power (relative to the reference) to maintain the same temperatures. This differential heat flow to the sample (endothermic) causes a drop in the DSC curve (Figure 6.3). At T_m, the sample crystals want to melt at constant temperature, so a sudden input of large amounts of heat is required to keep the sample temperature even with the reference temperature. This results in the characteristic endothermic melting peak. Crystallization, in which large amounts of heat are given off at constant temperature, gives rise to a similar but exothermic peak (although this peak is missing for many polymers that either do not crystallize appreciably or are already crystalline before heating). By measuring the net energy flow to or from the sample, heat capacities and heats of fusion can be determined. Decomposition (endothermic) and oxidation (exothermic) reactions can also be conveniently studied (as can exothermic polymerization reactions, as discussed later).

In Figure 6.3, the crystallization and melting peaks are shown as occurring over a temperature range, and the rates of heating (or cooling) can broaden (for higher dT/dt rates) or narrow (for slow heating or cooling) these peaks. However, even with a very slow temperature ramp, the melting transition will occur over a range of temperatures rather than being well defined at a single temperature (e.g., compared with a pure low molecular weight substance, say ice, polymer samples are polydisperse, having a range of crystallite sizes and having entanglements and complex secondary interactions that cause broad melting peaks).

A DSC is programmed to heat the sample at a constant rate. The higher the rate, the quicker the measurement, a practically desirable result. Unfortunately, because polymer chains cannot respond instantaneously to the changing temperature, the measurement is further from equilibrium. The dependence of the measured T_g or T_m on heating rate is at least partially responsible for the range of values observed in the literature. To approach the true equilibrium values, very low heating rates should be used or, better yet, several heating rates should be used and the results extrapolate back to zero heating rate. Because of the time and effort involved, this is rarely done. Newer DSC models have a temperature modulation program that allows fine-tuning of these measurements to more accurately measure T_g and T_m. Dynamic mechanical measurements (see Chapter 16) can also provide useful information on thermal transitions.

6.5 FACTORS THAT INFLUENCE T_g [4]

In general, the glass transition temperature depends on five factors.

1. The *free volume* of the polymer v_f, which is the volume of the polymer mass not actually occupied by the molecules themselves (think of a kitchen sponge, but on a molecular level), that is, $v_f = v - v_s$, where v is the specific volume of the polymer mass and v_s is the volume of the solidly packed molecules. The higher v_f, the more room the molecules will have to move around, resulting in a lower T_g. It has been estimated that for all polymers $v_f/v \approx 0.025$ at T_g.

Example 6.1 Glass transition temperatures have been observed to increase at pressures of several thousand psi. Why?

Solution. High pressures compress polymers, reducing v. Since v_s does not change appreciably, v_f is reduced.

---

2. The attractive forces between the molecules—the more strongly they are bound together, the more thermal energy will be required to produce motion. The solubility parameter δ (defined explicitly in Chapter 7) is a measure of intermolecular forces, thus T_g increases with δ. Polyacrylonitrile, because of the frequent, strong polar bonding between chains, has a T_g higher than its degradation temperature and therefore, though linear, is not thermoplastic.

$$\begin{array}{c} \text{H} \ \ \text{H} \\ | \ \ \ | \\ \text{+C--C+}_x \\ | \ \ \ | \\ \text{H} \ \ \text{C}\equiv\text{N} \end{array}$$

3. The internal mobility of the chains, that is, their freedom to rotate about bonds. Figure 6.4 shows potential energy as a function of rotation angle about a bond in a polymer chain. For a carbon–carbon bond, there are three other bonds to each carbon (assuming the backbone does not have double bonds): two will be pendent groups and the third will be the continuation of the polymer backbone. The minimum energy configuration, arbitrarily chosen as $\theta_f = 0$, is the position where the largest substituents (the rest of the chain) are far away from each other as possible. As the bond is rotated, the substituent groups are brought into juxtaposition, and energy is required to "push them over the hump." The maximum energy is needed to get the two chain substituents past one another, and this energy must be available if complete rotation is to be obtained. Rotation is necessary for type 1 and type 2 motions.

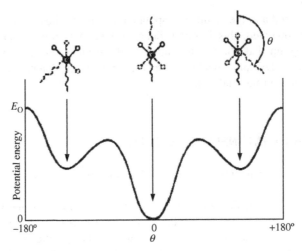

FIGURE 6.4   Rotation about a bond in a polymer chain backbone, viewed along the bond. The dotted substituents are on the rear carbon atom.

TABLE 6.1   Effect of Potential Energy on T_g for Selected Polymers [4]

Polymer	δ	T_g (°C)	E_o (kcal/mol)
Polydimethyl siloxane	7.3	−120	∼0
Polyethylene	7.9	−85(?)	3.3
Polytetrafluoroethylene	6.2	>20	4.7

Table 6.1 shows how T_g increases with E_o, the potential energy, for a series of polymers with approximately the same δ. Note how the ether oxygen "swivels" in the silicone chain permitting very free rotation.

Example 6.2 Poly(α-methyl styrene) has a higher T_g than PS. Why?

H H
+C−C+
H ϕ
Polystyrene

H CH$_3$
+C−C+
H ϕ
Poly(x-methylstyrene)

Solution. The methyl group introduces extra steric hindrance to rotation, giving a higher E_o.

---

4. The stiffness of the chains—chains that have difficulty coiling and folding will have higher T_g values. This stiffness usually goes hand-in-hand with high E_o, therefore it is difficult to separate the effects of 3 and 4.

Chains with parallel bonds in the backbone (ladder-type polymers), for example, polyimides (Example 2.4R), and those with highly aromatic backbones, such as aramids (see Chapter 4.8) have extremely stiff chains and, therefore, tend to have high T_g values. This makes these polymers mechanically useful at elevated temperatures but also very difficult to process.

5. The chain length—as do many mechanical properties of polymers, the glass transition temperature varies according to the empirical relation:

$$T_g = T_g^\infty - \frac{C}{x} \tag{6.1}$$

where C is a constant for the particular polymer, and T_g^∞ is the asymptotic value of the glass transition temperature for infinite chain length. This reflects the increased ease of motion for shorter chains. The decrease in T_g with x is only noticeable at relatively low chain lengths. For most commercial polymers, x is high enough so that $T_g \approx T_g^\infty$.

Example 6.3 Measurements such as those described above show that the addition of a small molecular weight chemical known as an external plasticizer (see Chapter 7) softens a polymer by reducing its glass transition temperature. Explain.

Solution. The plasticizer molecules pry apart the polymer chains, in essence increasing the free volume available to the chains (although not truly free, the small plasticizer molecules interfere with chain motions much less than would other chains). Also, by forming secondary bonds with the polymer chains themselves, type 1 and type 2 motions are easier.

6.6 THE EFFECT OF COPOLYMERIZATION ON T_g

The glass transition temperatures for random copolymers vary monotonically with composition between those of the homopolymers. They can be approximated fairly well from knowledge of the T_g values of the homopolymers, T_{g1} and T_{g2}, with the empirical relation:

$$\frac{1}{T_g} = \frac{w_1}{T_{g1}} + \frac{w_2}{T_{g2}} \tag{6.2}$$

where the w's are weight fractions of the monomers in the copolymer. This relation forms the basis for a method of estimating the T_g values of highly crystalline polymers, where the properties of the small amount of amorphous material are masked by the majority of crystalline material present. If a series of random copolymers can be produced in which the randomness prevents crystallization over a certain composition range, then Equation 6.2 can be used to extrapolate to $w_1 = 1$ or $w_2 = 1$, giving the T_g values of the homopolymers. This method is open to question because it assumes that the presence of major amounts of crystallinity does not restrict the molecular response in the amorphous regions. In fact, the T_g values of highly crystalline polymers (polyethylene, in particular) are still open to debate.

6.7 THE THERMODYNAMICS OF MELTING

The crystalline melting point T_m in polymers is a phase change similar to that observed in low molecular weight organic compounds, metals, and ceramics.

The Gibbs free energy of melting is given by

$$\Delta G_m = \Delta H_m - T \Delta S_m \tag{6.3}$$

At the equilibrium crystalline melting point, T_m, $\Delta G = 0$, therefore

$$T_m = \frac{\Delta H_m}{\Delta S_m} \tag{6.4}$$

Now ΔH_m is the energy needed to overcome the crystalline bonding forces at constant T and P, and is essentially independent of chain length of high polymers. For a given mass or volume of polymer, however, the shorter the chains are, the more randomized they become upon melting, giving a higher change in entropy on mixing, ΔS_m. (For a more detailed description of this in connection with solutions, see Chapter 7.) Thus, the crystalline melting point decreases with decreasing chain length and in a polydisperse polymer, the distribution of chain lengths give a distribution of melting points.

Equation 6.4 also indicates that chains that are strongly bound in the crystal lattice, that is, have a high ΔH_m, will have a high T_m, as expected. Also, the stiffer and less mobile chains, those that can randomize less upon melting and therefore have low ΔS_m, will tend to have higher T_m values.

Example 6.4 Discuss how the crystalline melting point varies with n in the "nylon n" series, where n can be varied.

$$\begin{matrix} H & O \\ | & || \\ \end{matrix}$$
$$\{ N-C(CH_2)_{n-1} \}_x$$

Solution. Increasing values of n dilute the nylon linkages that are responsible for interchain hydrogen bonding, and thus should lower ΔH_m and the crystalline melting point. As n goes to infinity, the structure approaches that of linear polyethylene. This should represent the asymptotic minimum T_m, with the chains held together only by van der Waal's forces.

Table 6.2 illustrates the variation in T_m and some other properties with n for some commercial members of the nylon series.

TABLE 6.2   Variation of Properties with n for Nylon n's

n	T_m (°C)	ρ (g/cm^3)	Tensile Strength (psi)	Water Absorption % in 24 h
6	216	1.14	12,000	1.7
11	185	1.04	8000	0.3
12	177	1.02	7500	0.25
∞ (PE)	135	0.97	5500	nil

Example 6.5 Consider the following classes of linear, aliphatic polymers:

$$\{O-\overset{\overset{\displaystyle O}{\|}}{C}-\underset{\underset{\displaystyle H}{|}}{N}-(CH_2)_n\}_x \qquad \{\overset{\overset{\displaystyle O}{\|}}{C}-\underset{\underset{\displaystyle H}{|}}{N}-(CH_2)_n\}_x \qquad \{\underset{\underset{\displaystyle H}{|}}{N}-\overset{\overset{\displaystyle O}{\|}}{C}-\underset{\underset{\displaystyle H}{|}}{N}-(CH_2)_n\}_x$$

Polyurethanes		Polyamides		Polyureas
T_m	$<$	T_m	$<$	T_m

For given values of n and x, the crystalline melting points increase from left to right, as indicated. Explain.

Solution. The polyurethane chains contain the $-O-$ swivel, thus they are the most flexible, having the largest ΔS_m, and having the lowest T_m. Hydrogen bonding, and thus ΔH_m, is roughly comparable in the polyurethanes and polyamides. The polyureas and polyamides should have chains of comparable flexibility (no swivel), but with the extra $-NH-$, the polyureas form stronger or more extensive hydrogen bonds, and therefore have a higher ΔH_m than the polyamides.

Example 6.6 Experiments show that uniaxial orientation (drawing) increases the crystalline melting point. Explain.

Solution. The entropy of melting, ΔS_m, is the difference between the entropies of the amorphous and the crystalline materials:

$$\Delta S_m = S_a - S_c$$

drawing align the amorphous chains in the direction of stretch, increasing order, reducing S_a and, thus, ΔS_m. According to Equation 6.4, this increases T_m.

Example 6.7 Liquid-crystalline polyesters (Section 4.8) have significantly higher T_m values than non-LC polyesters. For example, Xydar has a T_m value in the vicinity of 400 °C, whereas poly(ethylene terephthalate), shown in Example 2.4D, has a T_m of 267 °C and poly(butylene terephthalate), shown in Example 2.4C, has a T_m of 224 °C. Explain.

Solution. Here, one might expect the ordinary polyesters to have higher ΔH_m's, particularly the PET, because of the decreased spacing between the polar ester linkages, so the explanation must lie in the ΔS_m's. Because thermotropic LCPs by definition maintain considerable order in the molten state, they randomize less upon melting and therefore have lower S_m's than non-LC polyesters.

The crystalline melting point also increases a bit with the degree of crystallinity of a polymer. For example, low-density polyethylene (approximately 50% crystalline) has a T_m of about 115 °C, whereas high-density polyethylene (approximately 80% crystalline) melts at about 135 °C. This can be explained by treating the amorphous material as an impurity. It is well known that introducing an impurity lowers the melting point of common materials.

In a similar fashion, greater amounts of noncrystalline "impurities" lower the crystalline melting point of a polymer.

6.8 THE METASTABLE AMORPHOUS STATE

Since polymer chains are largely immobilized below T_g, if they are cooled rapidly through T_m to below T_g, it is sometimes possible to obtain a metastable amorphous state in polymers that would be crystalline at equilibrium. This rapid cooling effectively locks the chains in a random, amorphous state. As long as the material is held below T_g, this metastable amorphous state persists indefinitely. When annealed above T_g (and below T_m), the polymer crystallizes, as the chains gain the mobility necessary to pack into a lattice. This behavior leads to classifying polymers as materials that have a *thermal history,* indicating that the rate of cooling affects crystallinity (and other properties).

Poly(ethylene terephthalate), from Example 2.4D, because of its bulky chain structure, crystallizes sluggishly and is therefore relatively easy to obtain in the metastable amorphous state. When desired, crystallinity can be promoted by slow cooling from the molten state, annealing between T_g and T_m, or by the addition of nucleating agents. PET has important commercial applications both in the metastable amorphous state (soda bottles) and in the crystalline state (textile fibers, microwaveable food trays, molding resin). On the other hand, no one has yet succeeded in producing an amorphous polyethylene, with its much more flexible chains, although the degree of crystallinity can be reduced substantially by rapid cooling (Figure 6.2). Interestingly, metallurgists can also produce amorphous metals by rapidly cooling certain alloys.

Example 6.8 Poly(ethylene terephthalate), or PET (Mylar, Dacron), is cooled rapidly from 300 °C (state 1) to room temperature. The resulting material is rigid and perfectly transparent (state 2). The sample is then heated to 100 °C (state 3) and maintained at that temperature, during which time it gradually becomes translucent (state 4). It is then cooled down to room temperature and is again found to be rigid, but is now translucent rather than transparent (state 5). For this sample of PET, $T_{mf} = 267$ °C and $T_{gf} = 69$ °C. Sketch a general specific-volume versus temperature curve for a crystallizable polymer, illustrating T_g and T_m, and show the locations of states 1–5 for this PET sample.

Solution. Figure 6.5 illustrates the general v versus T curve for a crystallizable polymer. The dotted upper portion represents the metastable amorphous material obtainable by rapid cooling. The history of the PET sample is shown in the diagram. The metastable amorphous material (transparent, state 2) is obtained by rapid cooling to below T_g; however, when heated to state 3 (between the T_g and T_m), the sample undergoes the process of annealing, becoming more crystalline. The schematic DSC curve in Figure 6.3 illustrates what would be observed in a DSC measurement starting with a material in the metastable amorphous state (state 2) and heating it to above T_m (state 1).

Note the change in slope at the glass transition temperature. Also, as long as state 3 is above T_g, the chains will have enough mobility to rearrange (eventually) into a more stable semicrystalline state. Thus (comparing states 2 and 5), this information is not enough to simply state the molecular weight of PET and temperature to determine the properties; the thermal history influences crystallinity that can have a profound effect on the properties of the material (from density to optical properties and mechanical behavior). Thus, this is a

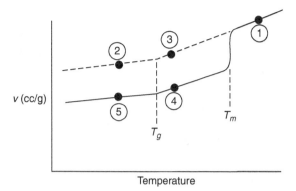

FIGURE 6.5   Specific volume–temperature relation for crystallizable polymers. Numbers apply to Example 6.8.

great example of why polymers have a *thermal history*. State 5 could also have been reached by slowly cooling the PET sample from state 1.

Polymer *blends*, which are mixtures of two or more different homopolymers, are also commonly used in commercial materials. Except in the rare case where interpolymer crystallites can form, blends exhibit multiple T_g's and T_m's, one for each component of the blend.

6.9 THE INFLUENCE OF COPOLYMERIZATION ON THERMAL PROPERTIES

Compared with polymer blends, copolymers have only one T_g and one T_m. The influence of random copolymerization on T_m and T_g is interesting and technologically important. Occasionally, if two repeating units are similar enough sterically to fit into the same crystal lattice, random copolymerization will result in copolymers whose crystalline melting points vary linearly with composition between those of the pure homopolymers. Much more common, however, is the case where the homopolymers form different crystal lattices because of steric differences. The random incorporation of minor amounts of repeating unit B with A will disrupt the A lattice, lowering T_m beneath that of homopolymer A, and vice versa. In an intermediate composition range, the disruption will be so great that no crystallites can form and the copolymers will be completely amorphous. A phase diagram for such a random copolymer system is shown in Figure 6.6.

The physical properties of random copolymers are determined from their composition and the temperature, as can be found in a diagram such as the one qualitatively shown in Figure 6.6. In region 1, the polymer is a homogeneous, amorphous, and, if pure, transparent material. The distinction between melt and rubbery behavior is not sharp; at higher temperatures, the material flows more easily and becomes less elastic in character (assuming that the polymer does not reach degradation temperatures). It should be kept in mind, though, that the viscosities of polymer melts, even well above T_g or T_m, are far greater than those encountered in nonpolymeric materials.

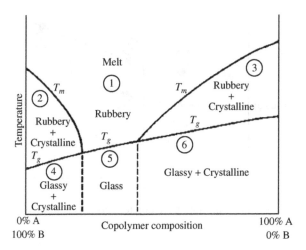

FIGURE 6.6 Phase diagram for a random copolymer system.

A copolymer in region 5 is a typical amorphous, glassy polymer: hard, rigid, and usually brittle. Again, if the polymer is pure, it will be perfectly transparent. PMMA (Lucite, Plexiglas) and PS are familiar examples of homopolymers with these properties.

Copolymers in regions 2 and 3 consist of rigid crystallites dispersed in a relatively soft, rubbery, amorphous matrix. Since the refractive indices of the crystalline and amorphous phases are in general different, materials in these regions will be translucent to opaque, depending on the size of the crystallites, the degree of crystallinity, and the thickness of the sample. Since the crystallites restrict chain mobility, the materials are not elastic, but the rubbery matrix confers flexibility and toughness. The stiffness depends largely on the degree of crystallinity; the more rigid crystalline phase present, the stiffer the polymer. Polyethylene (e.g., flexible squeeze bottles or rigid bleach bottles) is a good example of a homopolymer with these properties, typical of a polymer that is between its T_g and T_m at room temperature.

Copolymers in regions 4 and 6 consist of crystallites in an amorphous, glassy matrix. Since both phases are rigid, the materials are hard, stiff, and rigid. Again, the two phases impart opacity. Nylon 6/6 and nylon 6 are examples of homopolymers in a region below both T_g and T_m at room temperature.

6.10 EFFECT OF ADDITIVES ON THERMAL PROPERTIES

A number of different compounds are used as polymer additives to tailor the properties of materials for their end-use. As discussed in detail in Chapter 18, these additives include dyes, fillers, plasticizers, and a variety of other compounds. Because each of these additives gets between polymer chains (at least to some extent), they reduce the number of polymer–polymer interactions and cause an increase in chain mobility (when compared to the pure polymer at the same temperature). This causes a drop in T_g (which is exactly what a plasticizer is designed to do, Figure 6.7). This decrease in T_g depends on the thermodynamic compatibility of the additive with the polymer. For some fillers, this may be quite low, causing little change in T_g, but for chemical additives that interact with the polymer, T_g can be reduced significantly. The thermodynamics of polymer solutions and polymer

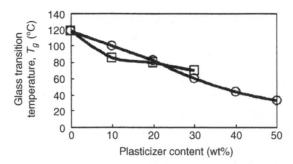

FIGURE 6.7   Effect of adding dioctyl phthalate (DOP) (a plasticizer), squares, or imidazolium ionic liquid additives (hmimPF6), circles, to PMMA on T_g [5]. Reproduced by permission of the Royal Society of Chemistry.

additives are discussed in detail in the next chapter. The melting point of a polymer with an additive will not be influenced greatly, but the degree of crystallinity (and crystal sizes) will drop markedly, as the additive will interfere in the formation of regular crystals.

6.11   GENERAL OBSERVATIONS ABOUT T_g AND T_m

Some other useful observations regarding T_g and T_m are that for polymers with a symmetrical repeating unit, such as polyethylene $-(CH_2-CH_2)-$ and poly(vinylidene chloride), or Saran®, $-(CH_2-CCl_2)-$, $T_g/T_m \approx 1/2$ (for absolute T's); for unsymmetrical repeating units, such as polypropylene $-(CH_2-CHCH_3)-$ and polychlorotrifluoroethylene $-(CF_2-CFCl)-$, $T_g/T_m \approx 2/3$. These are just rough estimates, but in all cases $T_g < T_m$.

6.12   EFFECTS OF CROSSLINKING

To this point, the discussion has centered on non-crosslinked (and generally linear, rather than branched) polymers. Light crosslinking, as in rubber bands, will not alter things appreciably. However, higher degrees of crosslinking, if formed in the amorphous molten state or in the solution, as is usually the case, will hinder the alignment of chains necessary to form a crystal lattice and will thus reduce or prevent crystallization. Similarly, cross-linking restricts chain mobility and causes an increase in the apparent T_g. When the crosslinks are more frequent than every 40–50 main chain atoms, the type 2 motions necessary to reach the rubbery state can never be achieved and the polymer will degrade before reaching T_g.

6.13   THERMAL DEGRADATION OF POLYMERS

As mentioned earlier in this chapter, several polymers start to degrade before reaching a melting point. All polymers eventually degrade given enough heating. The degradation process is sort of a depolymerization, except the bonds break randomly along the backbone, resulting in a smaller $\bar{M}_n$ and $\bar{M}_w$ (and higher PI). The reductions in the molecular weight also degrade the mechanical properties of the polymers. In many applications, additives are used to stabilize polymers that may be used at high temperatures. However, anyone who has microwaved a plastic bowl to the point of deformation understands that polymers can degrade appreciably even at moderate temperatures, especially when compared to glass or metal materials.

PVC is an example of a polymer that self-catalyzes its thermal degradation. Beginning at temperature around 160 °C, HCl is formed as a byproduct of thermal degradation, and, even worse, the acid catalyzes further degradation of PVC. This is why PVC is not used for high-temperature applications, and even for moderate temperatures, a thermal stabilizer is normally added.

Thermal degradation is one of the primary challenges in effective polymer recycling. Although thermoplastics can be melted and reused, uneven temperature distributions (or the heat evolved while grinding polymers) can cause some degree of thermal degradation, reducing $\bar{M}_n$ and $\bar{M}_w$. This is partly why recycled polymers are most often used in applications where high mechanical strength is less important, such as park benches and grocery bags.

6.14   OTHER THERMAL TRANSITIONS

Thermal transitions other than T_g and T_m are sometimes observed in polymers. Some polymers possess more than one crystal form, so there will be an equilibrium temperature of transition from one to another. Similarly, second-order transitions below T_g occur in some materials (T_g is then termed the α transition, the next lower is the β transition, and so forth). These are attributed to motions of groups of atoms smaller than those necessary to produce T_g (type 3 motions, Section 6.2). These transitions may strongly influence properties. For example, *tough* amorphous plastics (e.g., polycarbonate) have such a transition well below room temperature, while *brittle* amorphous plastics (e.g., PS and PMMA) do not.

The existence of another transition above T_g has been claimed, but is still the subject of considerable controversy. This T_{ll} (*liquid–liquid* transition) presumably represents the boundary between type 1 and type 2 motions. It has been observed in a number of systems [6–8], and it has been suggested that $T_{ll} \approx 1.2\, T_g$ (in absolute temperature) for all polymers [6]. For each article that reports T_{ll}, however, it seems that there is another that claims T_{ll} results from impurities (traces of solvent or unreacted monomer) in the sample or is an artifact of the experimental or data-analysis technique [9,10].

PROBLEMS

1   Pure nylon 6/6 (see Example 2.4E for the structure) is a white, opaque plastic at room temperature. If isophthalic acid (HOOC$-\phi-$COOH, with the acids in the *meta* positions, on the first and third carbons) is substituted for adipic acid in the polymerization, a rigid, transparent plastic is obtained. By varying the ratio of the two acids, one can get a series of polyamide copolymers. Sketch a phase diagram like Figure 6.6 for these copolymers. Be sure to show which repeating unit goes with each composition extreme and also show the room temperature.

2   Three DSC runs are made on a semicrystalline polymer sample starting at room temperature and passing through the glass-transition temperature and melting point.

Three different heating rates are used: 1, 5, and 20 °C/min. Sketch qualitatively the expected DSC thermograms to show how you think the observed T_g and T_m will vary with heating rate.

3 Injection molding consists of squirting a molten polymer into a cold metal mold. When thick parts are molded from a crystallizable polymer (e.g., polypropylene), they sometimes exhibit "sink marks," where the surface of the part has actually sunk away from the mold wall.

(a) Explain why. *Hint*: Polymers have very low thermal diffusivity.

(b) How would an amorphous (non-crystallizable) polymer perform in injection molding?

4 Two diols, ethylene glycol (Example 2.4D) and bisphenol-A (Example 2.4N), are commercially available at low cost. Which would you choose for polyesterification with a diacid if your objective was to:

(a) produce a transparent polyester and

(b) obtain the higher T_g.

5 A patent claims that a new polymer forms strong, highly crystalline parts when injection molded. Furthermore, it is claimed that $T_g > T_m$ for this material. Comment.

6 An amorphous emulsion copolymer consisting of 60 wt% methyl methacrylate (homopolymer $T_g = 105$ °C) and 40 wt% ethyl acrylate (homopolymer $T_g = -23$ °C) has been proposed as the basis of a latex paint formulation. A latex consists of tiny ($\approx 1-10\,\mu$m) polymer particles suspended in water. After application of the paint, these particles must coalesce to form a film upon evaporation of water.

(a) Would the copolymer be suitable for an outdoor paint?

(b) An actual paint formulation based on this copolymer contains some medium-volatility solvent dissolved in the latex particles. This solvent is designed to evaporate over a period of a few hours after application of the paint. What is it doing there?

7 High molecular weight linear polyesters from 1,4-butanediol (HO$-$(CH$_2$)$_4-$OH) and terephthalic acid (HOOC$-\phi-$COOH), with the acids in the *para* or opposite positions) are successful engineering plastics (materials with high mechanical strength and good thermal stability). They are, however, not used as a blister (sturdy, see-through) packaging material. A polymer for the blister-packaging market is made from the two monomers above plus isophthalic acid (*meta*$-\phi-$(COOH)$_2$). Explain the difference in the applications.

8 Polyethylene ($T_m = 135$ °C, $T_g < T_{\text{room}}$) may be lightly crosslinked by a chemical reaction with an organic peroxide at 175 °C. Heat-shrink tubing is made from such a crosslinked polyethylene. When heated at room temperature to about 150 °C, its diameter shrinks by a factor of three or four. Explain the thermal history required to make and use this material and the driving force for shrinkage when heat is applied.

9 A diagram like Figure 6.1 is prepared by heating a polymer well above its T_g and then rapidly cooling it to the desired temperature and holding it there until v is measured. Sketch v versus T curves for v measurements taken 1 min after cooling and 100 h after cooling. Illustrate how this would affect the value of T_g obtained.

10   Professor Irving Inept of MIT (Monongahela Institute of Technology) figures that he can pad his publication list by publishing a series of polymer tables. Similar to the steam tables from thermodynamics or the CRC Handbook of Chemistry and Physics, they will contain the thermodynamic properties of various polymers as functions of temperature and pressure. Discuss the difficulties associated with

(a)   obtaining the necessary data and

(b)   applying the published numbers in practice.

11   Sketch on a copy of Figure 6.5 the path of a DSC test on a materials starting in states 2 and 5 and heated to above T_m.

REFERENCES

[1]   Meares, P., *Trans. Faraday Soc.* **53**, 31 (1957).

[2]   Mandelkern, L., *Rubber Chem. Technol.* **32**, 1392 (1959).

[3]   Wunderlich, B., *Thermal Analysis*, Academic, San Diego, CA, 1990.

[4]   Tobolsky, A.B., *Properties and Structure of Polymers*, Wiley, New York, 1960, Chapter 2.

[5]   Scott, M.P., M. Rahman, and C.S. Brazel, *Eur. Polym. J.* **39**, 1947 (2003).

[6]   Kumar, P.L., *et al.*, *Org. Coat. Plast. Chem.* **44**, 396 (1981).

[7]   Ibar, J.P., *Polym. Prepr.* **22**(2), 405 (1981).

[8]   Boyer, R.F., *Macromolecules* **15**(6), 1498 (1982).

[9]   Plazek, D.J., *et al.*, *J. Polym. Sci., Polym. Phys. Ed.* **20**(9), 1533, 1551 1565, 1575 (1982).

[10]   Loomis, L.D. and P. Zollar, *J. Polym. Sci., Polym. Phys. Ed.* **21**(2), 241 (1983).

CHAPTER 7

POLYMER SOLUBILITY AND SOLUTIONS

7.1 INTRODUCTION

Most people associate polymers with solid materials—from rubber bands to car tires to Tupperware®— but what about acrylic and latex paints or ketchup and salad dressing or oil-drilling fluids? The thermodynamics and statistics of polymer solutions is an interesting and important branch of physical chemistry, and is the subject of many good books and large sections of books in itself. It is far beyond the scope of this chapter to attempt to cover the subject in detail. Instead, we will concentrate on topics of practical interest and try to indicate, at least qualitatively, their fundamental bases. Three factors are of general interest:

1. What solvents will dissolve what polymers?
2. How do the interactions between polymer and solvent influence the properties of the solution?
3. To what applications do the interesting properties of polymer solutions lead?

7.2 GENERAL RULES FOR POLYMER SOLUBILITY

Let us begin by listing some general qualitative observations on the dissolution of the polymers:

1. *Like dissolves like*, that is, polar solvents will tend to dissolve polar polymers and nonpolar solvents will tend to dissolve nonpolar polymers. Chemical similarity between

Fundamental Principles of Polymeric Materials, Third Edition. Christopher S. Brazel and Stephen L. Rosen.
© 2012 John Wiley & Sons, Inc. Published 2012 by John Wiley & Sons, Inc.

the polymer and the solvent is a fair indication of solubility; for example, poly(vinyl alcohol) or PVA, $-(CH_2CHOH)_n-$, will dissolve in water, $H-O-H$, and polystyrene or PS, $-(CH_2CH\phi)_n-$, will dissolve in toluene, $\phi-CH_3$, but toluene will not dissolve PVA and water will not dissolve PS (which is really good news for those who drink coffee out of foamed PS cups). Most polymers will also dissolve in their monomer.

2. In a given solvent at a particular temperature, the solubility of a polymer will decrease with increasing molecular weight.

3. (a) Crosslinking eliminates solubility (swelling is the best the polymer can do); only linear or branched polymers can be dissolved.

   (b) Crystallinity, in general, acts like crosslinking, but it is possible in some cases to find solvents strong enough to overcome the crystalline bonding forces and dissolve the polymer. Heating the polymer toward its crystalline melting point also leads to solubility in appropriate solvents. For example, nothing dissolves polyethylene at room temperature. At $100\,^{\circ}$C, however, it will dissolve in a variety of aliphatic, aromatic, and chlorinated hydrocarbons.

4. The rate of polymer dissolution in a solvent decreases with increasing molecular weight. For reasonably high molecular weight polymers, it can be orders of magnitude slower than for nonpolymeric solutes (and could take days for highly entangled polymer chains to eventually dissolve).

It is important to note here that items 1, 2, and 3 are *equilibrium* phenomena and can be described thermodynamically (at least in principle), while item 4 is a *rate* (or kinetic) phenomenon and is governed by the rates of disentanglement and diffusion of polymer chains in the solvent.

Example 7.1 The polymers of ω-amino acids are termed "nylon n," where n is the number of consecutive carbon atoms in the chain. These polymers are made from a single monomer, as these amino acids have two reactive functional groups. The general formula is

$$\begin{matrix} H & O \\ | & \| \end{matrix}$$
$$\leftidx{}{\!\!\!}{} \{N-C\{CH_2\}_{n-1}\}_x$$

The polymers are semicrystalline and will not dissolve in either water or hexane at room temperature. They will, however, reach an equilibrium level of absorption when immersed in each liquid. Describe how and why water and hexane absorption will vary with n.

Solution. Water is a highly polar liquid; hexane is nonpolar. The polarity of the nylons depends on the relative proportion of polar nylon linkages in the chains. As n increases, the polarity of the chains decrease (they become more hydrocarbon-like), so hexane absorption increases with n and water absorption decreases.

$$\begin{matrix} H & O \\ | & \| \end{matrix}$$
$$\{N-C\}$$

Example 7.2 Dry-cleaning companies use starch (a natural polymer) to remove wrinkles and impart stiffness to clothes (similar to ironing at home). The dry-cleaning process adds starch in a solution of perchloroethylene (C_2Cl_4), which is a good solvent for starch to the clothes during cleaning and before pressing with a hot iron.

(a) Explain how starch works in this process, based on the thermal and solution properties of polymers. Discuss the effects of a solvent and heat on the glass transition temperature.

(b) Because starch can be hydrolyzed in humid environments, and is not particularly mechanically tough, wrinkles begin to reappear in clothes as they are worn. A scientist at Crud Chemicals has come up with a new idea that will prevent clothes from rewrinkling. The new idea will replace the aqueous starch used in ironing with an aqueous solution of PS, which does not degrade and has a higher glass transition temperature. Why will the new idea not work?

Solution. (a) Starch is a polymer based on a hexose repeat unit (glucose, specifically) with the chemical formula: $C_6H_{12}O_6$.

Starch is made into a solution with perchloroethylene when it is sprayed onto clothing. At this point, it is above its T_g, which is lowered considerably by being in a good solvent. This means that the starch is flexible and can be worked into the clothing fibers. When heat is applied by an iron, the solvent evaporates, causing the T_g to rise. By pressing the iron over the surface, the starch aligns with the direction of motion, and since the T_g is now higher than room temperature, it becomes rigid and glassy, fixing the clothes without wrinkles.

(b) It is not a bad idea; many clothing companies stitch more rigid fibers into their "wrinkle-free" clothing lines, but the idea of ironing using a dispersion of PS in water has one serious flaw: *PS is insoluble in water.* PS will form globules or hard spheres in water (and the T_g will be unchanged since the water is not a solvent and cannot decrease T_g as it does with starch) and when ironed into a fabric, the microspheres of styrene will simply be pressed into spots on the fabric, but it would be ineffective at pressing the shirt to remove wrinkles.

7.3 TYPICAL PHASE BEHAVIOR IN POLYMER–SOLVENT SYSTEMS

Figure 7.1 shows schematically a phase diagram for a typical polymer–solvent system, plotting temperature versus the polymer fraction in the system (essentially a T-x diagram from thermodynamics). At low temperatures, the typical polymer does not dissolve well in many solvents (although the dissolution process can take time, since polymer chains must be disentangled to form a uniform solution). The diagram shows a two-phase system at low temperatures but a completely miscible system at higher temperatures. In the lower temperature phase envelope, the dotted tie lines connect the compositions of phases in equilibrium: a solvent-rich phase (dilute solution) on the left and a polymer-rich (swollen polymer or gel) phase on the right. As the temperature is raised, interactions between the solvent and the polymer increase, and the compositions of the phases come closer together, until at the upper critical solution temperature (UCST) they are identical. Above the UCST, the system forms homogeneous (single-phase) solutions across the entire composition range. The location of the phase boundary depends on the interaction between the polymer and the solvent as well as the molecular weight of the polymer.

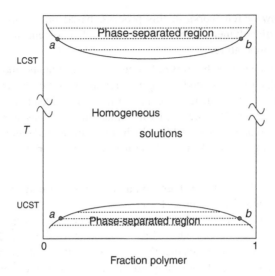

FIGURE 7.1 Schematic phase diagram for polymer–solvent system: (a) dilute solution phase; (b) swollen polymer or "gel" phase.

In recent years, a number of systems have been examined that also exhibit a lower critical solution temperature (LCST), as shown at the top of Figure 7.1. Here, a more unusual thermodynamic property is exhibited when the polymer phase separates from the solvent after being heated. (One might question the nomenclature the puts the LCST above the UCST, but that is the accepted definition.) LCSTs are more difficult to observe experimentally because they often lie well above the normal boiling points of the solvents.

When we talk about a polymer being soluble in a particular solvent, we generally mean that the system lies between its LCST and UCST, that is, it forms a homogeneous solution over the entire composition range. Keep in mind, however, that homogeneous solutions can still be formed toward the extremes of the composition range below the UCST and above the LCST.

7.4 THE THERMODYNAMIC BASIS OF POLYMER SOLUBILITY

"To dissolve or not to dissolve. That is the question." (with apologies to W.S.). The answer is determined by the sign of the Gibbs free energy. Consider the process of mixing pure polymer and pure solvent (state 1) at constant pressure and temperature to form a solution (state 2):

$$\Delta G = H - T\Delta S \tag{7.1}$$

where

$\Delta G =$ the change in Gibbs free energy upon mixing

$\Delta H =$ the change in enthalpy due to mixing

$T =$ absolute temperature

$\Delta S =$ the change in entropy due to mixing

If only ΔG is negative, will the solution process be thermodynamically feasible. The absolute temperature must be positive by definition and the change in entropy for a solution process is generally positive (the chains become more disordered when dissolved in solution). One possible exception is for lyotropic liquid-crystal materials, where the chains organize into crystals when in solution. The positive product ($T\Delta S$) is preceded by a negative sign. Thus, the third term ($-T\Delta S$) in Equation 7.1 favors solubility. The change in enthalpy (ΔH) may be either positive or negative. A positive ΔH means that the solvent and polymer "prefer their own company," that is, the pure materials are in a lower energy state than the mixed solution, while a negative ΔH indicates that the solution is in the lower energy state. If the latter situation is true, a solution is assured. Negative ΔH's usually arise where specific interactions such as hydrogen bonds are formed between the solvent and polymer molecules. But, if ΔH is positive, then $\Delta H < T\Delta S$ must be true for the polymer to be soluble (and linking this to the discussion of UCSTs, raising the temperature would be one way to achieve solubility).

One thing that makes polymers unusual is that the entropy change in forming a polymer solution is generally much smaller than that which occurs on dissolution of equivalent masses or volumes of low molecular weight solutes. The reasons for this are illustrated qualitatively on a two-dimensional lattice model in Figure 7.2. With the low molecular weight solute (say, styrene monomer, $CH_2{=}CH\phi$), the solute molecules may be distributed randomly throughout the lattice, the only restriction being that a lattice slot cannot be occupied simultaneously by two (or more) molecules. This gives rise to a large number of configurational possibilities, that is, high entropy. In a polymer solution, however, each chain segment is confined to a lattice site adjacent to the next chain segment, greatly reducing the configurational possibilities. This also gives a reasonable illustration of the challenge of dissolving polymers, as the chains must disentangle from a solid coil or blob to float freely in a good solvent. Returning to the lattice, note that for a given number of chain segments (equivalent masses or volumes of polymer), the more chains they are split up into, that is, the lower their molecular weight, the higher the entropy upon dissolution. This directly explains observation 2 in Section 7.2, the decrease in solubility with molecular weight. But in general, for high molecular weight polymers, because the $T\Delta S$ term is so small, if ΔH is positive then it must be even smaller if the polymer is to be soluble. So in the absence of specific interactions, predicting polymer solubility largely boils down to minimizing ΔH.

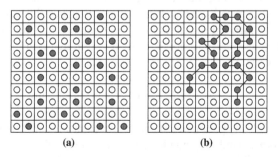

FIGURE 7.2 Lattice model of solubility: (a) low molecular weight solute; (b) polymeric solute. ○ = Solvent; ● = Solute.

7.5 THE SOLUBILITY PARAMETER

How can ΔH be estimated? Well, for the formation of *regular* solutions (those in which the solute and the solvent do not form specific interactions), the change in *internal energy* per unit volume of solution is given by

$$\Delta H \approx \Delta E = \phi_1 \phi_2 (\delta_1 - \delta_2)^2 \, [=] \text{cal/cm}^3 \text{ soln} \qquad (7.2)$$

where

$\Delta E =$ the change in internal energy *per unit volume of solution*
$\phi_i =$ component volume fractions
$\delta_i =$ *solubility parameters*

The subscripts 1 and 2 usually (but not always) refer to solvent and solute (polymer), respectively. The *solubility parameter* is defined as follows:

$$\delta = (\text{CED})^{1/2} = (\Delta E_v / v)^{1/2} \qquad (7.3)$$

where

CED $=$ *cohesive energy density*, a measure of the intermolecular forces holding the molecules together in the liquid state
$\Delta E_v =$ molar change in internal energy on vaporization
$v =$ molar volume of liquid (cm^3/mol)

Traditionally, solubility parameters have been given in (cal/cm^3)$^{1/2}$ = hildebrands (in honor of the originator of regular solution theory), but they are now more commonly listed in (MPa)$^{1/2}$, where 1 hildebrand = 0.4889 (MPa)$^{1/2}$.

Now, for a process that occurs at constant volume and constant pressure, the changes in internal energy and enthalpy are equal. Since the change in volume on solution is usually quite small, this is a good approximation for the dissolution of polymers under most conditions, so Equation 7.2 provides a means of estimating enthalpies of solution if the solubility parameters of the polymer and the solvent are known.

Note that regardless of the magnitudes of δ_1 and δ_2 (they must be positive), the predicted ΔH is always positive, because Equation 7.2 applies only in the absence of the specific interactions that lead to negative ΔH's. Inspection of Equation 7.2 also reveals that ΔH is minimized and the tendency toward solubility is therefore maximized by matching the solubility parameters as closely as possible. As a very rough rule-of-thumb (or heuristic principle, if you prefer),

$$|\delta_1 - \delta_2| < 1 \ (\text{cal/cm}^3)^{1/2} \quad \text{for solubility} \qquad (7.4)$$

Measuring the solubility parameter of a low molecular weight solvent is not a problem. Polymers, on the other hand, degrade long before reaching their vaporization temperatures, making it impossible to evaluate ΔE_v directly. Fortunately, there is a way around this impasse. The greatest tendency of a polymer to dissolve occurs when its solubility

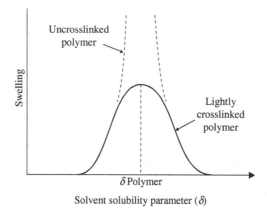

FIGURE 7.3 Determination of polymer solubility parameter by swelling lightly crosslinked samples in a series of solvents.

parameter matches to that of the solvent. If the polymer is crosslinked lightly, it cannot dissolve but will only swell. The maximum swelling will be observed when the solubility parameters of the polymer and the solvent are matched. So polymer solubility parameters are determined by soaking lightly crosslinked samples in a series of solvents of known solubility parameters. The value of the solvent's δ at which maximum swelling is observed is taken as the solubility parameter of the polymer (Figure 7.3).

Solubility parameters of solvent mixtures can be readily calculated from

$$\delta_{\text{mix}} = \frac{\Sigma y_i v_i \delta_i}{\Sigma y_i v_i} = \Sigma \phi_i \delta_i \tag{7.5}$$

where

$y_i =$ mole fraction of component i
$v_i =$ molar volume of component i
$\phi_i =$ volume fraction of component i

Equation 7.5 has often been used to prepare a series of mixed solvents for establishing the solubility parameter of a polymer as described above. Care must be exercised in this application, however, because the liquid that winds up inside the swollen polymer is not necessarily what you mixed up. In general, the crosslinked polymer will preferentially absorb the better (closer δ) solvent component, a phenomenon known as *partitioning*. In the absence of specific data on solvents, a group-contribution method is available for estimating both the solubility parameters and the molar volumes of liquids [1].

While the solubility-parameter concept has proved useful, there are unfortunately many exceptions to Equation 7.4. First, regular solution theory that leads to Equation 7.2 has some shortcomings in practice. Second, polymer solubility is too complex a phenomenon to be described quantitatively with a single parameter. Several techniques have been proposed that supplement solubility parameters with quantitative information on hydrogen bonding and dipole moments [2,3]. One of the simplest of these classifies solvents into three categories according to their hydrogen-bonding ability (poor, moderate, and strong).

Three different δ ranges are then listed for each polymer, one for each solvent category. Presumably, a solvent that falls within the δ range for its hydrogen-bonding category will dissolve the polymer. Another technique that has achieved widespread practical application is discussed in the next section.

7.6 HANSEN'S THREE-DIMENSIONAL SOLUBILITY PARAMETER

According to Hansen [4–7], the total change in internal energy on vaporization, ΔE_v, may be considered as the sum of three individual contributions: one due to hydrogen bonds, ΔE_h, another due to permanent dipole interactions, ΔE_p, and a third from dispersion (van der Waals or London) forces, ΔE_d:

$$\Delta E_v = \Delta E_d + \Delta E_p + \Delta E_h \tag{7.6}$$

Dividing by the molar volume v gives:

$$\frac{\Delta E_v}{v} = \frac{\Delta E_d}{v} + \frac{\Delta E_p}{v} + \frac{\Delta E_h}{v} \tag{7.7}$$

or

$$\delta^2 = \delta_d^2 + \delta_p^2 + \delta_h^2 \tag{7.8}$$

where $\quad \delta_j = (\Delta E_j/v)^{1/2}, \quad j = d, p, h$

Thus, the solubility parameter δ may be thought of as a vector in a three-dimensional d, p, and h space. Equation 7.8 gives the magnitude of the vector in terms of its components. A solvent, therefore, with given values of δ_{p1}, δ_{d1}, and δ_{h1} is represented as a point in space, with δ being the vector from the origin to this point.

A polymer (being denoted as component 2) is then characterized by δ_{p2}, δ_{d2}, and δ_{h2}. Furthermore, it has been found on a purely empirical basis that if δ_d is plotted on a scale twice the size as that used for δ_p and δ_h, then all solvents that dissolve that polymer fall within a sphere of radius R surrounding the point (δ_{p2}, δ_{d2}, and δ_{h2}).

Solubility judgments for the determination of R are usually based on visual observation of 0.5 g polymer in 5 cm^3 of solvent at room temperature. Given the concentration and temperature dependence of the phase boundaries in Figure 7.1, this is somewhat arbitrary, but it seems to work out pretty well in practice, probably because the boundaries are fairly "flat" for polymers of reasonable molecular weight.

The three-dimensional equivalent of Equation 7.4 is obtained by calculating the magnitude of the vector from the center of the sphere surrounding the solubility parameters for the polymer (δ_{p2}, δ_{d2}, and δ_{h2}) to the point representing the solvent (δ_{p1}, δ_{d1}, and δ_{h1}). If this is less than R, the polymer is deemed soluble:

$$\left[(\delta_{p1} - \delta_{p2})^2 + (\delta_{h1} - \delta_{h2})^2 + 4(\delta_{d1} + \delta_{d2})^2\right]^{1/2} < R \quad \text{for solubility} \tag{7.9}$$

(The factor 4 arises from the empirical need to double the δ_d scale to achieve a spherical solubility region.)

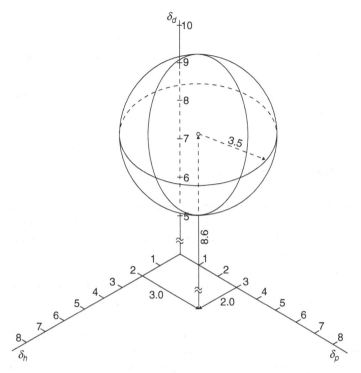

FIGURE 7.4 The Hansen solubility sphere for PS ($\delta_d = 8.6$, $\delta_p = 3.0$, $\delta_h = 2.0$, $= 3.5$) [5].

Figure 7.4 shows the solubility sphere for PS ($\delta_p = 3.0$, $\delta_d = 8.6$, and $\delta_h = 2.0$, $R = 3.5$, all in hildebrands) [4]. Note that parts of the PS sphere extend into regions of negative δ. The physical significance of these areas is questionable, at best.

The range of δ_d's spanned by typical polymers and solvents is rather small. In practice, therefore, the three-dimensional scheme is often reduced to two dimensions, with polymers and solvents represented on δ_h–δ_p coordinates with a polymer solubility circle of radius R.

Values of the individual components of δ_p, δ_d, and δ_h have been developed from measured δ values, theoretical calculations, studies on model compounds, and plenty of computer fitting. They are extensively tabulated for solvents [4–9]. They, along with R, are less readily available for polymers, but have been published [4,5,9]. Mixed solvents are handled by using a weighted average for the individual δ_j components according to Equation 7.5.

Despite its semiempirical nature, the three-dimensional solubility parameter has proved to be of great practical utility, particularly in the paint industry, where the choice of solvents to meet economic, ecological, and safety constraints is of critical importance. It is capable of explaining those cases in which the solvent and the polymer δ's are almost perfectly matched, yet the polymer will not dissolve (the δ vectors have the same magnitudes, but different directions), or where two nonsolvents can be mixed to form a good solvent (the individual solvent components lie on opposite sides outside the sphere, the mixture within). Inorganic pigments may also be characterized by δ vectors. Pigments whose δ vectors closely match to those of a solvent tend to form stable suspensions in that solvent.

Example 7.3 A polymer has a solubility parameter $\delta = 9.95$ ($\delta_p = 7.0$, $\delta_d = 5.0$, $\delta_h = 5.0$) and a solubility sphere of radius $R = 3.0$ (all numbers in hildebrands). Will a solvent with $\delta = 10$ ($\delta_p = 8$, $\delta_d = 6$, $\delta_h = 0$) dissolve it?

Solution. No. The solvent point lies in the δ_p–δ_d plane (i.e., $\delta_h = 0$). The closest approach the polymer solubility sphere makes to this plane is $5.0 - 3.0 = 2$. Thus, despite nearly identical δ's the solvent will not dissolve the polymer.

7.7 THE FLORY–HUGGINS THEORY

Theoretical treatment of polymer solutions was initiated independently and essentially simultaneously by Flory [10] and Huggins [11] in 1942. The *Flory–Huggins theory* is based on the lattice model shown in Figure 7.2. For the case of the low molecular weight solute (Figure 7.2a), it is assumed that the solute and the solvent molecules have roughly the same volumes; each occupies one lattice site. With the polymeric solute (Figure 7.2b), a *segment* of the polymer molecule (which corresponds roughly but not necessarily exactly to a repeat unit) has the same volume as a solvent molecule and also occupies one lattice site.

By statistically evaluating the number of arrangements possible on the lattice, Flory and Huggins obtained an expression for the (extensive) configurational entropy changes (those due to geometry alone), ΔS^*, in forming a solution from n_1 moles of solvent and n_2 moles of solute:

$$\Delta S^* = -R(n_1 \ln \phi_1 + n_2 \ln \phi_2) \tag{7.10}$$

where the ϕ's are volume fractions,

$$\phi_1 = \frac{x_1 n_1}{x_1 n_1 + x_2 n_2} \tag{7.10a}$$

$$\phi_2 = \frac{x_2 n_2}{x_1 n_1 + x_2 n_2} \tag{7.10b}$$

and the x's are the number of segments in the species. For the usual monomeric solvent, $x_1 = 1$. For a polydisperse polymeric solute, strictly speaking, a term must be included in Equation 7.10 for each individual species in the distribution, but x_2 is usually taken as $\bar{x}_n$, the number-average degree of polymerization, with little error. (Writing the volume fractions in terms of moles implies equal molar segmental volumes.) Note that while ϕ_1, ϕ_2, and n_1 are the same in Figures 7.2a and b, $n_2 = 20$ molecules for the monomeric solute, but only 1 molecule for the polymeric solute.

Example 7.4 Estimate the configurational entropy changes that occur when

    (a)   500 g of toluene (T, ϕ–CH_3) is mixed with 500 g of styrene monomer (S, ϕ–$CH=CH_2$)

(b)   500 g T is mixed with 500 g of PS, $\overline{M}_n = 100,000$

(c)   500 g of PS, $\overline{M}_n = 100,000$ is mixed with 500 g of polyphenylene oxide (PPO), (see Example 2.4K), $\overline{M}_n = 100,000$. (This is one of the rare examples where two high molecular weight polymers are soluble in one another.)

Solution. $M_T = 92$, $M_S = 104$. Using these values and those given for the polymers, we get $n_i = 500 \text{ g}/M_i$. In the absence of other information, we must assume that the number of segments equals the number of repeat units for the polymers. Therefore, $x_i = \overline{M}_{ni}/m_i$, where m_i is the molecular weight of the repeat unit, $m_{PS} = 104$, $m_{PPO} = 120$. These quantities may now be inserted in Equations 6.10, 6.10a, and 6.10b. The results are summarized below ($R = 1.99 \text{ cal/mol·K}$):

i	$n_i \text{(mol)}$	x_i	ϕ_i
(a) Toluene	5.44	1	0.531
Styrene	4.81	1	0.469
	$\Delta S^* = 14.1 \text{ cal/K}$		
(b) Toluene	5.44	1	0.531
PS	0.005	962	0.469
	$\Delta S^* = 6.85 \text{ cal/K}$		
(c) PS	0.005	962	0.536
PPO	0.005	833	0.464
	$\Delta S^* = 0.0138 \text{ cal/K}$		

The result for (c) illustrates why polymer–polymer solubility essentially *requires* a negative ΔH.

An expression for the (extensive) enthalpy of mixing, ΔH, was obtained by considering the change in adjacent-neighbor (molecules or segments) interactions on the lattice, specifically the replacement of [1,1] and [2,2] interactions with [1,2] interactions upon mixing:

$$\Delta H = RT \chi \phi_2 n_1 x_1 \quad [=]\text{cal} \tag{7.11}$$

where χ is the *Flory–Huggins polymer–solvent interaction parameter*. Initially, χ was interpreted as the enthalpy of interaction per mole of solvent divided by RT. By equating Equations 7.2 and 7.11 (keeping in mind that the enthalpy in Equation 7.2 is based on a unit volume of solution, while that in Equation 7.11 is an extensive quantity) and making use of Equation 7.10a, it may be shown that the Flory–Huggins parameter is related to the solubility parameters by

$$\chi = \frac{v(\delta_1 - \delta_2)^2}{RT} \tag{7.12}$$

where v is the molar segmental volume of species 1 and 2 (assumed to be equal). For the dissolution of a polymer in a monomeric solvent (e.g., PS in styrene), v is taken as the molar volume of the solvent, v_1. Based on our knowledge of solubility parameters, we see that Equation 7.12 predicts $\chi \geq 0$. Actually, negative values have been observed.

If it is assumed that the entropy of solution is entirely configurational, substitution of Equations 7.10 and 7.11 into Equation 7.1 gives

$$\Delta G = \text{RT}(n_1 \ln \phi_1 + n_2 \ln \phi_2 + \chi \phi_2 n_1 x_1) \tag{7.13}$$

Again, for the usual monomeric solvent, $x_1 = 1$. For a polydisperse solute, the middle term on the right side of Equation 7.13 must be replaced by a summation over all the solute species; however, treatment as a single solute with $x_2 = \overline{x}_n$ usually suffices.

In terms of the Flory–Huggins theory, the criterion for complete solubility of a high molecular weight polymer across the composition range is

$$\chi \leq 0.5 \quad \text{for solubility} \tag{7.14}$$

It is now recognized that there is an interactive as well as a configurational contribution to the entropy of solution, that is also included in the χ term, so χ is now considered to be a ΔG (rather than strictly a ΔH) of interaction per mole of solvent divided by RT. The first two terms on the right of Equation 6.13 therefore represent the configurational entropy contribution to ΔG, while the third term is the interaction contribution and includes both enthalpy and entropy effects.

The Flory–Huggins theory has been used extensively to describe phase equilibria in polymer systems. It can, for example, qualitatively describe the lower phase boundary (UCST) in Figure 7.1, although it rarely gives a good quantitative fit of experimental data. Partial differentiation of Equation 7.13 with respect to n_1 (keeping in mind that ϕ_1 and ϕ_2 are functions of n_1) gives the chemical potential of the solvent. This is, of course, a key quantity in phase equilibrium that also makes χ experimentally accessible. Further development is beyond the scope of this chapter, but the subject is well treated in standard works on polymer solutions [12–15].

The limitations of the Flory–Huggins theory have been recognized for a long time. It cannot predict an LCST (Figure 7.1). It is perhaps not surprising that χ depends on temperature, but it unfortunately turns out to be a function of concentration and molecular weight as well, limiting practical application of the theory. These deficiencies are thought to arise because the theory assumes no volume change upon mixing and the statistical analysis on which it is based is not valid for very dilute solutions, particularly in poor solvents. There has been considerable subsequent work done to correct these deficiencies and extend lattice-type theories [16,17].

Experimental values for χ have been tabulated for a number of polymer–solvent systems, both single values [9] and even as a function of composition [8]. They may be used with Equation 7.14 to predict solubility.

7.8 PROPERTIES OF DILUTE SOLUTIONS

Well, now that the thermodynamics have been covered, we know how to get the polymer into solution, but how the polymer chains are organized in the solution is also important. Let us initially assume that the polymer is in a fairly dilute solution, therefore we can assume that there are very few entanglements between the polymer chains. Entanglements begin to set in when the dimensionless Berry number (Be), the product of intrinsic viscosity and concentration, exceeds unity [18]:

$$\text{Be} = [\eta]c > 1 \quad \text{for entanglements} \tag{7.15}$$

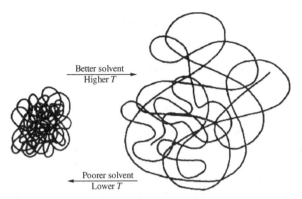

FIGURE 7.5 The effects of solvent power and temperature on a polymer molecule in solution.

For typical polymer–solvent systems, this usually works out to a few percent polymer.

In a "good" solvent (one whose solubility parameter closely matches that of the polymer), the secondary forces between polymer segments and solvent molecules are strong, and the polymer molecules will assume a spread-out conformation in solution. In a "poor" solvent, the attractive forces between the segments of the polymer chain will be greater than those between the chain segments and the solvent; in other words, the chain segments "prefer their own company," and the chain will ball up tightly (as in Figure 7.5).

Imagine a polymer in a good solvent, for example, PS ($\delta = 9.3$) in chloroform ($\delta = 9.2$). A nonsolvent is now added, say methanol ($\delta = 14.5$). (Despite the large difference in solubility parameters, chloroform and methanol are mutually soluble in all proportions because of the large ΔS of solution for low molecular weight compounds.) Ultimately, a point is reached where the mixed solvent becomes too poor to sustain solution, and the polymer precipitates out because the attractive forces between polymer segments become much greater than that between the polymer and the solvent. At some point, the polymer teeters on the brink of solubility, $\Delta G = 0$ and $\Delta H = T\Delta S$. This point obviously depends on the temperature, polymer molecular weight (mainly through its influence on ΔS), and the polymer–solvent system (mainly through its influence on ΔH). Adjusting either the temperature or the polymer–solvent system allows fractionation of the polymer according to molecular weight, as successively smaller molecules precipitate upon lowering the temperature or going to poorer solvent.

In the limit of infinite molecular weight (the minimum possible ΔS), the situation where $\Delta H = T\Delta S$ is known as the θ or Flory condition. For a polymer of infinite molecular weight in a particular solvent, the θ temperature equals the UCST, indicating that the solution is on the brink of phase separation. Under these conditions, the polymer–solvent and polymer–polymer interactions are equal, and the solution behaves in a so-called ideal fashion with the second virial coefficient equal to 0, etc., and the MHS exponent (Eq. 5.18) $a = 0.5$.

For a given polymer, θ conditions can be reached at a fixed temperature by adjusting the solvent to give a θ solvent or with a particular solvent by adjusting the temperature to reach the θ or Flory temperature. Any actual polymer will still be soluble under θ conditions, of course, because of its lower-than-infinite molecular weight and consequently larger ΔS.

Getting back to our example, we might ask what happens to the viscosity of the solution upon moving from a good solvent to a poor solvent (to make a fair comparison, imagine that as soon as a nonsolvent is added, an equivalent amount of good solvent is removed, maintaining a constant polymer concentration). This question can be answered qualitatively by imagining the polymer molecules to be rigid spheres (they really are not) and applying an equation derived by Einstein [19] for the viscosity of a dilute suspension of rigid spheres:

$$\eta = \eta_s(1 + 2.5\phi) \tag{7.16}$$

where η is the viscosity of the suspension (solution), η_s is the viscosity of the solvent, and ϕ is the volume fraction of the polymer spheres. (We further assume that the viscosities of the low molecular weight liquids are comparable, i.e., that η_s does not change.) Going from good to poor solvent, the polymer molecules ball up, giving a smaller effective ϕ, *lowering* the solution viscosity. Thus, solution viscosity can be controlled by adjusting solvent power.

Example 7.5 Indicate how solvent "power" (good solvent versus poor solvent) will influence the following:

(a) The intrinsic viscosity of a polymer sample at a particular T and the molecular weight, $\overline{M}_v$, estimated by viscometry.

(b) The molecular weight of a polymer sample as determined by osmometry.

Solution. (a) The measured viscosities of solutions will be less in a "poor" solvent to that in a good one, leading to lower intrinsic viscosity. The lower viscosity will translate to a lower $\overline{M}_v$ when calculated using the MHS Equation 5.18.

(b) If the membrane were ideal, solvent power would make no difference, since osmometry measures only the number (moles) of solute per unit volume, regardless of geometry. Some of the smallest particles can sneak through any real membrane, however. The poor solvent, causing the molecules to ball up tightly, will allow passage of more molecules through the membrane. This will lower the observed osmotic pressure causing the calculated $\overline{M}_n$ to move on the high side.

Consider now the effects of temperature on the viscosity of a solution of polymer in a relatively poor solvent. As is the case with all simple liquids, the solvent viscosity, η_s, decreases with increasing temperature. Increasing temperature, however, imparts more thermal energy to the segments of the polymer chains, causing the molecules to spread out and assume a larger effective ϕ in solution. Thus, the effects of temperature on η_s and ϕ tend to compensate, giving a solution that has a much smaller change in viscosity with temperature than that of the solvent alone. In fact, the additives that are used to produce the so-called multiviscosity motor oils (such as 10W-40) are polymers with compositions adjusted so that the base oil is a relatively poor solvent at the lowest operating temperatures. As the engine heats up, the polymer molecules uncoil, providing a much greater resistance to thinning than is possible from oil alone (and preventing, at least as the commercials say, "thermal breakdown").

Viscosities (at low shear rates) of dilute solutions of polymers with known molecular weights may be calculated using Equations 5.15, 5.16, and 5.18, reversing the procedure for obtaining molecular weights from viscosity measurements.

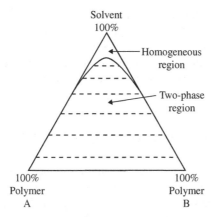

FIGURE 7.6   Typical ternary phase diagram for a polymer–polymer-common solvent system. The dotted tie lines connect the composition of phases in equilibrium.

7.9 POLYMER–POLMYER-COMMON SOLVENT SYSTEMS

As was illustrated in Example 7.3c, ΔS for the dissolution of one polymer in another is extremely small. For this reason, the true solubility of one polymer in another is relatively rare, although many more examples have come to light in recent years. When such solubility does occur, it generally results from strong interactions (e.g., a large, negative ΔH), most often hydrogen bonds, between the polymer pairs. Polymer–polymer miscibility has been extensively reviewed [20,21].

For the far more common case where two polymers are insoluble in one another, even when a common solvent is added (one that is infinitely soluble with each polymer alone), the two polymers usually cannot coexist in a homogeneous phase beyond a few percent concentration. A schematic phase diagram for such a system is shown in Figure 7.6. Beyond a few percent polymer (the exact value depends on the chemical nature of the polymers and the solvent and molecular weights of the polymers), two phases in equilibrium are formed, each phase containing a nearly pure polymer. These results extend to more than two polymers. In general, each polymer will coexist in a separate phase. This fact was the basis of an early proposal for separating and recycling mixed plastic waste [22].

7.10 POLYMER SOLUTIONS, SUSPENSIONS, AND EMULSIONS

Given a good solvent and a temperature between the UCST and LCST for a polymer–solvent system, a homogeneous solution will form. The amount of polymer added will certainly increase viscosity, but in any well-solvated polymer, the chains will expand, uncoil, and allow many polymer–solvent interactions. For thermodynamically bad polymer–solvent combinations, polymer–polymer interactions are preferred, and each polymer chain will be tightly coiled, leading to a two-phase system with a solid polymer precipitated out. A *suspension* can be made by introducing energy of mixing, but this only spreads the insoluble polymer out in a nonsolvent. When mixing is stopped, the polymer will settle out again (think oil and water as in traditional Italian salad dressing). Polymer *emulsions* can overcome this settling out through the introduction of *surfactants*—molecules that have two distinct regions—one that is solvent-philic and the other polymer-philic. The surfactants line up along the interface between the polymer phase and the solvent phase, allowing a stable emulsion to form (think emulsified salad dressings that require no shaking to mix). Both polymer suspensions and emulsions are important in both reactions and final products and are discussed in later chapters.

7.11 CONCENTRATED SOLUTIONS: PLASTICIZERS

Up to this point, we have considered relatively dilute polymer solutions. Now let us look at the other end of the spectrum, where the polymer is the major constituent of the solution. A pure, amorphous polymer consists of a tangled mass of polymer chains. The ease with which this mass can deform depends on the ability of the polymer chains to untangle and slip past one another. One way of increasing this ability is to raise the temperature, increasing segmental mobility of the polymer chains. Another way is to add a low molecular weight (generally low volatility) liquid to the polymer as an *external plasticizer*. By forming secondary bonds to the polymer chains and spreading them apart, the plasticizer reduces secondary bonds in polymer–polymer interactions. The plasticizer provides more room for the polymer molecules to move around, yielding a softer, more easily deformable mass. This corresponds well with one of the major functions of plasticizers, lowering T_g.

Example 7.6 If you have ever heard baseball announcers comment that "On these humid nights, the ball just isn't hit as hard" or "A pitcher throws a heavier ball on these humid nights," you may have wondered if there was a scientific basis for these statements. They cannot be justified in terms of the properties of the humid air (fluid dynamics). Can you justify them in terms of the properties of the baseball? A baseball is made largely of tightly wound wool yarn. Wool, a natural polymer, is similar, physically and chemically, to nylons (see Example 7.1).

Solution. The polar linkages in wool with hydrogen bond to water. The higher the humidity, the greater the equilibrium moisture content of the wool, accounting for the heavier ball. The absorbed water will also act as an external plasticizer for the wool, softening it and decreasing its resiliency. Just think what would happen if the baseballs were dunked in water before a game!

Quantitative confirmation of Example 7.6 was provided by a noted technical journal, *Sports Illustrated*, which, between swimsuit photos, reported on the effects of moisture on the mass and resilience of baseballs [23]:

- Balls baked in a 212 °F oven for 24 h lost 12 g and gained 5.8% in vigor.
- 24 h in a steam-filled moist room caused them to gain 11 g and lose 17.4 % of their bounce.

A similar reversible plasticization of fibers by heat and moisture is applied in the process of steam-ironing fabrics. Using a sizing agent (such as starch) in ironing also begins by plasticizing the clothing, but becomes rigid as the water evaporates (see Example 7.2).

A thermodynamically poor plasticizer should be more effective than a good one, giving a lower viscosity at a given level (smaller ϕ, fewer entanglements), but since it is less strongly bound to the polymer, it will have a greater tendency to exude out over a period of time, leaving the stiff polymer mass. This was a familiar problem with some early imported cars. When parked in the sun, a greasy plasticizer film would condense on the windows and, over a period of time, the upholstery and dashboard would crack from loss of plasticizer. (Much of that new-car smell is due to plasticizers, which evaporate slowly, particularly in hot weather.) Thus, a balance must be struck between plasticizers' efficiency and permanence. Also from the standpoint of permanence, it is necessary that the plasticizers have low volatility. Plasticizers therefore, generally, have higher molecular weights than solvents (400–600 versus 100 or less for typical organic solvents), but are still well below the high polymer range in this respect.

A polymer may be *internally plasticized* by random copolymerization with a monomer whose homopolymer is very soft (has a low T_g, see Chapter 6). There are obviously never any permanence problems when this is done. The composition of the copolymer can be adjusted to give the desired properties at a particular temperature. Above this temperature, the copolymer will be softer than intended and below this temperature, it will be harder. With external plasticizers, the same sort of temperature compensation as in the multi-viscosity motor oils is obtained, giving materials that maintain the desired flexibility over a broader temperature range.

The most common externally plasticized polymer is poly(vinyl chloride) (PVC) ($-CH_2CHCl-$). A typical plasticizer for PVC is dioctyl phthalate (DOP), the diester of phthalic acid and *iso*-octyl alcohol (*note*: DOP is the more common term in industry, but the IUPAC name for this chemical is di(2-ethyl hexyl) phthalate or DEHP):

PVC, because of its high chlorine content, is inherently fire resistant (see Chapter 18 for other methods to make polymers fire resistant). Dilution of the PVC with a hydrocarbon plasticizer increases flammability (especially important since many plasticizers are included at 20 wt% or higher). Therefore, chlorinated waxes and phosphate esters, for example, tricresyl phosphate (TCP), are used as plasticizers that maintain fire resistance. So-called polymeric plasticizers, which are actually oligomers with molecular weights on the order of 1000, provide low volatility and good permanence. It might also be noted that a low molecular weight fraction in a pure polymer behaves as a plasticizer often in an undesirable fashion.

Epoxidized plasticizers also perform the important function of stabilizing PVC. When PVC degrades, HCl is given off that catalyzes further degradation. Not only that, HCl attacks metallic molding machines, molds, extruders, etc. The epoxy plasticizers are made by epoxidizing polyunsaturated vegetable oils:

$$\text{\char`\~\char`\~}\overset{\overset{H}{|}}{C}=\overset{\overset{H}{|}}{C}\text{\char`\~\char`\~} + H_2O_2 \xrightarrow{\text{Catalyst}} \text{\char`\~\char`\~}\underset{\diagdown \! O \! \diagup}{\overset{\overset{H}{|}}{C}-\overset{\overset{H}{|}}{C}}\text{\char`\~\char`\~} + H_2O$$

Oxirane ring

The *oxirane rings* soak up HCl and minimize further degradation (this is an admittedly simplified view of a very complex phenomenon) [24]:

$$\text{\char`\~\char`\~}\underset{\diagdown \! O \! \diagup}{\overset{\overset{H}{|}}{C}-\overset{\overset{H}{|}}{C}}\text{\char`\~\char`\~} + HCl \longrightarrow \text{\char`\~\char`\~}\underset{\underset{OH}{|}}{\overset{\overset{H}{|}}{C}}-\underset{\underset{Cl}{|}}{\overset{\overset{H}{|}}{C}}\text{\char`\~\char`\~}$$

Having their origin in vegetable oils, these epoxidized plasticizers can often obtain FDA clearance for use in food-contact applications. Further advances in plasticizers have been reviewed [25].

Unplasticized (or nearly so) PVC is a rigid material used for pipe and fittings, house siding, window frames, bottles, etc. The properties of plasticized PVC vary considerably depending on the plasticizer level. Plasticized PVC is familiar as a gasket material, a leather-like upholstery material (Naugahyde®), wire and cable covering, shower curtains, water hoses, and packaging film.

Plastisols are an interesting and useful technological application of plasticized PVC. A typical formulation might consist of 100 parts DOP phr (per hundred parts polymer resin) plus some stabilizers, pigments, etc. Although the PVC is thermodynamically soluble in the DOP at room temperature, the rate of dissolution of high molecular weight PVC at room temperature is extremely low. Therefore, initially the plastisol is a milky suspension of finely divided PVC particles in the plasticizer. As a suspension rather than a solution, the viscosity is of the order of magnitude of that of the plasticizer itself (Eq. 6.16), and it can be applied to substrates or molds by brushing, dipping, rolling, etc. When heated to about 350 °F (175 °C), the increased thermal agitation of the polymer molecules speeds up the dissolution process greatly. If there is no filler or pigment present, the solution process can be observed as the plastisol becomes transparent. When cooled back to room temperature, the viscosity of the solution is so high that for all practical purposes, it may be considered as a flexible solid. Examples of plastisols include doll "skin," toy rubber duckies, and the covering of wire dish racks. Vinyl foam is also made by whipping air into a plastisol prior to fusion.

The viscosity of the initial suspension may be lowered by incorporating a volatile organic diluent for the plasticizer, giving an *organsol*, or by whipping it in water to form a *hydrosol*. In both cases, the diluent vaporizes upon heating.

Further discussion of the plasticized polymers is included in Chapter 18.

PROBLEMS

1   Crud Chemicals is using GPC for quality-control checks on their PS ($\delta = 9.3$) product. Normally, the column runs at 25 °C with THF solvent ($\delta = 9.1$). Unbeknownst to

them, a disgruntled former employee contaminated their THF supply with methanol ($\delta = 14.5$) before leaving. Being fairly sophisticated technically, he has loaded it up pretty well without causing the polymer to precipitate out. How and why will this affect the calculated average molecular weights?

2   Crud's analytical laboratory corrects its solvent problem (see above), but the temperature controller on their GPC column goes haywire, allowing the column temperature to rise to 45 °C without their knowledge. How and why will this affect their calculated average molecular weights for PS?

3   Crud's laboratory now wants to use the GPC column (from Problem 7.1) to analyze copolymers of styrene (75%) with acrylonitrile (25%). Not having monodisperse calibration standards for the copolymer, the scientists at Crud go ahead and use the existing PS calibration. How and why will this affect their calculated average molecular weights?

4   According to Hansen's three-dimensional/solubility parameter scheme, which of the following solvents will dissolve PS (Figure 7.4)? How well do these predictions agree with Equation 7.4?

Solvent	δ_d	δ_p	δ_h
n-Hexane	7.28	0	0
Styrene	9.09	0.49	2.0
Methyl chloride	7.48	3.0	1.9
Acetone	7.58	5.08	3.4
Ethylacetate	7.72	2.6	3.5
Isopropanol	7.72	3.0	8.02
Phenol	8.80	2.9	7.28
All δ's are in $(cal/cm^3)^{1/2} = $ hildebrands			

5   Apply the general thermodynamic criteria for phase equilibria and derive equations that would allow you to describe the lower-phase envelope in Figure 7.1 in terms of the Flory–Huggins theory.

6   Show the derivation of Equation 7.12.

7   Sketch qualitatively how you would expect the molecular weight of the polymer to affect the phase diagram in Figure 7.1.

8   Sketch qualitatively how you would expect the ternary phase diagram (Figure 7.6) to be affected by: (a) polymer molecular weight and (b) temperature. Assume that the polymer has UCST behavior in the solvent system.

9   Figure 7.6 illustrates a ternary phase diagram for a polymer–polymer-common solvent system. Sketch qualitatively a ternary phase diagram for a polymer–solvent-non-solvent system. Include tie lines and illustrate the effects of temperature and polymer molecular weight on two additional diagrams. Assume that the solvent and nonsolvent are soluble in all proportions, and that the polymer has UCST behavior.

10   Polymer solubility parameters can be determined through viscosity measurements. Describe how you would do this.

11 A small amount of finely divided polymer is put into a beaker containing a liquid. The beaker is covered and the contents are magnetically stirred with no change in the milky appearance over several hours. The beaker is then uncovered, with the stirring continued. The contents of the beaker gradually become transparent over a period of several hours. Explain (there are at least two possibilities).

12 It has been proposed that a mixture of waste thermoplastics could be separated by extraction with a single solvent by using different temperatures. That is, polymer A would be extracted from the mixture at the lowest temperature, T_A, polymer B at a higher temperature T_B, and so on, leading to nearly pure recovered polymers. Comment on the thermodynamic feasibility of this proposal.

13 You are getting ready to challenge Tiger Woods to a day on the golf course, but to have any shot to compete with Tiger, you decide to stack the odds in your favor by soaking Tiger's golf balls in a good solvent to reduce the distance the balls will fly off of the golf tee. Choose an appropriate solvent that will swell the internal material of the golf ball to reduce the momentum transfer from Tiger's tee shots to help you "fix" the game. (Check also to see if the solvent you choose will penetrate the Surlyn® coating; you may assume that the exposure time is small enough for the coating to remain intact, but allows some solvent to penetrate.) *Note*: You will need to determine what the core of a golf ball is made of and determine at least one solvent that will swell it; you should also visit DuPont's website and search for chemical resistance data for Surlyn. Will the solvent increase or decrease the glass transition temperature of the golf ball core?

REFERENCES

[1] Fedors, R.F., *Polym. Eng. Sci.* **14**, 147, 472 (1974).

[2] Beerbower, A., L.A. Kaye, and D.A. *Pattison, Chem. Eng.* **74**(26), 118 (1967).

[3] Seymour, R.B., *Modern Plastics* **48**(10), 150 (1971).

[4] Hansen, C.M., *J. Paint. Technol.* **39**(505), 104 (1967).

[5] Hansen, C.M., *The Three Dimensional Solubility Parameter and Solvent Diffusion Coefficient*, Danishi Technical Press, Copenhagen, Denmark, 1967.

[6] Hansen, C.M. and A. Beerbower, Solubility parameters, in *Encyclopedia of Chemical Technology*, Suppl. Vol., 2nd ed., Wiley, New York, 1971.

[7] Hansen, C.M., *Ind. Eng. Chem. Prod. Res. Dev.* **8**, 2 (1969).

[8] Brandrup, J. and E.H. Immergut (eds), *Polymer Handbook*, 3rd ed., Wiley-Interscience, New York, 1989.

[9] Barton, A.F.M., *Handbook of Polymer-Liquid Interaction Parameters and Solubility Parameters*, CRC Press, Boca Raton, FL, 1990.

[10] Flory, P.J., *J. Chem. Phys.* **10**, 51 (1942).

[11] Huggins, M.L., *Ann. N.Y. Acad. Sci.* **43**, 1 (1942).

[12] Flory, P.J., *Principles of Polymer Chemistry*, Cornell University Press, Ithaca, NY, 1953.

[13] Morawetz, H., *Molecules in Solution*, Wiley-Interscience, New York, 1975.

[14] Tompa, H., *Polymer Solutions*, Butterworths, London, 1956.

[15] Kwei, T.K., Macromolecules in solution, Chapter 4 in *Macromolecules: An Introduction to Polymer Science*, F. A. Boveyand F.H. Winslow (eds), Academic, New York, 1979.

[16] Sanchez, I.C. and R.H. Lacombe, *Macromolecules* **11**(6), 1145 (1978).

[17] Pesci, A.I. and K.F. Freed, *J. Chem. Phys.* **90**(3), 2017 (1989).

[18] Hager, B.L. and G.C. Berry, *J. Polym. Sci.: Polym. Phys. Ed.* **20**, 911 (1982).

[19] Einstein, A., *Ann. Phys.* **19**, 289 (1906); **34**, 591 (1911).

[20] Olabisi, O., L.M. Robeson, and M.T. Shaw, *Polymer-Polymer Miscibility*, Academic, New York, 1979.

[21] Coleman, M.M., J.F. Graf, and P.C. Paniter, *Specific Interactions and the Miscibility of Polymer Blends*, Technomic, Lancaster, PA, 1991.

[22] Sperber, R.J. and S.L. Rosen, *Polym. Eng. Sci.* **16**(4), 246 (1976).

[23] *Sports Illustrated,* Time, Inc., July 20, 1970, p. 22.

[24] Anderson, D.F. and D.A. McKenzie, *J. Polymer Sci.: A-1*, **8**, 2905 (1970).

[25] Rahman, M. and C.S. Brazel, *Prog. Polym. Sci.*, **29**, 1223 (2004).

PART II

POLYMER SYNTHESIS

In Part I, molecular organization of polymers and functional groups required for their formation have been introduced. It also included some of the important chemical characteristics needed to understand crystallinity, thermal behavior, and solubility. This part discusses the reaction mechanisms and factors influencing the formation of polymers. We have already discovered in Chapter 2 that the primary polymerization mechanisms are condensation (or step-growth) polymerization (Chapter 8), and free-radical (or addition) polymerization (Chapter 9). Some advanced techniques for polymerizations that result in more specifically tailored structures are covered in Chapter 10. Chapter 11 introduces the concept of reactivity ratios for copolymerization, as the growth of polymer chains is not entirely random. Finally, Chapter 12 discusses industrial methods of polymer synthesis, the equipments required, and how some of the challenges in working with high molecular weight viscous materials can be circumnavigated. This part of the book may be thought of as polymerization reaction engineering.

Fundamental Principles of Polymeric Materials, Third Edition. Christopher S. Brazel and Stephen L. Rosen
© 2012 John Wiley & Sons, Inc. Published 2012 by John Wiley & Sons, Inc.

CHAPTER 8

STEP-GROWTH (CONDENSATION) POLYMERIZATION

8.1 INTRODUCTION

Polymerization reactions may be divided into two categories according to the mechanism by which the chains grow. In *step-growth polymerization*, also known as condensation polymerization, chains of *any* lengths x and y can combine to form longer chains:

$$x\text{-mer} + y\text{-mer} \rightarrow (x + y)\text{-mer} \qquad \text{(step growth)} \qquad (8.1a)$$

In *chain-growth* (also known as addition or free-radical) polymerization, a chain of any length x can only add a monomer molecule to continue its growth:

$$x\text{-mer} + \text{monomer} \rightarrow (x + 1)\text{-mer} \qquad \text{(chain growth)} \qquad (8.1b)$$

These two mechanisms generally require different organic functional groups, carboxylic acids, amines, alcohols, etc., for step-growth and alkenes for chain-growth. Another distinction that you may have picked up on in Chapter 2 is that the molecular weight of the monomer and the repeat unit are generally the same for chain-growth polymers, but the repeat unit is smaller (due to the condensation of water) for most step-growth polymers. This is not always true, since some condensation polymerizations do not split out a water molecule (see Examples 2.4F and 2.4Q). This chapter covers step-growth polymerization, while chain-growth polymerization is considered in Chapters 9 and 10.

Regardless of the type of polymerization reaction, quantitative treatments are usually based on the assumption that the reactivity of the functional group at a chain end is independent of the length of the chain, x and y in Equations 8.1a and 8.1b. Experimentally, this is an excellent assumption for x's greater than five or six. Since most polymers must develop x's on the order of a hundred (or much greater) to be of practical value, this

Fundamental Principles of Polymeric Materials, Third Edition. Christopher S. Brazel and Stephen L. Rosen.
© 2012 John Wiley & Sons, Inc. Published 2012 by John Wiley & Sons, Inc.

assumption introduces little error while enormously simplifying the mathematical treatment. With these basic concepts in mind, we proceed to a more detailed and quantitative treatment of polymerization reactions.

8.2 STATISTICS OF LINEAR STEP-GROWTH POLYMERIZATION

Consider the two equivalent linear step-growth reactions, assuming difunctional monomers and stoichiometric equivalence:

$$\frac{x}{2}ARA + \frac{x}{2}BR'B \rightarrow ARA[BR'B - ARA]_{(\frac{x}{2}-1)}BR'B* \qquad (8.2a)$$

$$xARB \rightarrow ARB[ARB]_{x-2}ARB \qquad (8.2b)$$

where A and B represent the complementary functional groups. It is important to note that x is used here to denote the number of *monomer residues* or *structural units* in the chains, rather than repeat units (which would be either [BR'B-ARA] or [ARB][1]).

Each of the polymer molecules above contains a total of x A groups:

> *one unreacted* A group (on the end) and
> $x - 1$ *reacted* A groups,

so x still represents the degree of polymerization with either repeat unit listed above.

Considering a volume of the reaction mixture at a certain time during polymerization, we can use statistical analysis to find the polymer chain length and average molecular weight. For this analysis, we will let

p = probability of finding a reacted A group
  = number or mole fraction of reacted A groups present
  = *conversion* or extent of reaction of A groups
$(1 - p)$ = the probability of finding an unreacted A group
N = the total number of molecules present in the reaction mass (of all sizes)
n_x = the number of molecules containing x A groups, both reacted and unreacted

$$\left(N = \sum_{x-1}^{\infty} n_x \right)$$

The total probability of finding a molecule with x A groups (reacted and unreacted) is equal to the mole or number fraction of those molecules present in the reaction mass n_x/N. This, in turn, is equal to the probability of finding a molecule with $x - 1$ reacted A groups and one unreacted A group. Since the overall probability is the product of the individual probabilities,

---

[1] Some of the molecules formed in this type of reaction will be capped with two A groups. For each of these, however, stoichiometric equivalence requires that there should be one capped with two B groups, so Equation 8.2a represents the *average* reaction.

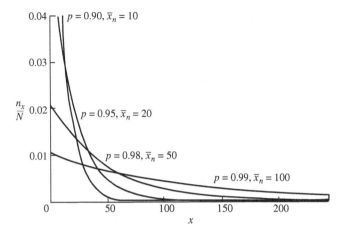

FIGURE 8.1   Number- (mole-) fraction distributions for linear step-growth polymers at different conversions (Eq. 8.3).

$$\frac{n_x}{N} = p^{(x-1)}(1-p) = \text{mole (number) fraction } x\text{-mer} \tag{8.3}$$

This is the so-called *most probable distribution*. It results from the random nature of the reaction between chains of different lengths, which will produce polymers of various lengths at any time during the reaction. As the reaction proceeds with time, the conversion increases to yield longer chains. The distribution is plotted in Figure 8.1 for several conversions, p. Note that at any conversion, the shorter chains are always more numerous, that is, the longer the chain, the fewer the chain. Also, the fraction of short chains decreases with conversion, while the fraction of long chains increases with conversion, because increasing conversion hooks together shorter chains to form longer chains.

Because each growing polymer chain has two functional groups (one at each end), step-growth polymerizations can involve reactions between monomers, oligomers, or polymers of any length, x:

$$
\begin{array}{ccccc}
\text{1-mer} & + & \text{1-mer} & \rightarrow & \text{2-mer} \\
\text{2-mer} & + & \text{3-mer} & \rightarrow & \text{5-mer} \\
\text{5-mer} & + & \text{95-mer} & \rightarrow & \text{100-mer}
\end{array}
$$

This is an unique characteristic of step-growth polymerizations and contrasts with chain-growth polymerizations (Chapter 9) that add only monomers (1-mers) to growing chains (Eq. 8.2*b*).

8.3 NUMBER-AVERAGE CHAIN LENGTHS

Since a distribution of chain lengths always arises in polymerization reactions, the average chain length is often of interest. Each molecule in the reaction mass has, on an average, one unreacted A group. If N_o is the original number of molecules present in the reaction mass, there are N_o unreacted A groups present at the start of the reaction. At some time later during the reaction there are fewer molecules (N), and therefore N unreacted A groups are present, so $N_o - N$ of the A groups have reacted. Thus,

$$p = \frac{N_0 - N}{N_0} = \text{fraction of reacted A groups present} \qquad (8.4)$$

Because the N_o original monomer molecules are distributed among the N molecules present in the reaction mass, the average number of monomer residues per molecules, $\overline{x}_n$, is given by

$$\overline{x}_n = \frac{N_o}{N} = \text{average number of monomer residues per molecule} \qquad (8.5)$$

Combining Equations 8.4 and 8.5 gives

$$\overline{x}_n = \frac{1}{1 - p} \quad \text{(Carothers' equation)} \qquad (8.6)$$

This rather simple conclusion was reached by W.H. Carothers, the discoverer of nylon and one of the founders of polymer science, in the 1930s. (The same result may be obtained in a more laborious fashion by inserting Equation 8.3 into Equation 5.24.) Its importance becomes obvious when it is realized that typical linear polymers must have $\overline{x}_n$ values on the order of at least 100 to achieve useful mechanical properties. This requires a conversion of at least 99%, assuming difunctional monomers in perfect stoichiometric equivalence. Such high conversions are almost unheard in most organic reactions, but are necessary to achieve high molecular weight condensation polymers.

Since condensation reactions are reversible, many step-growth polymerizations would reach equilibrium at low conversions if the low molecular weight product of the reaction (usually condensed water) were not efficiently removed (e.g., by heat and vacuum or by a second reaction) to drive the reaction to high conversions.

Example 8.1 Crud Chemicals is producing a linear polyester in a batch reactor by the condensation of a hydroxy acid HORCOOH. It has been proposed to follow the progress of the reaction by measuring the amount of water removed from the reaction mass. Assuming pure monomer, and that all the water of condensation can be removed and measured, derive for them an equation relating $\overline{x}_n$ of the polymer to N_o, the moles of monomer charged to the reactor, and M, the total moles of water evolved since the start of the reaction.

Solution. In this polycondensation reaction, each reacted −OH (or −COOH) group produces one molecule of water. Therefore, the moles of reacted −OH (or −COOH) groups is M. The probability of finding a reacted group (of either kind) is

$$p = \frac{M}{N_0} = \frac{\text{moles reacted} - \text{OH groups}}{\text{total moles} - \text{OH groups}}$$

Plugging into Carothers' Equation 8.6 yields

$$\overline{x}_n = \frac{1}{1 - (M/N_0)}$$

So far, we have only discussed polymerizations where there is a stoichiometric balance of A and B groups (which is always true for monomers such as hydroxy acids, as shown in Example 8.1). However, if the number of A and B groups is not equal, the length of the chains formed will be reduced (even at 100% conversion!). As an analogy to explain the result of an imbalanced stoichiometric feed, assume that each person in a classroom is a monomer—girls are AA's and boys are BB's. They have to link hands girl–boy–girl, etc. If there are an equal number of boys and girls, a complete ring can be made with all hands connected. However, if there are fewer girls in the class, then there will be several B hands that are left unlinked, even when all of the girls' A hands are connected. This causes a serious drop in the molecular weight for condensation polymers and is of utmost importance in designing reactors for step-growth polymerizations. We can use similar (but more involved) reasoning for Example 8.1 to derive

$$\bar{x}_n = \frac{1+r}{2r(1-p)+(1-r)} = \frac{1+r}{1+r-2rp} \tag{8.7}$$

where $r = N_{A0}/N_{B0}$ is the stoichiometric ratio of the functional groups present (A is taken to be the limiting reactant, so r is always less than or equal to 1, and p represents the fraction of A groups reacted).

To examine the effect of stoichiometric imbalance, consider the limiting case of complete conversion, $p = 1$, for which Equation 8.7 reduces to

$$(\bar{x}_n)_{\max} = \frac{1+r}{1-r} \quad (\text{for } p = 1) \tag{8.8}$$

When A and B groups are supplied in stoichiometric equivalence ($r = 1$), all the reactant molecules are combined in a single molecule of essentially infinite molecular weight. Reducing r to 0.99 reduces $(\bar{x}_n)_{\max}$ to 199, and for $r = 0.95$, all the way down to 39. This is an extremely important concept! To reach the large chain lengths needed for useful materials, step-growth polymerizations must have functional groups supplied at or very close to stoichiometric equivalence. Considering the usual industry purity levels and precision of weight techniques, this is not always easy to achieve. One way to avoid this problem is to use a difunctional monomer of the type ARB, with the stoichiometric equivalence built into the monomer. This is not always feasible or possible, so many step-growth polymerization reactions also require careful reactor design. For the production of nylon 6/6, for example, where adipic acid (ARA) and hexamethylene diamine (BR'B) are used as the monomers (see Example 2.4E), the monomers form an ionic salt of a perfect 1:1 ratio ("nylon salt"), which is carefully purified by crystallization before polymerization! Another option for forming nylons is to use interfacial polymerization, with each monomer in a separate liquid phase (see Chapter 12). Also, chain length may sometimes be pushed up despite an initial stoichiometric imbalance by eliminating excess amount of a reactant. For example, if a polyester is made from a diacid plus and an excess amount of glycol, the reaction must stop at a point when all the chains are capped by OH groups. If the glycol is volatile enough to be driven off by applying heat and vacuum, however, the chains may combine further with the elimination of excess amount of glycol:

$$HO-R-O \overbrace{\underset{\substack{\| \\ O}}{C}-R'-\underset{\substack{\| \\ O}}{C}-O-R-O}^{}_x H \ + \ H \overbrace{O-R-O-\underset{\substack{\| \\ O}}{C}-R'-\underset{\substack{\| \\ O}}{C}}^{}_y O-R-OH$$

$$\longrightarrow \ HO-R-O \overbrace{\underset{\substack{\| \\ O}}{C}-R'-\underset{\substack{\| \\ O}}{C}-O-R-O}^{}_{(x+y)} H \ + \ HO-R-OH$$

On the other hand, reducing r by the deliberate introduction of an excess amount of one of the monomers or of some monofunctional material provides a convenient means of limiting chain length when desired.

8.4   CHAIN LENGTHS ON A WEIGHT BASIS

The previous development of the distribution of chain lengths (Chapter 5) in a linear step-growth polymer, although perfectly legitimate, is in some ways misleading, because it describes the *number* of molecules of a given chain length present and equally counts both monomer units and chains containing many hundreds of monomer units, that is, each is one molecule. For example, in a mixture consisting of one monomer molecule and one 100-mer, the number or mole fraction of each is 1/2. Another way of looking at it is to inquire about the relative weights of the various chain lengths present. On this basis, the weight fraction of monomer in the mixture is only 1/101. (This assumes that the monomer and the repeat unit have the same molecular weight.) By neglecting the weight of the small molecule that splits out in condensation reactions, we can most simply obtain the weight-fraction distribution of chain lengths by combining Equations 5.33, 8.3 and 8.6 to get

$$\frac{w_x}{W} = x p^{(x-1)} (1-p)^2 \tag{8.9}$$

This "most-probable" weight fraction distribution is shown in Figure 8.2. Although the smaller molecules (monomers, dimers, trimers, etc.) are the most numerous, their

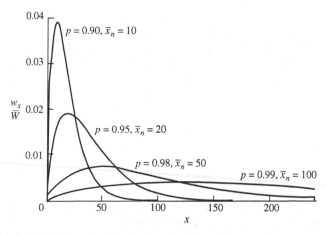

FIGURE 8.2   Weight-fraction distributions for linear step-growth polymers (Eq. 8.9).

combined weight is an insignificant portion of the total weight. It should be noted that step-growth polymerization typically results in very few monomers, as most of them react to form at least dimers or trimers. This is an important point of contrast with chain-growth polymerization (Chapter 9), where only large molecular weight polymers and monomers exist as the conversion increases. In the figure, the peak is at $x = -1/(\ln p) \approx 1/(1-p)$, that is, the x-mer present in greatest weight is approximately $\bar{x}_n$. The neglect of a molecule of condensation (reacted and unreacted A groups do not have the same weight, but each of them is part of one monomer residue) in this derivation can lead to significant errors at low x's. The exact solution is quite complex [1].

The weight-average chain length, $\bar{x}_w$, can be obtained by inserting the number distribution, Equation 8.3 into Equation 6.23 and integrating at constant p:

$$\bar{x}_x = \frac{\int_0^\infty (n_x/N)x^2\,dx}{\int_0^\infty (n_x/N)x\,dx} = \frac{\int_0^\infty x^2 p^{(x-1)}dx}{\int_0^\infty x p^{(x-1)}dx} = \frac{1+p}{1-p} \tag{8.10}$$

(Evaluation of the above integrals is a good exercise for mathematical masochists!) Combining Equations 8.10 and 8.6 gives an expression for the polydispersity index in terms of conversion:

$$\frac{\bar{x}_w}{\bar{x}_n} = 1 + p \quad (0 \leq p \leq 1) \tag{8.11}$$

8.5   GEL FORMATION

If a step-growth polymerization batch contains some monomer with a functionality $f > 2$ and if the reaction is carried to a high enough conversion, a cross-linked network or *gel* may be formed (Figure 8.3). In this context, a gel is defined as a molecule of essentially infinite molecular weight, extending throughout the reaction mass. In the production of thermosetting polymers, the reaction must be terminated short of the conversion at which a gel is formed or

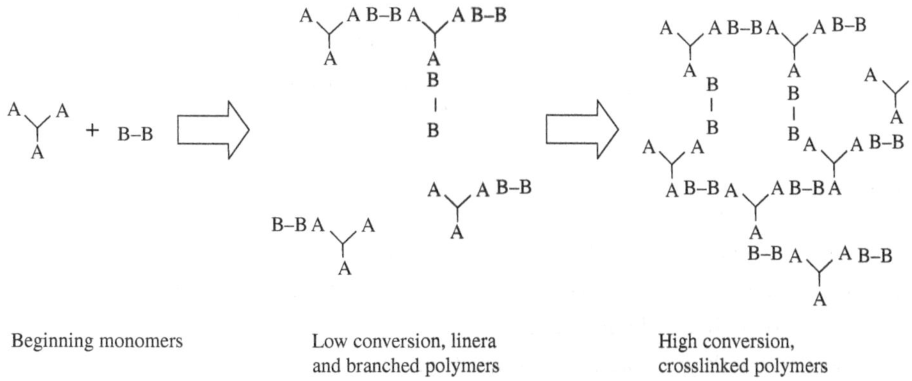

Beginning monomers   Low conversion, linera   High conversion,
                          and branched polymers   crosslinked polymers

FIGURE 8.3   Time course of reaction of multifunctional A to difunctional B monomer. First small polymers are formed that are not crosslinked, but as enough A and B groups react, a tight network forms.

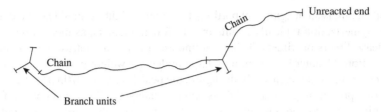

FIGURE 8.4 A growing polymer chain with two branch points indicating the position of the trifunctional monomers that have been added.

the product could not be molded or processed further (crosslinking is later completed in the mold). Hence, the prediction of this *gel point* conversion is of great practical importance.

The multifunctional monomer is represented by A_f for example,

$$f = 3 \qquad\qquad f = 4$$

while the reaction mixture usually also consists of difunctional monomers of the form ARA and BR'B. In a gel network, the multifunctional monomer acts as a *branch unit*. These branch units are connected to *chains*, which are linear segments of difunctional units, that lead either to another branch unit or to an unreacted end (Figure 8.4).

The branching coefficient α is defined as the probability that a given functional group on a branch unit is connected to another branch unit by a chain (rather than to an unreacted end).

Further analysis is based on two assumptions. (1) All functional groups are equally reactive. This might not always be true. For example, the secondary (middle) hydroxyl group in glycerin ($HOCH_2-CH(OH)-CH_2OH$) is probably not as reactive as the primary (end) hydroxyls, due to steric hindrances. (2) Furthermore, intramolecular condensations (or polymer back-biting, i.e., reactions between A and B groups *on the same molecule*) do not occur. In other words, each reaction between an A and a B reduces the number of molecules in the reaction mass by one.

Consider the reaction:

$$\underset{\substack{\uparrow\\1}}{AA} + BB + A_f \rightarrow A_{f-1}A-[\underset{\substack{\uparrow\\2}}{BB}-\underset{\substack{\uparrow\\3}}{AA}-]_i\underset{\substack{\uparrow\\4}}{BB}-AA_{f-1}$$

$$\overset{\text{Branch}}{\underset{\text{unit}}{}} \quad \overset{\text{Linear}}{\underset{\text{chain}}{}} \quad \overset{\text{Branch}}{\underset{\text{unit}}{}}$$

where

$\rho =$ fraction of the *total* A groups in the branch units A_f

$(1 - \rho) =$ fraction of the total A groups in the difunctional AA molecules

$p_A =$ probability of finding a reacted A group

$p_B =$ probability of finding a reacted B group

The probabilities of finding each of the numbered types of bonds in the polymer shown on the right of the above reaction are tabulated below.

Bond	Probability of Formation	
1	p_A	probability of A reacting with B
2	$p_B(1 - \rho)$	probability of B reacting with A on AA
3	p_A	probability of A reacting with B
4	$p_B \rho$	probability of B reacting with A on A_f

Therefore, the probability of finding the chain shown, which, counting from left to right, contains 1-type one bond, i-type two bonds, i-type three bonds, and 1-type four bond is:

$$p_A[p_B(1 - \rho)p_A]^i p_B \rho$$

From the standpoint of connecting branch units, i may have any value between zero and infinity. Thus, to obtain the probability of finding a chain of any i that connects the branch units, we must sum over all i to find the branching coefficient:

$$\alpha = \sum_{i=0}^{\infty} p_A[p_B(1 - \rho)p_A]^i p_B \rho = \frac{p_A p_B \rho}{1 - p_A p_B(1 - \rho)} \tag{8.12}$$

Now,

$$r = \frac{N_{A0}}{N_{B0}} \tag{8.13}$$

$$p_A = \frac{N_{A0} - N_A}{N_{A0}} \tag{8.14}$$

$$p_B = \frac{N_{B0} - N_B}{N_{B0}} \tag{8.15}$$

From the stoichiometry of the reaction, the number of reacted A groups must be equal to the number of reacted B groups:

$$N_{A0} - N_A = N_{B0} - N_B \tag{8.16}$$

Therefore,

$$p_B = \frac{N_{A0}}{N_{B0}} p_A = r p_A \tag{8.17}$$

and

$$\alpha = \frac{r p_A^2 \rho}{1 - r p_A^2(1 - \rho)} = \frac{p_B^2 \rho}{r - p_B^2(1 - \rho)} \tag{8.18}$$

Each of the terminal branch units in the polymer molecule on the right above has $(f-1)$ unreacted functional groups. If at least one of these unreacted A groups is then connected to another branch unit, a gel is formed. Since all the unreacted A groups on branch units are statistically identical, all the branch units in the reaction mass will be connected together by chains, which is our definition of a gel. (In general, there will still be some monomer and linear chains floating throughout the gel network, i.e., the reaction mass is not a single molecule at the gel point.) This occurs when $\alpha\,(f-1) \geq 1$, that is, when the product of the probability of a chain connecting a given functional group to another branch unit and the number of remaining unreacted functional groups becomes a certainty. Thus, the critical value of the branching coefficient for gelation is:

$$\alpha_c = \frac{1}{f-1} \qquad (8.19)$$

Combining Equations 8.18 and 8.19 gives the gel-point conversion, p_c.

Example 8.2 Calculate the gel-point conversion of a reaction mass consisting of 2 mol of glycerin plus 3 mol of phthalic anhydride.

Solution. By convention, A should be chosen as the functional group on the multifunctional monomer. Here, $A = OH$, $f = 3$, and $\alpha_c = 0.5$ (Eq. 8.19). Note that there is no difunctional AA molecule in this problem. Also, $N_{A0} = N_{B0}$, so $r = 1$. Since all the A $(-OH)$ groups are on the multifunctional (glycerin) molecules, $\rho = 1$. Plugging these values into Equation 8.18 and solving gives $p_A = 0.707$.

Experimentally, the gel-point conversion for the system in Example 8.2 was found to be $p_A = 0.765$. This was done by noting the conversion at which bubbles ceased to rise in the reaction mass. The difference between the experimental and the theoretical results can be accounted for by some intramolecular condensation (back-biting) and the lower reactivity of the secondary hydroxyls at the center of the glycerin molecules. Also, Bobalek *et al.* [3] showed that as large molecules grew within the reaction mass, their molecular weights became high enough to cause them to precipitate from the solution before a true, infinite gel network was formed (a particle just 1 μm in diameter has a molecular weight in the order of 10^{11} g/mol). Higher conversions were needed experimentally to link these precipitated particles together to prevent the rise of bubbles.

Regardless of the reasons, however, the theoretical calculation usually gives the *lower limit*, or most conservative value, for the gel-point conversion, giving a margin of safety in practice. An apparent exception can sometimes occur. In concentrated solutions, physical entanglements of chains can cause a polymer to behave as a crosslinked gel at conversions lower than that predicted by this theory. Also, in the preparation of unsaturated polyester resins (see Example 2.3), the double bond that would be assumed inert in the condensation reaction participates in crosslinking reactions during condensation polymerization. This, in effect, makes the value of f much greater than that for condensation alone, which results in premature gelling. Additional crosslinking is normally counteracted by incorporating an inhibitor (see Chapter 9) for addition reactions during the polycondensation (step-growth) reaction.

Example 8.3 Calculate the minimum number of moles of ethylene glycol (I) needed to produce a gel when reacted with 1 mol of phthalic anhydride (II) and 1 mol of BTDA (III).

HO–CH$_2$–CH$_2$–OH

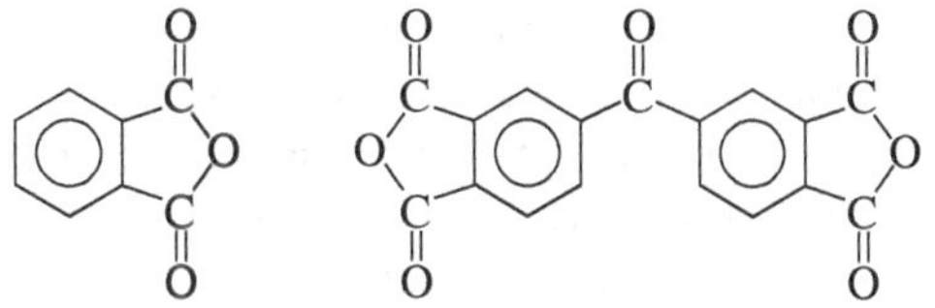

(I)   (II)   (III)

Solution. First, recognize that BTDA (III) is a dianhydride and, therefore, can be considered a tetrafunctional acid. Again, A is by convention the functional group on the multifunctional monomer. Here, A = COOH, and (I) = BB, (II) = AA, and (III) = A$_4$.

The tendency toward gelation increases with conversion, and with the minimum moles of (I) present, A groups will be in stoichiometric excess so that B is the limiting reactant. Therefore,

$$(p_B)_{max} = 1.0 \quad \text{for minimum (I)}$$

and from Equation 8.19:

$$\alpha_c = \frac{1}{f-1} = \frac{1}{4-1} = \frac{1}{3}$$

Here,

$$\rho = \frac{\text{moles A on (III)}}{\text{moles A on [(II) + (III)]}} = \frac{4}{6} = \frac{2}{3}$$

At the gel point, with $p_B = 1$ and $\rho = 2/3$, using Equation 8.18,

$$\frac{1}{3} = \frac{r p_A^2 \rho}{1 - r p_A^2 (1-\rho)} = \frac{p_B^2 \rho}{r - p_B^2 (1-\rho)}$$

$$r = \frac{7}{3} = \frac{N_{A0}}{N_{B0}}$$

$$N_{B0} = \frac{N_{A0}}{r} = 6\left(\frac{3}{7}\right) = \frac{18}{7}$$

But, $N_{B0} = 2 N_I$, that is, there are two moles of hydroxyl per mole of glycol. Therefore,

$$N_I = \frac{N_{B0}}{2} = \frac{18}{14} = \frac{9}{7}\text{mol}$$

8.6 KINETICS OF POLYCONDENSATION

We now know something about the chemistry of forming step-growth polymers and about the relation of the length of the polymer chains with the reaction conversion; but how long do these reactions occur? The kinetics of polycondensation reactions is similar to that of ordinary condensation reactions. Since the average chain length is related to conversion in linear polycondensation by Equation 8.7, and the conversion is given as a function of time by the kinetic expression, $\overline{x}_n$ is directly related to the reaction time and can thus be controlled by limiting the reaction time. Similarly, the time to reach a gel point is related by the rate expression and Equations 8.18 and 8.19. The rate of reaction depends on several parameters, including the concentrations of monomer species, as demonstrated in Example 8.4, and the temperature (through usual Arrhenius behavior, in that an increase in temperature raises kinetic rate constants and speeds up the reaction rate, see Chapter 9).

Example 8.4 For a second-order, irreversible polycondensation reaction with rate proportional to the concentrations of reactive A and B groups, obtain expressions for conversion and number-average chain length as a function of time for a stoichiometrically equivalent batch.

Solution. For a second-order reaction, assuming constant volume,

$$\frac{-d[A]}{dt} = k[A][B] \tag{8.20}$$

where k is the reaction-rate constant and the brackets indicate concentrations. For $[A]_o = [B]_o$, $[A] = [B]$ at all times, and

$$\frac{-d[A]}{dt} = k[A]^2 \tag{8.21}$$

Separating variables and integrating between the limits when $t = 0$, $[A] = [A]_o$ and when $t = t$ and $[A] = [A]$ gives

$$\frac{1}{[A]} - \frac{1}{[A]_0} - kt \tag{8.22}$$

Since

$$p = ([A]_0 - [A])/[A]_0 = ([A]_0 kt)/(1 + [A]_0 kt). \tag{8.23}$$

combining Equations 8.6 and 8.23 gives

$$\overline{x}_n = 1 + [A]_0 kt \qquad (8.24)$$

Therefore, the number-average chain length increases linearly with time.

Two cautions are in order about the preceding example. First, by writing an *irreversible* rate expression, we have assumed that any molecule of condensation is being continuously and efficiently removed from the reaction mass so that there is no depolymerization. Second, not all step-growth reactions are of second order. Some polyesterifications, for example, are catalyzed by their own acid groups and are, therefore, first order in hydroxyl concentration, second order in acid, and third order overall. The rate may also be proportional to the concentration of an added catalyst (usually acids or bases for polycondensations), if used.

References 4 and 5 provide a more comprehensive and detailed look at step-growth polymerizations, extending and applying many of the concepts introduced here.

PROBLEMS

1   Crud Chemicals wishes to limit chain length in their linear polycondensation reaction by adding monofunctional B to the equimolar AA, BB reactant mix. Obtain an expression for the maximum number-average chain length possible at 100% conversion, $(\overline{x}_n)_{max}$, when N_B moles of B are added per mole of AA or BB.

2   One mole of hydroxybutyric acid (HO−(CH$_2$)$_3$−COOH) is placed in a reactor that reaches 50% conversion.

   (a)   How many moles of unreacted COOH groups are left in the reactor? How many moles of unreacted monomer are left in the reactor?

   (b)   What are the average $\overline{M}_n$, $\overline{M}_w$, and PDI for the reaction mixture?

3   Polycarbonate is made by mixing 20 kg bisphenol A with 15 kg phosgene (see Example 2.4N) in a batch reaction. What are the maximum number-average chain length, x_n, and maximum number-average molecular weight, M_n, of this polymer?

4   (a)   Starting with Example 8.3, calculate the maximum number of moles of ethylene glycol that could be added to the reaction mixture to form a gel when reacted with 1 mol of phthalic anhydride and 1 mol of BTDA.

   (b)   Sketch qualitatively a triangular ternary diagram for the A4, AA, BB system at maximum conversion that shows the gel and no-gel regions. (A sample ternary diagram is shown in Figure 7.6.) Locate the solution to part (a) and the solution to Example 8.3 in your sketch.

   (c)   Describe what happens to prevent gel formation "just across the border" in each case.

5   Calculate the gel point conversion (based on the limiting reactant) for reacting 3 mol ethylene glycol (Example 2.4D) with 1 mol of tetrafunctional butyl dianhydride (Example 2.4R, but replace the "R" group with (CH$_2$)$_4$).

6   (a)   Modify Equation 8.18 to handle a condensation reaction batch that contains monofunctional B in addition to the A$_f$, A−A and B−B present in the original

analysis. You will need to include one additional stoichiometric parameter:

$$\beta = \text{fraction of total B's on the B–B molecules.}$$

(b) Oil-based paints are made by condensing glycerin, phthalic anhydride, and a drying oil. The drying oil is an unsaturated monoorganic acid R-COOH (e.g., linseed oil). The paint film is crosslinked after application of an addition reaction involving the double bonds in R. Since the polymer must be soluble for application, it cannot be allowed to form gel during the condensation reaction. With a 2/3 ratio of glycerin/phthalic anhydride, what is the minimum amount of drying oil that must be incorporated in the reaction mass to prevent gel formation during the condensation reaction?

(c) For the type of A_f, B–B, B system in (b), sketch qualitatively a ternary diagram similar to the one in Figure 7.6 showing the gel/no-gel composition regions at maximum conversion. Locate the solution to (b) in the diagram.

7 (a) How many moles of monofunctional A must be added to a mixture of 1 mol of BB and 1 mol of A_3 to prevent gelation at maximum conversion?

(b) Sketch a ternary diagram for the A_3, BB, A system showing the gel/no-gel regions at maximum conversion. Locate the solutions to part (a) in your sketch.

8 Modify the bond probability of the formation table in Section 8.5 to handle a reaction mass that contains monofunctional A in addition to the AA, BB, and A_f present in the original analysis. You will need one additional stoichiometric parameter:

$$a = \text{fraction of A groups on the monofunctional A's.}$$

9 See Example 8.2. Given 1 mol of glycerin, over what range may the moles of phthalic anhydride be varied with the possibility of forming a gel?

10 Calculate the gel/no-gel boundaries and plot the ternary diagram showing the gel/no-gel regions for the A_3, BB, B system at maximum conversion.

11 Crud Chemicals is producing a casting resin that consists of an AB-type monomer to which a catalyst is added to promote polymerization in the mold. The material has many useful properties but is deficient in heat and solvent resistance. A proposal is made to improve heat and solvent resistance by adding a tetrafunctional monomer to the mix to provide crosslinks. Unfortunately, no one in-house knows if this will work or how to calculate the minimum amount of A_4 needed to form a gel; Therefore, they call you in as a consultant. Evaluate the proposal for them.

12 A step-growth polymerization is started with the following initial composition:

A_3	1 mol
AA	1 mol
B	1 mol
BB	2 mol

At $p_A = 0.5$, what is the numerical probability of finding one branch unit connected to another by a single BB?

13 Calculate the minimum number of moles of phthalic anhydride needed to form a gel when reacted with 1 mol each of glycerin, ethylene glycol, and propylene glycol.

14 Obtain the expression that relates $\bar{x}_n$ to time for a linear polyesterification that is catalyzed by its own acid groups, that is, second order in COOH and first order in OH. Assume an irreversible reaction at constant volume and $[A]_o = [B]_o$.

15 Assume that the reaction in Example 8.2 follows the kinetic scheme outlined in Section 8.6. Obtain an expression (in terms of the rate constant k) for the time to reach the gel point t_{gel}.

REFERENCES

[1] Grethlein, H.E., *Ind. Eng. Chem. Fund.* **8**, 206 (1969).

[2] Flory, P.J., *Principles of Polymer Chemistry*, Cornell University Press, Ithaca, NY, 1953.

[3] Bobalek, E.G., *et al.*, *J. Appl. Polym. Sci.* **8**, 625 (1964).

[4] Gupta, S.K. and A. Kumar, *Reaction Engineering of Step-Growth Polymerizations*, Plenum, New York, 1987.

[5] Odian, G., *Principles of Polymerization*, Wiley, New York, 1991.

CHAPTER 9

FREE-RADICAL ADDITION (CHAIN-GROWTH) POLYMERIZATION

9.1 INTRODUCTION

One of the most important types of addition polymerization is initiated by the action of *free-radicals*, electrically neutral species with an *unshared electron*. Free-radicals are highly reactive and these types of polymerizations begin with the formation of free-radicals by an initiation reaction. In the developments to follow, a dot (·) will represent a single electron or free-radical. The single bond, a pair of shared electrons, will be denoted by a double dot (:) or, where it is not necessary to indicate electronic configurations, by the usual covalent bond sign ($-$). A double bond, two shared electron pairs, is indicated by :: or $=$.

Free-radicals for the initiation of addition polymerization are usually generated by the breakdown of a chemical initiator. These initiators are classified by the method used to catalyze the breakdown: *thermal* initiators, *redox* initiators, or *photo* initiators.

Thermal initiators form free-radicals by the thermal decomposition of compounds such as organic peroxides or azo compounds. Two common examples of thermal initiators are benzoyl peroxide and azobisisobutyronitrile:

$$\phi-\overset{\overset{\displaystyle O}{\|}}{C}-O{:}O-\overset{\overset{\displaystyle O}{\|}}{C}-\phi \quad \longrightarrow \quad 2\,\phi-\overset{\overset{\displaystyle O}{\|}}{C}-O{\cdot} \quad \longrightarrow \quad 2\,\phi{\cdot} \ + \ 2CO_2$$

$$(CH_3)_2-\underset{\underset{\displaystyle N}{\overset{\displaystyle |}{\underset{\displaystyle \|}{C}}}}{C}{:}N{=}N{:}\underset{\underset{\displaystyle N}{\overset{\displaystyle |}{\underset{\displaystyle \|}{C}}}}{C}-(CH_3)_2 \quad \longrightarrow \quad 2(CH_3)_2-\underset{\underset{\displaystyle N}{\overset{\displaystyle |}{\underset{\displaystyle \|}{C}}}}{C}{\cdot} \ + \ N_2$$

To start these reactions, moderately high temperatures (70–100 °C) are needed.

Redox initiators form free-radicals by an oxidation–reduction reaction using chemicals such as ammonium persulfate, usually in combination with an accelerator, such as N,N,N′,N′-

Fundamental Principles of Polymeric Materials, Third Edition. Christopher S. Brazel and Stephen L. Rosen.
© 2012 John Wiley & Sons, Inc. Published 2012 by John Wiley & Sons, Inc.

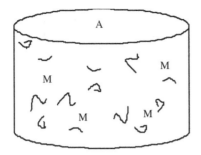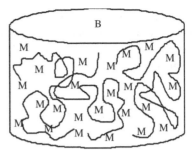

FIGURE 9.1 Comparison of reactor contents at 50% conversion (a) step-growth or condensation polymerization and (b) addition (or free-radical) polymerization. M = monomer.

tetramethyl ethylene diamine. These reactions can begin a polymerization at lower temperatures. Photo initiators are used less frequently, but can form free-radicals upon exposure to (usually) ultraviolet light. In each case, one initiator molecule forms two free-radicals, often with the formation of a gas. Once the free-radical is formed from the initiator, the polymerization can begin.In contrast to step-growth polymerizations, where condensation can occur between two units of any chain length, addition polymerizations add monomers one at a time to a growing polymer chain (Figure 9.1). This has an important consequence for forming large chains: in addition polymerizations, at any given time during the reaction, the remaining monomer can be easily separated (by evaporation), leaving behind high molecular weight polymers that are mechanically useful (even at somewhat low conversions); recall that for step-growth polymers, conversions topping 99% are required to get even moderate molecular weight materials.

9.2 MECHANISM OF POLYMERIZATION

The initiator molecule, represented in a generic sense by I, undergoes a first-order decomposition with a rate constant k_d to give two free-radicals, $2\,R\cdot$:

$$I \xrightarrow{k_d} 2R \cdot \quad \text{Decomposition} \tag{9.1}$$

This reaction has a rate constant k_d. The radical then adds a monomer by grabbing an electron from the electron-rich double bond, forming a single (R−C) bond with the monomer, but leaving an unshared electron at the other end:

$$
\begin{array}{cc}
& \text{H H} & \text{H H} \\
& | \ | & | \ | \\
\text{R}\cdot + & \text{C::C} \rightarrow \text{R:C:C}\cdot \\
& | \ | & | \ | \\
& \text{H X} & \text{H X}
\end{array}
$$

Addition (Activation)

This may be abbreviated by

$$\text{R}\cdot + M \xrightarrow{k_a} P_1 \cdot \quad \text{Addition} \tag{9.2}$$

where M is the monomer, and $P_1\cdot$ represents a growing polymer with a single repeat unit with a free-radical attached to the other carbon that used to be part of the double bond. The rate constant for addition is k_a. Note that the initiator radical actually combines with the monomer, so when a polymer is formed by this method, the end group is normally different from the normal repeat units of the polymer. Fortunately, due to the large size of most polymers, the end group supplied by the initiator is normally insignificant and is usually ignored. The other important thing to notice from Equation 9.2 is that the product of the addition reaction is still a free-radical; thus, it proceeds to propagate the chain by adding another monomer unit:

$$P_1 \cdot + M \xrightarrow{k_p} P_2 \cdot$$

again maintaining the unshared electron at the chain end, which adds another monomer unit:

$$P_2 \cdot + M \xrightarrow{k_p} P_3 \cdot$$

and so on. In general, this propagation step is written as

$$P_x \cdot + M \xrightarrow{k_p} P_{(x+1)} \cdot \quad \text{Propagation} \tag{9.3}$$

where k_p is the rate constant for the propagation step. We have again assumed that reactivity is independent of chain length by using the same k_p for each propagation reaction.

The chains continue to grow as long as monomers continue to be available and accessible in the reaction mixture. The reaction ends by a termination step using one of two methods. Two chains can bump together and stick, with their unshared electrons combining to form a single bond between them (combination):

$$P_x \cdot + P_y \cdot \xrightarrow{k_{te}} P_{(x+y)} \quad \text{Combination} \tag{9.4a}$$

where $P_{(x+y)}$ is a dead polymer chain of $(x+y)$ repeating units. Or, one can abstract a proton from the penultimate carbon on the other forming two dead chains with lengths x and y (disproportionation):

$$P_x \cdot + P_y \cdot \xrightarrow{k_{td}} P_x + P_y \quad \text{Disproportionation} \tag{9.4b}$$

The relative proportion of each termination mode depends on the particular polymer and the reaction temperature, but in most cases, one or the other predominates. Note that another potential method for terminating a polymerization reaction is by chain transfer, where the free-radical is transferred to another species in the reaction mixture, such as a solvent, as discussed in Section 9.6.

9.3 GELATION IN ADDITION POLYMERIZATION

A significant portion of Chapter 8 was devoted to describing gelation and the formation of network polymers by step-growth polymerization, including the gel point conversion, α_c, for such reactions. In free-radical polymerizations, all monomers have the same reactive functional group (the alkene double bond). Monomers that have only one C=C produce linear polymers, but those with two or more alkenes normally result in branch points and the formation of crosslinked networks. Such multifunctional monomers are termed *crosslinking agents*, and need only be added at 1 mol% (or even less) to successfully form a network. Because free-radical polymers add monomers to long growing chains, these crosslinking agents are added just as the other monomers and begin to form crosslinked networks, even at low conversions (particularly if the amount of crosslinking agent is high). The three-part mechanism (initiation, propagation, and termination) is largely the same as above. The most important parameters in determining the structure of these polymers are the fraction of crosslinking agents added and their functionality (number of double bonds). The resulting structures are characterized by $\overline{M}_c$, the molecular weight between crosslinks. Lightly crosslinked gels will swell appreciably in a good solvent, while those made with high concentrations of crosslinking agents are often brittle and easily crack. Because a gel point conversion is much less important for networks made by free-radical polymerization, no further discussion is included in this chapter.

9.4 KINETICS OF HOMOGENEOUS POLYMERIZATION

Free-radical polymerizations can take place in either homogeneous (single phase) or heterogeneous (multiple phase) mixtures. In this section, we will consider only homogeneous reactions, where the monomer is either reacted in bulk or a good solvent is used so that the monomer, initiator, and polymers of all lengths are soluble in a single phase. To determine the rate of polymerization, we must consider the rates of each of the steps (initiation, addition, propagation, and termination) involved in the mechanism.

In practice, not all the radicals generated in Equation 9.1 actually initiate chain growth as in Reaction (9.2). Some recombine through the reverse reactions, or others are used up by side reactions. Thus, f, the fraction of radicals generated in the initiation step that actually initiate chain growth through Reactions 9.1 and 9.2, is important in developing rate equations:

$$r_i = \left(\frac{1}{V}\right)\left(\frac{dP_1 \cdot}{dt}\right) = 2fk_d[I] \quad \text{Initiation rate} \tag{9.5}$$

where V is the volume of the reaction mass, $P_1 \cdot$ is the moles of chain radicals with $x = 1$, and $[I]$ is the initiator concentration (moles/volume). In the notation used here, the quantity Q of a species without brackets represents the moles of the species and when enclosed in square brackets, $[Q]$, the *molar concentration* of the species. They are related by

$$[Q] = \frac{Q}{V} \tag{9.6}$$

Although it is rarely mentioned explicitly, Equation 9.5 is based on the generally valid assumption that the rate of Reaction (9.2) is much greater than that of Reaction (9.1), that is, the initiator decomposition is *rate controlling*. Thus, as soon as an initiator radical is formed, it grabs a monomer molecule, starting chain growth, so k_a does not appear in the rate expression.

According to Reaction (9.3), the rate of monomer removal in the propagation step is

$$r_p = -\left(\frac{1}{V}\right)\left(\frac{dM}{dt}\right)_p = k_p[M][P\cdot] \quad \text{Propagation rate} \tag{9.7}$$

where M is the moles of monomer and $[P\cdot]$ is the total concentration of growing chain radicals of all lengths:

$$[P\cdot] = \sum_{x=1}^{\infty}[P_x\cdot] \tag{9.8}$$

The rate of removal of chain radicals is the sum of the rates of the two termination reactions. Since both are second order

$$r_l = -\left(\frac{1}{V}\right)\left(\frac{dP\cdot}{dt}\right) = 2k_t[P\cdot]^2 \quad \text{Termination rate} \tag{9.9}$$

where the termination rate constant, k_t, adds the rate constants for termination by combination (k_{tc}) and by disproportionation (k_{td}):

$$k_t = k_{tc} + k_{td} \tag{9.10}$$

Unfortunately, the unknown quantity $[P\cdot]$ is present in the equations. It cannot be measured during a reaction, but can be removed from the equations by making the standard kinetic assumption of a steady-state concentration of a transient species, in this case, the chain radicals, such that $d[P\cdot]/dt = 0$ during the reaction. For $[P\cdot]$ to remain constant, chain radicals must be generated at the same rate at which they are removed. Thus, we assume that the rates of initiation and termination must be equal ($r_i = r_t$), or

$$2fk_d[I] = 2k_t[P\cdot]^2 \tag{9.11}$$

This gives the chain-radical concentration as

$$[P\cdot] = \left(\frac{fk_d[I]}{k_t}\right)^{1/2} \tag{9.12}$$

which, when inserted into Equation 9.7 provides an expression for the rate of monomer consumption in the propagation reaction:

$$r_p = -\left(\frac{1}{V}\right)\left(\frac{dM}{dt}\right)_p = k_p\left(\frac{fk_d}{k_t}\right)^{1/2}[I]^{1/2}[M] \tag{9.13}$$

One molecule of monomer is consumed in the addition step (Eq. 9.2) also, but for long chains this is insignificant, so that Equation 9.13 may be taken as the overall rate of polymerization, the rate at which monomer is converted to polymer.

Equation 9.13 is the classical rate expression for a homogeneous, free-radical polymerization. It has been an extremely useful approximation over the years, particularly in fairly dilute solutions, but deviations from it are sometimes observed. It is now evident that many deviations are due to the fact that the termination reaction is diffusion controlled, so k_t is not really a constant but decreases in time as the polymerization proceeds. In certain cases of practical interest, these deviations can be of major significance. They are discussed in Chapter 12 on polymerization practice.

It is often more convenient to work in terms of *monomer conversion X*. By making use of the definition of conversion:

$$X \equiv \frac{M_0 - M}{M_0} \tag{9.14}$$

and Equation 9.6 for M, we can rewrite Equation 9.13 as

$$\frac{dX}{dt} = k_p \left(\frac{fk_d}{k_l}\right)^{1/2} [I]^{1/2}(1 - X) \tag{9.13a}$$

Note that the conversion X used here is a very different quantity than the p used in Chapter 8 for step-growth polymers. Here, X is the fraction of *monomer molecules* that have reacted to form polymer. The quantity p represents the fraction of *functional groups* that have reacted (whether on a monomer or part of a growing polymer chain). In step-growth polymerization, each monomer molecule has several functional groups that react independently, so p is not the fraction of reacted monomer. (Compare to free-radical addition, where both sides of the double bond are reacted in sequence.)

Integration of Equation 9.13a gives conversion as a function of time. A form occasionally used in studying the initial stages of an isothermal, batch reaction is obtained by assuming a constant initiator concentration $[I]_o$ with the initial condition $X = 0$ at $t = 0$:

$$X = 1 - \exp\left\{-k_p \left(\frac{fk_d}{k_l}\right)^{1/2} [I]_0^{1/2}t\right\} \quad \text{(constant } [I]) \tag{9.15}$$

The assumption of constant $[I]$ may not be realistic at high conversions for two reasons. First, the initiator concentration can decrease significantly with time. Second, the volume of the reaction mass will, in general, change with conversion. Most liquid-phase addition polymerizations undergo a density increase of 10–20% from pure monomer ($X = 0$) to polymer ($X = 1$) (carrying out the reaction in an inert solvent reduces this, of course).

We can include the first-order initiator decay:

$$\left(\frac{1}{V}\right)\left(\frac{dI}{dt}\right) = -k_d[I] = -k_d\left(\frac{I}{V}\right) \tag{9.16a}$$

and integrate from $t = 0$ with $I = I_o$ to get

$$I = I_0 \exp(-k_d t) \tag{9.16b}$$

We then assume that the volume of the reaction mass is linear with conversion, as outlined by Levenspiel [1]:

$$V = V_0(1 + \varepsilon X) \tag{9.17}$$

where ε is the fractional change in volume from $X = 0$ to $X = 1$ and V_o is the volume of the reaction mass at $X = 0$. Combining Equations 9.6 (for I), 9.13, 9.16b, and 9.17 gives

$$\frac{dX}{dt} = k_p \left(\frac{fk_d}{k_t}\right)^{1/2} [I]_0^{1/2} \frac{1 - X}{(1 + \varepsilon X)^{1/2}} \exp\left(\frac{-k_d t}{2}\right) \tag{9.18}$$

The variables in Equation 9.18 may be separated and it may be integrated analytically with respect to time, at least, to give

$$\int_0^x \frac{(1 + \varepsilon X)^{1/2}}{1 - X} dX = \left(\frac{2k_p}{k_d}\right)\left(\frac{fk_d}{k_t}\right)^{1/2} [I]_0^{1/2} \left[1 - \exp\left(\frac{-k_d t}{2}\right)\right] \tag{9.19}$$

Unfortunately, the left side of Equation 9.19 cannot be integrated analytically by normal human beings. An explicit expression for X as a function of t is extremely difficult to determine, so calculation of $X(t)$ is normally done by numerical or by computational methods.

If we can neglect volume change ($\varepsilon = 0$), we can solve the integral explicitly to get

$$X = 1 - \exp\left\{\left(\frac{2k_p}{k_d}\right)\left(\frac{fk_d}{k_t}\right)[I]_0^{1/2} \left[\exp\left(\frac{-k_d t}{2}\right) - 1\right]\right\} \quad (\varepsilon = 0) \tag{9.20}$$

This expression has some interesting and important implications when compared with Equation 9.15. Setting $t = \infty$ reveals that there is a maximum attainable conversion that depends on $[I]_o$:

$$X_{\max} = 1 - \exp\left\{-\left(\frac{2k_p}{k_d}\right)\left(\frac{fk_d}{k_t}\right)^{1/2}[I]_0^{1/2}\right\} \quad (\varepsilon = 0) \tag{9.21}$$

This "dead-stop" situation is basically a matter of the initiator being used up before the monomer, but it is not revealed by Equation 9.15, which always predicts complete conversion in the limit of infinite time. Thus, the use of Equation 9.15 instead of Equation 9.19 or 9.20 can result in considerable error at high conversions, although the results approach one another at low conversions.

Example 9.1 For the homogeneous polymerization of pure styrene with azobisisobutyronitrile at 60 °C in an isothermal, batch reactor,

(a) compare Equations 9.15, 9.19, and 9.20 by plotting conversion versus time, using the data given below,

(b) calculate and compare the times needed to reach 90% conversion according to Equations 9.15, 9.19, and 9.20, and

(c) determine the minimum $[I]_o$ needed to achieve 90% conversion. Note that according to Equation 9.15, there is *no* minimum.

> Data: $k_p^2/k_t = 1.18 \times 10^{-3}$ L/mol,
> $k_d = 0.96 \times 10^{-5}$ s
> $[I]_o = 0.05$ mol/L
> $f = 1$
> $\varepsilon = -0.136$(13.6% shrinkage from monomer to polymer)
> Assume $\xi = 1$ (termination exclusively by disproportionation)

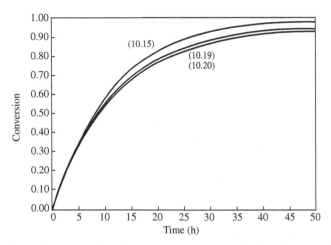

FIGURE 9.2   Conversion versus time for an isothermal, free-radical batch polymerization (data of Example 9.2): Equation 9.15, $[I] = $ constant; Equation 9.19, $\varepsilon = -0.136$; Equation 9.20, $\varepsilon = 0$.

Solution. (a) Equations 9.15 ($[I]$ constant) and 9.20 (V constant) may be solved analytically, but Equation 9.19 (both $[I]$ and V changing) requires a numerical solution. The results are shown in Figure 9.2. Graphically, Equations 9.19 and 9.20 are almost indistinguishable. Note that volume shrinkage increases concentrations and speeds up the reaction slightly.

(b) Numerical solution of Equation 9.19 gives a time of 34.0 h to reach 90% conversion. Neglecting initiator decay and solving Equation 9.15 for t gives 26.9 h (21% low!) and neglecting volume change (Eq. 9.20) gives 36.1 h (6.2% too high). In many cases, the error introduced by neglecting volume change is probably less than the precision with which the rate constants are known. The error will, however, increase with higher conversions.

(c) Numerical solution of Equation 9.19 with $t = \infty$ gives a minimum $[I]_o$ of 0.00989 mol/L for 90% conversion. Solving Equation 9.21 with $X_{max} = 0.90$ gives $[I]_o = 0.0108$, a 9.07% error from neglecting volume change.

Side reactions cause deviations from the classical kinetics. Agents that cause these reactions are generally categorized as *inhibitors* or *retarders* (though *accelerators* can also be used). An inhibitor delays the start of the reaction, but once begun, the reaction proceeds at the normal rate. Liquid vinyl monomers are usually shipped with a few parts per million of inhibitor to prevent polymerization in transit. A retarder slows down the reaction rate, which can be used to allow a reacting mixture to be worked with before setting. Some chemicals combine both effects. These are illustrated in Figure 9.3. Oxygen is a free-radical scavenger, so it behaves as an inhibitor for free-radical polymerization; thus, these reactions are normally carried out under a blanket of nitrogen.

9.5   INSTANTANEOUS AVERAGE CHAIN LENGTHS

As in step-growth polymerization, a distribution of chain lengths is always obtained in a free-radical addition polymerization because of the inherently random nature of the termination reaction with regard to chain length. Expressions for the number-average

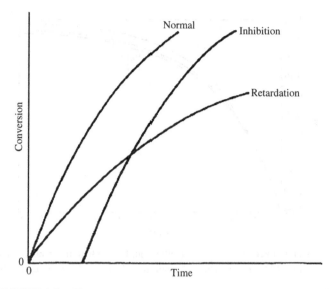

FIGURE 9.3   Change in reaction behavior due to inhibitor or retarder.

chain length are usually couched in terms of the kinetic chain length, ν, which is the rate of monomer addition to growing chains over the rate at which chains are started by initiator radicals, that is, the average number of monomer units per growing chain radical at a particular instant.[1] It thus expresses the efficiency of the initiator radicals in polymerizing the monomer:

$$\nu = \frac{r_p}{r_i} = \frac{k_p[M]}{2(fk_dk_t)^{1/2}[I]^{1/2}} \tag{9.22}$$

If the growing chains terminate exclusively by disproportionation, they undergo no change in length in the process, but if combination is the exclusive mode of termination, the growing chains, on average, double in length upon termination. Therefore,

$$\bar{x}_n = \nu \quad \text{(termination by disproportionation)} \tag{9.23a}$$

$$\bar{x}_n = 2\nu \quad \text{(termination by combination)} \tag{9.23b}$$

The average chain length may be expressed more generally in terms of a quantity ξ, the average number of dead chains produced per termination reaction, which equals the ratio of the rate of dead chain formation to the rate of termination reactions. Since each disproportionation reaction produces *two* dead chains and each combination *one*,

$$\text{Rate of dead chain formation} = (2k_{td} + k_{tc})[P\cdot]^2 \tag{9.24}$$

---

[1] Note that ν only includes the growing polymer chains, ignoring the monomer ($X = 1$) remaining in the reactor. This is different from the calculations of $\bar{x}_n$ in Chapter 8 for step-growth polymers, but is reflective of the length of polymers formed in free-radical polymerizations, assuming that the monomer remaining at the end of the reaction is separated from the polymer.

$$\text{Rate of termination reactions} = (k_{tc} + k_{td})[P \cdot]^2 \tag{9.25}$$

$$\xi = \frac{k_{tc} + 2k_{td}}{k_{tc} + k_{td}} = \frac{k_{tc} + 2k_{td}}{k_t} \tag{9.26}$$

The instantaneous number-average chain length is the rate of addition of monomer units to all chains (r_p) over the rate of dead chain formation:

$$\bar{x}_n = \frac{k_p[P\cdot][M]}{(2k_{td} + k_{tc})[P\cdot]^2} = \frac{k_p[M]}{\xi(fk_dk_t)^{1/2}[I]^{1/2}} \tag{9.27a}$$

When $k_{td} \gg k_{tc}$, $\xi = 2$, and when $k_{tc} \ll k_{td}$, $\xi = 1$, duplicating the previous result, but Equation 9.27a can also handle various degrees of mixed termination.

Keeping in mind that $\bar{x}_n$ is one of the most important factors in determining certain mechanical properties of polymers, what is the significance of Equation 9.27a A growing chain may react with another growing chain and terminate, or it may add another monomer unit and continue its growth. The more monomer molecules that are in the vicinity of the chain radical, the higher the probability of another monomer addition, hence the proportionality to [M]. On the other hand, the more initiator radicals that are present competing for the available monomer, the shorter the chains will be, on an average, causing the inverse proportionality to the square root of [I]. Equation 9.13 shows that the rate of polymerization can be increased (always an economically desirable goal) by increasing both [M] and [I], the former being more efficient than the latter. However, there is an upper limit to [M] set by the density of the pure monomer at the reaction conditions. So it is often tempting to increase [I] to attain higher rates. But according to Equation 9.27a, this unavoidably lowers $\bar{x}_n$. You cannot have your cake and eat it, too.

9.6 TEMPERATURE DEPENDENCE OF RATE AND CHAIN LENGTH

We may assume that the temperature dependence of the individual rate constants in each of the steps of the reaction are given by the Arrhenius expression:

$$k_i = A_i \exp\left(\frac{-E_i}{RT}\right) \tag{9.28}$$

where k_i is the rate constant for a particular elementary reaction, A_i is its frequency factor, and E_i is its activation energy. This expression is common in chemical kinetics, and results in the rule of thumb that for every 10 °C increase in temperature, the rate approximately doubles (which holds for temperatures around 300 K). Because most addition polymerizations are exothermic, the temperature of the reaction mixture increases during the polymerization, which causes an increase in the kinetic rate constants, which raises the temperature even faster, and so on, causing a run-away reaction. This behavior is related to the *Tromsdorff effect*, whereby chain termination occurs more quickly than expected, leading to lower molecular weight chains. This subject is covered in more detail in Chapter 12, *Polymerization Practice*.

If we neglect the temperature dependence of f, Equation 9.13 becomes

$$r_p = C \exp\left[\frac{-(E_p + E_d/2 - E_t/2)}{RT}\right] \qquad (9.29)$$

where C combines temperature-independent quantities into a single constant and $E_p + E_d/2 - E_t/2$ is the *effective* activation energy for polymerization. Therefore, between two (absolute) temperatures T_1 and T_2:

$$\ln \frac{r_p(T_2)}{r_p(T_1)} = \frac{-(E_p + E_d/2 - E_t/2)}{R}\left(\frac{1}{T_2} - \frac{1}{T_1}\right) \qquad (9.30)$$

Similar treatment of Equation 9.27a, with the assumption that ξ is independent of temperature, results in:

$$\ln \frac{\bar{x}_n(T_2)}{\bar{x}_n(T_1)} = \frac{-(E_p - E_d/2 - E_t/2)}{R}\left(\frac{1}{T_2} - \frac{1}{T_1}\right) \qquad (9.31)$$

where $E_p - E_d/2 - E_t/2$ is the *effective* activation energy for the number-average chain length.

Example 9.2 For a typical free-radical addition reaction, $E_p - E_t/2 = 5\,\text{kcal/mol}$ and $E_d = 30\,\text{kcal/mol}$. Estimate the changes in r_p and $\bar{x}_n$ for a typical homogeneous, free-radical addition polymerization when the temperature is raised from 60 to 70 °C.

Solution. The effective activation energy for polymerization is $+20\,\text{kcal/mol}$. Inserting this into Equation 9.30 along with $T_1 = 273 + 60 = 333\,\text{K}$, $T_2 = 273 + 70 = 343\,\text{K}$, and $R = 1.987\,\text{cal/mol-K}$ gives:

$$\ln \frac{r_p(70\,°\text{C})}{r_p(60\,°\text{C})} = 0.879 \quad \text{and} \quad \frac{r_p(70\,°\text{C})}{r_p(60\,°\text{C})} = 2.41 (!)$$

that is, the rate increases by 141% (or more than double) over the given 10 °C temperature interval. This rather spectacular (and sometimes dangerous) sensitivity is mainly due to the large activation energy for the rate-controlling initiator decomposition.

The effective activation energy for $\bar{x}_n$ is $-10\,\text{kcal/mol}$. With Equation 9.31, we get

$$\ln \frac{\bar{x}_n(70\,°\text{C})}{\bar{x}_n(60\,°\text{C})} = -0.439 \quad \text{and} \quad \frac{\bar{x}_n(70\,°\text{C})}{\bar{x}_n(60\,°\text{C})} = 0.644$$

so $\bar{x}_n$ is subject to a 35.6% *decrease* over the same temperature range.

9.7 CHAIN TRANSFER AND REACTION INHIBITORS

In practice, another type of reaction sometimes occurs in free-radical addition polymerizations. These *chain-transfer* reactions kill a growing chain radical and can start a new one in its place (as long as the radical transfers to another monomer):

$$R' : H + P_x \cdot \xrightarrow{k_{tr}} P_x + R' \cdot \qquad (9.32a)$$

$$R' \cdot + M \xrightarrow{k_a'} P_1 \cdot \quad \text{etc.} \qquad (9.32b)$$

Thus, chain transfer results in shorter chains, and if the reactions in Equation 9.32a are not too frequent compared to the propagation reaction and do not have very low rate constants, chain transfer will not change the overall rate of polymerization appreciably.

The compound $R':H$ in the above reaction is known as a *chain-transfer agent*. They are also sometimes referred to as chain terminating agents, because the free-radical on the growing polymer P_x is removed. Under appropriate conditions, almost anything in the reaction mass may act as a chain-transfer agent, including initiator, monomer, solvent, and dead polymer.

Reaction inhibitors can also change the polymerization rate. Inhibitors for addition polymerizations are typically free-radical scavengers—chemicals that can gobble up free-radicals quickly and prevent or slow the initiation and propagation steps of the polymerization mechanism. Oxygen is one such scavenger, and in some addition polymerizations, even oxygen dissolved in the reaction mixture must be removed to obtain appreciable molecular weight polymers. Other inhibitors are often added to monomers being shipped so that they do not autopolymerize before a reaction is desired. Quinone-type chemicals are often used in this capacity.

Example 9.3 Show, using dots to represent the electrons involved, how chain transfer to a dead polymer leads to long-chain branching in polyethylene.

Solution.

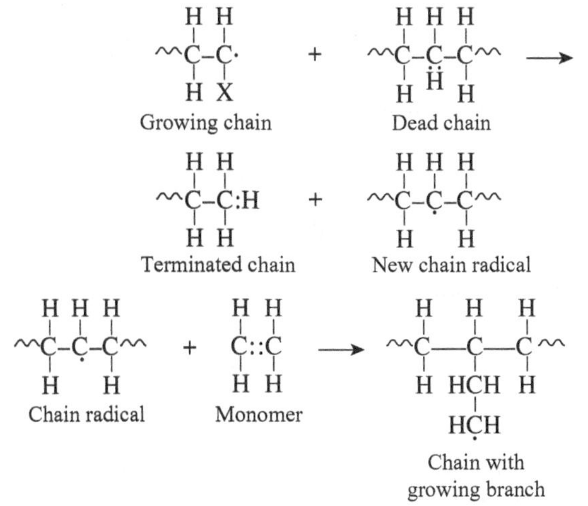

Example 9.4 Show, as above, how short branches arise in polyethylene when a growing chain "bites its own back," that is, the radical transfers to an atom a few [5–8] carbon atoms down the chain.

Solution.

Most frequently, mercaptans (also known as thiols), the sulfur analogs of alcohols (R′S:H), are added to the reaction mass as effective chain-transfer agents to lower the average chain length.

The rate of dead chain formation by chain transfer

$$r_{\text{tr}} = -\left(\frac{1}{V}\right)\left(\frac{dP\cdot}{dt}\right)_{tr} = k_{tr}\left[R' : H\right][P\cdot] \tag{9.33}$$

must be added to the denominator of Equation 9.27a to give the total rate of dead chain formation:

$$\bar{x}_n = \frac{k_p[P\cdot][M]}{(2k_{td} + k_{tc})[P\cdot]^2 + k_{tr}[R' : H][P\cdot]} \tag{9.34}$$

Using Equations 9.12 and 9.26 gives

$$\bar{x}_n = \frac{k_p[M]}{\xi(fk_dk_t[I])^{1/2} + k_{tr}[R' : H]} \tag{9.35}$$

Taking the reciprocal of Equation 9.35 gives

$$\frac{1}{\bar{x}_n} = \frac{1}{(\bar{x}_n)_0} + C\frac{[R' : H]}{M} \tag{9.36}$$

where C is the chain-transfer constant $= k_{tr}/k_p$ and $(\bar{x}_n)_0$ is simply the average chain length in the absence of chain transfer from Equation 9.27a. Thus, a plot of $1/(\bar{x}_n)$ versus [R′:H]/[M] is linear with a slope of C and intercept of $1/(\bar{x}_n)_0$.

Example 9.5 Prove that the use of a chain-transfer agent with $C = 1$ will maintain the ratio [R':H]/[M] constant in a batch reaction.

Solution.

$$\frac{r_{tr}}{r_p} = \frac{-(1/V)(d\text{R}' : \text{H}/dt)}{-(1/V)(dM/dt)} = \frac{d[\text{R}' : \text{H}]}{d[M]} = \frac{k_{tr}[\text{R}' : \text{H}][P\cdot]}{k_p[M][P\cdot]} = C\frac{[\text{R}' : \text{H}]}{[M]}$$

Separating variables and integrating gives

$$\int_{[\text{R}':\text{H}]_o}^{[\text{R}':\text{H}]} \frac{d[\text{R}' : \text{H}]}{[\text{R}' : \text{H}]} = C \int_{[M]_o}^{[M]_o} \frac{d[M]}{[M]}$$

$$\frac{[\text{R}' : \text{H}]}{[\text{R}' : \text{H}]_o} = \left(\frac{[M]}{[M]_o}\right)^C$$

(9.37)

Therefore, only if $C = 1$, will $([\text{R}':\text{H}]/[M]) = ([\text{R}':\text{H}]_o/[M]_o) = \text{constant}$.

Values of C greater than one use up chain-transfer agent too quickly, leaving nothing to modify the polymer formed at high conversions, while low values leave a lot of unreacted chain-transfer agent around toward the end of the reaction when enough is used to be effective at the beginning. If the agent happens to be a mercaptan, and if you have ever smelled a mercaptan (rotten eggs), you can appreciate what a problem leftover agent can be. These problems are circumvented by (1) adding an active ($C > 1$) agent over the course of the reaction, and (2) using mixtures of active ($C > 1$) and sluggish ($C < 1$) chain-transfer agents in the initial charge.

Example 9.6 Generalize Equation 9.36 to incorporate transfer to a variety of different possible transfer agents, S_i:

$$P_x\cdot + S_i \xrightarrow{k_{tr,i}} P_x + S_i\cdot$$

Solution. The rate of dead chain formation from transfer to agent S_i must be added to the denominator of Equation 9.27a:

$$\overline{x}_n = \frac{k_p[P\cdot][M]}{(2k_{td} + k_{tc})[P\cdot]^2 + \sum_i k_{tr,i}[S_i][P\cdot]}$$

Invoking Equations 9.12 and 9.26 gives

$$\overline{x}_n = \frac{k_p[M]}{\xi(fk_dk_t[I])^{1/2} + \sum_i k_{tr,i}[S_i]}$$

or

$$\frac{1}{\bar{x}_n} = \frac{1}{(\bar{x}_n)_0} + \sum_i C_i \frac{[S_i]}{[M]}$$

where

$$C_i = \frac{k_{tr,i}}{k_p}$$

9.8 INSTANTANEOUS DISTRIBUTIONS IN FREE-RADICAL ADDITION POLYMERIZATION

A growing polymer chain $P_x \cdot$ can either propagate, terminate, or transfer. The propagation probability, q, is the probability that it will propagate rather than transfer or terminate and is given by

$$q = \frac{\text{Rate of propagation}}{\text{Rate of propagation } + \text{ Rate of transfer } + \text{ Rate of termination}}$$

$$q = \frac{k_p[M][P \cdot]}{k_p[M][P \cdot] + k_{tr}[R' : H][P \cdot] + 2k_t[P \cdot]^2} \tag{9.38}$$

which becomes, with the aid of Equation 9.12,

$$q = \frac{k_p[M]}{k_p[M] + k_{tr}[R' : H] + 2(fk_d k_t[I])^{1/2}} \tag{9.39}$$

To get chain-length distributions from q, we must separately consider termination by disproportionation and combination because the two mechanisms give rise to different distributions. First, consider only those chains whose growth is terminated by either disproportionation ($k_t = k_{td}$, $\xi = 2$) or chain transfer. The resulting distributions are the same, because a growing chain does not know if it has been killed by an encounter with another growing chain or with a chain-transfer agent. Either way, its growth stops with no change in length (that is not the case for combination). By taking the reciprocal of Equation 9.39, we get

$$\frac{1}{q} = 1 + \underbrace{\frac{k_{tr}[R' : H]}{k_p[M]} + \frac{2(fk_d k_{td}[I])^{1/2}}{k_p[M]}}_{1/\bar{x}_n \quad \text{for } \xi = 2 (9.35)} \tag{9.40}$$

Therefore,

$$\bar{x}_n = \frac{q}{(1 - q)} \tag{9.41}$$

Now, a polymer molecule containing x units, P_x, has been formed by $x - 1$ propagation steps (the remaining unit was incorporated in the addition step), each with a probability q, and one disproportionation or transfer step with a probability $1 - q$. The probability of finding such a molecule is equal to its number (mole) fraction; therefore,

$$\frac{n_x}{N} = (1 - q)q^{(x-1)} \tag{9.42}$$

which, lo and behold, is the "most probable" distribution (Eq. 8.3) again. But, although Equations 9.42 and 8.3 appear the same, p and q are totally different quantities. In batch step-growth polymerization, p increases monotonically from 0 to 1 as the reaction proceeds. In free-radical addition polymerization, however, q depends only indirectly on conversion, since $[I]$, $[M]$, and $[R':H]$ will, in general, vary with conversion. In the usual case, q is always close to 1, as indeed it must be to form high molecular weight polymer chains.

Example 9.7 Show the structures represented by $x = 1$ in Equation 9.42 for the polymerization of a vinyl monomer $H_2C{=}CHX$. Assume no chain transfer and initiation according to

$$I : I \rightarrow 2I \cdot$$

Solution. Since chains here are killed only by disproportionation, dead chains of length $x = 1$ would be

$$\begin{array}{ccc}
\text{H H} & & \text{H H} \\
| \ | & & | \ | \\
\text{R–C–C–H} & \text{and} & \text{R–C=C} \\
| \ | & & | \\
\text{H X} & & \text{X}
\end{array}$$

(With its double bond, the molecule on the right above could conceivably participate as a comonomer in subsequent polymerization. The mechanism here, however, ignores that possibility.)

This points out another important difference between the seemingly similar Equations 9.42 and 8.3. The distributions for free-radical addition include only terminated chains, not unreacted monomer. In a step-growth reaction, there are no terminated chains short of complete conversion, and $x = 1$ in Equation 8.3 represents unreacted monomer.

Note that by analogy to Equation 8.6, $\bar{x}_n$ for the most-probable distribution should be

$$\bar{x}_n = \frac{1}{1 - q} \tag{9.43}$$

which is greater by one than the value given by Equation 9.41. This discrepancy arises from the fact that the kinetic derivation of Equation 9.41 neglects the monomer molecule added to the chain in the addition step, while that molecule is specifically considered in the derivation of Equation 9.42. From a practical standpoint, this makes little difference, because in the usual free-radical addition polymerization, $q \rightarrow 1$ and $\bar{x}_n \gg \bar{x}_n 1$.

By analogy to Equations 8.9–8.11 for step-growth polymerization:

$$\frac{w_x}{W} = xq^{(x-1)}(1 - q)^2 \tag{9.44}$$

$$\overline{x}_w = \frac{1+q}{1-q} \tag{9.45}$$

$$\frac{\overline{x}_w}{\overline{x}_n} = 1 + q \tag{9.46}$$

Here, however, unlike the case in step growth, the approximations $q \to 1$ and $\overline{x}_n \gg 1$, and $\ln q \approx q - 1$ are almost always applicable, and Equations 9.42, 9.44, and 9.46 simplify to (recall Example 5.5)

$$\frac{n_x}{N} \approx \frac{1}{\overline{x}_n} \exp\left(\frac{-x}{\overline{x}_n}\right) \tag{9.47}$$

$$\frac{w_x}{W} \approx \frac{1}{\overline{x}_n^2} \exp\left(\frac{-x}{\overline{x}_n}\right) \tag{9.48}$$

$$\frac{\overline{x}_w}{\overline{x}_n} \approx 2 \tag{9.49}$$

Now, let us consider chains that *exclusively* terminate by combination ($k_t = k_{tc}$; $\xi = 1$):

$$q = \frac{k_p[M]}{k_p[M] + 2(fk_dk_{tc}[I])^{1/2}} \tag{9.50}$$

$$\frac{1}{q} = 1 + \underbrace{\frac{2(fk_dk_{tc}[I])^{1/2}}{k_p[M]}}_{(2/\overline{x}_n) \quad \text{for } \xi = 1} \tag{9.51}$$

$$(2/\overline{x}_n) \quad \text{for } \xi = 1 \tag{9.27b}$$

from which

$$\overline{x}_n = \frac{2q}{1-q} \tag{9.52}$$

A dead chain consisting of x units is formed by the combination of two growing chains, one containing y units and the other containing $x - y$ units:

$$P_y\cdot + P_{(x-y)}\cdot \to P_x \tag{9.53}$$

One growing chain has undergone $y - 1$ propagation steps, each of probability q. The other chain has undergone $x - y - 1$ propagation steps, each of probability q. Two growing chains are terminated, each with a probability of $1 - q$. But, each dead chain consisting of x total units could have been formed in $x - 1$ different ways; for example,

x Possible combinations of y and $(x - y)$
2 $1 + 1$
3 $2 + 1; 1 + 2$
4 $3 + 1; 2 + 2; 1 + 3$
5 $4 + 1; 3 + 2; 2 + 3; 1 + 4$

and all possible ways to form a chain of x units total must be counted. Therefore,

$$\frac{n_x}{N} = (x - 1)q^{(y-1)}q^{(x-y-1)}(1 - q)^2 = (x - 1)(1 - q)^2 q^{(x-2)} \tag{9.54}$$

For this distribution,

$$\bar{x}_n = \frac{2}{1 - q} \tag{9.55}$$

which is greater by two than the value given by Equation 9.52, since the first monomer incorporated in the addition step is now considered for each of the chains, and

$$\bar{x}_w = \frac{2 + q}{1 - q} \tag{9.56}$$

$$\frac{\bar{x}_w}{\bar{x}_n} = \frac{2 + q}{2} \tag{9.57}$$

For the usual free-radical case, where $q \to 1$, $\bar{x}_n \gg 1$, and $\ln q \approx q - 1$:

$$\frac{n_x}{N} \approx \frac{4x}{\bar{x}_n^2} \exp\left(-\frac{2x}{\bar{x}_n}\right) \tag{9.58}$$

$$\frac{w_x}{W} \approx \frac{4x^2}{\bar{x}_n^3} \exp\left(-\frac{2x}{\bar{x}_n}\right) \tag{9.59}$$

$$\frac{\bar{x}_w}{\bar{x}_n} \approx 1.5 \tag{9.60}$$

These distributions are plotted in reduced form in Figure 9.4. Note that termination by combination sharpens the distributions because of the low probability of two very short or two very long macroradicals combining.

As shown above, chains terminated by combination follow a different distribution than those terminated by disproportionation or chain transfer. If a material contains chains formed according to both distributions (as would happen, e.g., if a chain-transfer agent were added to a system that terminates inherently by combination), the distributions must

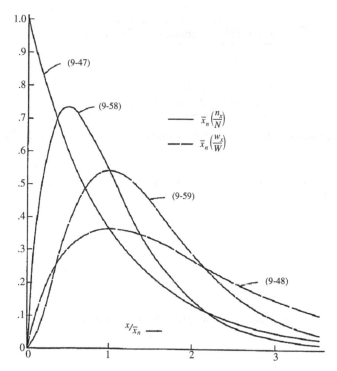

FIGURE 9.4 Instantaneous number- and weight-fraction distributions of chain lengths in free-radical addition polymerization. Equations 9.47 and 9.48 are for termination by disproportionation and/or chain transfer; Equations 9.58 and 9.59 are for termination by combination.

be summed according to the proportion (mole or mass fraction) of each, as detailed in Kenat *et al.* [2] To solve this situation, let ψ be the mole (or number) fraction of the chains that have been terminated by combination. The weight fraction of chains terminated by combination is $2\,\psi/(\psi+1)$. In terms of kinetic parameters, ψ is given by

$$\psi = \frac{2-\xi}{[\xi + \{k_{tr}[R' : H]/k_t f k_d [I]^{1/2}\}]} \tag{9.61}$$

where ξ is defined by Equation 9.26. In terms of ψ, the general distributions become

$$\bar{x}_n\left(\frac{n_x}{N}\right) = \left[\psi(\psi+1)^2\left(\frac{x}{\bar{x}_n}\right) + (1-\psi^2)\right]\exp\left[-(\psi+1)\left(\frac{x}{\bar{x}_n}\right)\right] \tag{9.62}$$

$$\bar{x}_n\left(\frac{w_x}{W}\right) = \left[\psi(\psi+1)^2\left(\frac{x}{\bar{x}_n}\right) + (1-\psi^2)\right]\left(\frac{x}{\bar{x}_n}\right)\exp\left[-(\psi+1)\left(\frac{x}{\bar{x}_n}\right)\right] \tag{9.63}$$

$$\bar{x}_w/\bar{x}_n = \frac{4\psi+2}{(\psi+1)^2} \tag{9.64}$$

For termination entirely disproportionation and/or chain transfer, $\psi=0$ and Equations 9.62–9.64 reduce to Equations 9.42–9.49; for termination exclusively by combination, $\psi = 1$, they reduce to Equations 9.58–9.60, but these new equations are also able to describe various degrees of mixed termination.

Example 9.8 Calculate q and $\bar{x}_n$ for the styrene polymerization in Example 9.1 *for conditions at the start of the reaction*. Neglect chain transfer. Styrene terminates by combination at 60 °C so that $k_t = k_{tc}$ ($\xi = 1$). $[M]_o = 8.72 \, \text{mol/L}$.

Solution. Using the rate constants from Example 9.1 along with $[M] = [M]_o$ and $[I] = [I]_o$ from above in Equation 9.39 gives $q = 0.9954$. This illustrates the point made above that $q \rightarrow 1$ right from the beginning of a typical free-radical addition polymerization.

Because termination is by combination, we use Equation 9.55 to get $\bar{x}_n = 434$ ($\bar{M}_n = 45{,}200$), so we are getting high molecular weight polymer right off the bat, too.

9.9 INSTANTANEOUS QUANTITIES

It is obvious from Equation 9.13 that the rate of polymerization is an instantaneous quantity; it depends on the particular values of $[M]$, $[I]$, and T (through the temperature dependence of the rate constants) that exist as a *particular instant* (and location, for that matter) in a reactor. In a uniform, isothermal batch reaction (Example 9.1), the rate of polymerization decreases monotonically because of the decreases in both $[M]$ and $[I]$ with time. In a similar fashion, $\bar{x}_n$ according to Equation 9.35 is a function of $[M]$, $[I]$, T, and $[R':H]$, all of which may vary with time (and/or location) in a reactor. But is the concept of an instantaneous $\bar{x}_n$ valid? How much do these quantities change during the lifetimes of individual chains? This important point is clarified in the following example.

Example 9.9 Consider the uniform, isothermal batch polymerization of Example 9.1. For conditions *at the start of the reaction*, calculate:

(a) the average lifetime of a growing chain, $\bar{t}$, and compare it with the half lives of monomer and initiator,
(b) the percent decrease in initiator concentration during the time interval $\bar{t}$,
(c) the conversion during the time interval $\bar{t}$, and
(d) the number of growing polymer chains per liter of reaction mass.

Additional necessary data: $k_p = 176 \, \text{L/mol. s}$

Solution. (a) This is always conceptually difficult. Think of it as calculating the time required for a car traveling at a constant velocity to cover a specified distance. Here, the distance corresponds to the kinetic chain length, ν (Eq. 9.22), the average number of monomer units added to a chain during its lifetime (note that the mode of termination is immaterial). The velocity corresponds to the rate at which the monomer is added to a single chain. Monomer is added to all chains (in a unit volume of reaction mass) at a rate r_p. In that unit volume of reaction mass, there are $[P\cdot]$ growing chains. Therefore, the rate of monomer addition to a single chain $= r_p/[P\cdot]$ and

$$\bar{t} = \frac{\nu[P\cdot]}{r_p}$$

With Equations 9.12, 9.13, and 9.22, this becomes

$$\bar{t} = \frac{1}{2(fk_d k_t [I])^{1/2}}$$

Inserting the given rate constants and $[I] = [I]_o = 0.05$ mol/L gives $\bar{t} = 0.141$ s (!). From Figure 9.1, the time to reach 50% conversion (monomer half life) is about 8 h. From equation 9.16b, the time to reach $(I/I_o) = 0.5$ is $-(\ln 0.5)/k_d$ or 20 h.

(b) We can confidently neglect any volume change over the extremely short average chain lifetime so that $(I/I_0) = ([I]/[I]_0) = \exp(-k_d \bar{t})$ (9.16b).

With $\bar{t} = 0.141$ /s

$$\% \text{ decrease in } [I] = 100\left(1 - \frac{[I]}{[I]_0}\right) = 1.35 \times 10^{-4}$$

(c) The results of (b) justifies the use of Equation 9.15. With $\bar{t} = 0.141/$ s, X $= 3.36$ $\times 10^{-6}$ of $3.36 \times 10^{-4}\%$.

(d) From equation 9.12, we get $[P\cdot] = 1.35 \times 10^{-7}$ mol growing chains/L. Or, multiplying by Avogadro's number, 6.02×10^{23} growing chains/mol growing chains, it is 8.14×10^{16} growing chains L.

Some important conclusions can be drawn from the preceding example. In a typical, homogeneous, free-radical addition polymerization, there are lots of chains growing at any instant. The average lifetime of a growing chain, however, is extremely short, many orders of magnitude smaller than the half-lives of either monomer or initiator. Once a chain has been initiated by the decomposition of an initiator molecule (recall that this is the slow step), it grows and dies in a flash, and once terminated, it plays no further role in the reaction (unless it happens to act as a chain-transfer agent, as in Example 9.6) and merely sits around inertly as more chains are formed. This is to be contrasted with step-growth polymerization, in which the chains always maintain their terminal reactivity and continue to grow throughout the reaction (see Table 9.1).

The lifetime of free-radical chains is, in fact, so short that changes in concentrations are entirely negligible during a chain lifetime. Hence, it is perfectly proper to characterize chains formed at any instant when $[M]$, $[I]$, T, and $[R':H]$ have a particular set of values. All the quantities defined to this point ($\bar{x}_n$, $\bar{x}_w$, q) and, therefore, the distributions are just such instantaneous quantities. For this reason, it was necessary to specify conditions "at the start of the reaction" in earlier examples to permit their calculations.

9.10 CUMULATIVE QUANTITIES

Unfortunately (for the sake of simplicity), as conditions vary within a polymerization reactor, so do the instantaneous quantities, making it very difficult to accurately determine $\bar{x}_n$ and $\bar{x}_w$ as functions of time. The polymer in the reactor is a mixture of material formed under varying conditions of temperature and concentrations, and therefore must be characterized by *cumulative* quantities, which are integrated averages of the instantaneous quantities of the material formed up until the reactor is sampled. The cumulative number-

TABLE 9.1   Comparison of Step-Growth (Chapter 8) and Addition (Chapter 9) Polymerizations

Characteristic	Step-Growth Polymerization	Addition Polymerization
Typical monomer structure	Bifunctional monomers with two organic functional groups (amines, alcohols, acids, etc.)	Monomers have a polymerizable double bond (such as in vinyl or acrylate monomers)
Reactor contents during polymerization at 50% conversion	$p = .5$; 50% of functional groups reacted, $\bar{x}_n = 2$. Only small chains present.	$X = 0.5$; large polymer chains and monomer are in the reactor, $\bar{x}_n$ depends on initiator concentration
Reaction speed (kinetics)	Generally slow; depends on reactivity of functional groups; dimers, trimers, etc., start to form from the start, making the solution syrupy and viscous.	Generally fast; solution contains primarily monomers until high conversions are reached, so the solution retains relatively low viscosity
Initiator required	No	Usually
Effect of temperature	Arrhenius behavior	Arrhenius behavior
Networks and gels	Requires tri- (or higher) functional monomers (A_3) and high enough conversions to exceed the critical gelation point.	Multiple polymerizable double bonds on a monomer result in network structures; a gel starts to form as soon as reaction propagates (even at low conversion)

average chain length, $<\bar{x}_n>$, is simply the total moles of monomer polymerized over the total moles of dead chains formed. In a batch reactor, for example,

$$\text{Moles of monomer in dead chains} = M_0 - M(t) = M_0 X(t)$$

$$\text{Moles of dead chains formed} = f\xi\{I_0 - I(t)\} + \{R' : H_0 - R' : H(t)\}$$

(For termination by disproportionation, each successful initiator reaction results in two dead chains, and for combination, one; hence, the need for ξ. Each reacted molecule of chain-transfer agent results in one dead chain.) Therefore,

$$< \bar{x}_n > (t) = \frac{M_0 X(t)}{f\xi\{I_0 - I(t)\} + \{R' : H_0 - R' : H(t)\}} \tag{9.65}$$

(If you wish to neglect volume change, all the molar quantities can be enclosed in square brackets to give concentrations.) Note that the preceding expression is indeterminate at $t = 0$. At this point, however, $\bar{x}_n = <\bar{x}_n> = 1$ (just monomers), both are given by Equation 9.35.

A good analogy for visualizing the relation between $\bar{x}_n$ and $<\bar{x}_n>$ is the "well-stirred, adiabatic bathtub." $\bar{x}_n$ corresponds to the temperature of the water leaving the tap, and $<\bar{x}_n>$ to the temperature of the water in the tub. So, the newly formed polymer entering from the top must be averaged with all that is been made earlier (the stuff already in the bathtub). As long as the tub is not full, the analogy applies to a batch reactor. If the tub is allowed to fill and overflow, it applies to a continuous stirred-tank reactor as well.

Example 9.10 For the reaction in Examples 9.1 and 9.9, calculate and plot $\bar{x}_n$ and $<\bar{x}_n>$ versus X.

Solution. For this homogeneous, isothermal, batch reaction, X is given as a function of time by Equations 9.19 (incorporating volume change) or 9.20 (neglecting volume change) as illustrated in Example 9.1. $I(t)$ is given by Equation 9.16b. By combining Equations 9.27, 9.14, 9.16b, 9.6, and 9.17 we get

$$\bar{x}_n(t) = \frac{k_p[M]_0\{1 - X(t)\}}{\xi(fk_dk_t)^{1/2}[I]_0^{1/2}\{1 + \varepsilon X(t)\}^{1/2}\exp(-k_dt/2)} \tag{9.27'}$$

Had it been necessary to include chain transfer, we would have used Equation 9.35 with [R':H] related to [M] with Equation 9.37. Similar treatment of Equation 9.65 gives

$$<\bar{x}_n>(t) = \frac{[M]_0X(t)}{f\xi[I]_0\{1 - \exp(-k_dt)\}} \tag{9.65'}$$

Calculations are facilitated by selecting time as the independent variable and relating X, $\bar{x}_n$, and $<\bar{x}_n>$ to t through Equations 9.19 or 9.20, 9.27' and 9.65'.

Alternatively, X may be chosen as the independent variable and Equation 9.19 or 9.20 solved for the corresponding times, which are then used in Equations 9.27' and 9.65'. The results are shown in Figure 9.5. Note that in this case, $\bar{x}_n$ goes through a minimum at high conversions. In fact, it ultimately goes to infinity, because in this example of "dead-stop" polymerization, [I] goes to zero while [M] remains finite. Not so obvious on this scale is the fact that $<\bar{x}_n>$ must also go through a minimum. As long as $\bar{x}_n$ is less than $<\bar{x}_n>$, the former will continue to drag the latter down (adding colder water from the tap cools off the water in the tub). When $\bar{x}_n$ is greater, it must pull $<\bar{x}_n>$ up (you warm up the water in the tub by adding hotter water from the tap). Therefore, there must be a minimum in $<\bar{x}_n>$

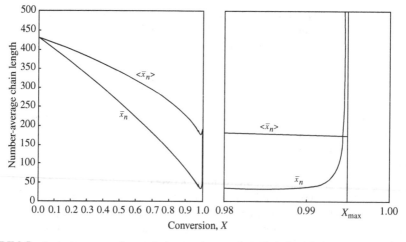

FIGURE 9.5 Instantaneous and cumulative number-average chain lengths versus conversion for an isothermal, free-radical, batch polymerization (data of Example 9.10).

where the two curves cross, although $<\overline{x}_n>$ does reach a finite value of 173.5 at the maximum conversion of 99.48%. (The numbers quoted apply for $\xi = -0.136$. For $\xi = 0$, they are 173.2 and 99.30%.) Even though the last tiny drop from the tap may be very hot, it is not going to significantly raise the temperature of the 50 gallons already in the tub.

It is instructive to compare the results of Example 9.10 with Equation 8.6 for step-growth polymerization. There, $\overline{x}_n$ increases monotonically with conversion (p), and high conversions are necessary for high chain lengths. This is not true for free-radical addition, where chains formed early in the reaction have fully developed chain lengths.

Also keep in mind that if samples are removed from the reactor and analyzed for number-average chain length, the quantity determined is $<\overline{x}_n>$. The cumulative quantity is what characterizes the reactor contents. The only way to measure $\overline{x}_n$ would be to sample at a very low conversion, where $\overline{x}_n \approx <\overline{x}_n>$.

9.11 RELATIONS BETWEEN INSTANTANEOUS AND CUMULATIVE AVERAGE CHAIN LENGTHS FOR A BATCH REACTOR

Given any instantaneous or cumulative average chain length as a function of conversion in a batch reactor, the others may be calculated as follows. Recall that conversion is defined as

$$X \equiv \frac{M_0 - M}{M_0} \tag{9.14}$$

The moles of monomer polymerized *up to* conversion X is

$$M_0 = M = M_0 X$$

while the moles of monomer polymerized in a conversion *increment* dX is

$$dM = M_0 dX$$

The moles of *polymer chains* formed in the conversion increment dX is:

$$dN = \frac{dM}{\overline{x}_n} = \frac{M_0 dX}{\overline{x}_n}$$

The total moles of polymer chains formed *up to* conversion X is obtained by integrating dN:

$$N = \int_0^x dN = \int_0^x \frac{M_0 dX}{\overline{x}_n}$$

But, by definition of $<\overline{x}_n>$, N is also given by

$$N = \frac{\text{moles monomer polymerized to conversion } X}{<\overline{x}_n>} = \frac{M_0 X}{<\overline{x}_n>}$$

Equating the two expressions for N gives

$$< \overline{x}_n >= \frac{X}{\int_0^x (dX/\overline{x}_n)} \tag{9.66}$$

Differentiating Equation 10.66 reverses the result:

$$\overline{x}_n = \frac{dX}{d(X/ < \overline{x}_n >)} \tag{9.67}$$

Equation 5.6 gives the weight average of a mixture in terms of a summation of the weight averages of the finite components of the mixture. For a mixture of *differential* components of weight-average chain length $\overline{x}_w$, this may be generalized to

$$< \overline{x}_w >= \frac{\int_0^x \overline{x}_w dW_p}{\int_0^x dW_p} = \frac{1}{W_p} \int_0^x \overline{x}_w dW_p$$

where

dW_p = weight of polymer formed in conversion increment of dX and

W_p = total weight of polymer formed up to conversion X.

Now $W_p = W_o X$ and $dW_p = W_o dX$, where W_o is the weight of monomer fed; so

$$< \overline{x}_w >= \frac{1}{X} \int_0^x \overline{x}_w dX \tag{9.68}$$

and by differentiating Equation 9.68, we get

$$\overline{x}_w =< \overline{x}_w > +X \left(\frac{d < \overline{x}_w >}{dX} \right) \tag{9.69}$$

The relation between $\overline{x}_n$ and $\overline{x}_w$ is, of course, determined on a microscopic scale by the nature of the instantaneous distribution, which fixes the instantaneous polydispersity index, $\overline{x}_w/\overline{x}_n$, given by Equation 9.64. However, it is not the instantaneous polydispersity index that characterizes the reactor product but rather the cumulative polydispersity index $<\overline{x}_w>/<\overline{x}_n>$. Even though the instantaneous polydispersity index may remain constant, if the instantaneous averages vary during the reaction (as in Example 9.10), the cumulative distribution must be broader than the instantaneous, increasing the cumulative polydispersity index (Example 5.2 provides a simple quantitative illustration of this). This is one of the reasons why commercial polymers often have polydispersity indices much greater than the minimum value given by Equation 9.64.

The "road map" below summarizes how to get between the various averages when one is known as a function of conversion:

$$\bar{x}_n(X) \quad \underset{(9.67)}{\overset{(9.66)}{\rightleftarrows}} \quad \langle \bar{x}_n \rangle (X)$$

$$\uparrow (9.64) \downarrow$$

$$\bar{x}_w(X) \quad \underset{(9.69)}{\overset{(9.68)}{\rightleftarrows}} \quad \langle \bar{x}_w \rangle (X)$$

Remember that $\langle \bar{x}_n \rangle$ and $\langle \bar{x}_w \rangle$ can be experimentally determined by sampling the reactor, but $\bar{x}_n$ and $\bar{x}_w$ can be obtained only by calculation at finite conversions.

Example 9.11 Which type of isothermal reactor would produce the narrowest possible distribution of chain lengths in a free-radical addition polymerization: continuous stirred tank reactor (CSTR, or backmix), batch (assume perfect stirring in each of the previous), plug-flow tubular, or laminar-flow tubular?

Solution. A CSTR has a large well-mixed tank with continuous inlet of reactant and outlet of product (polymer). In an ideal continuous stirred tank reactor, [M] and [I] in the tank are constant, and therefore $\bar{x}_n$ is too. This will give the narrowest possible distribution, due only to the microscopically random nature of the reaction. With an ideal CSTR, $\langle \bar{x}_w \rangle / \langle \bar{x}_n \rangle = \bar{x}_w / \bar{x}_n$. In a batch reactor, [$M$] and [$I$] decrease as the polymers are produced. In both of the tubular reactors, where the reaction occurs as the reactants flow through a tube, the [M] decreases as the reacting mixture moves from the inlet to the exit. In each of the latter three reactors, $\bar{x}_n$ changes with conversion, broadening the distribution so that $\{\langle \bar{x}_w \rangle / \langle \bar{x}_n \rangle\} > \bar{x}_w / \bar{x}_n$. Note that changes in temperature (which are not uncommon for exothermic polymerizations) can cause the rate constants to vary, and thus also contribute to broadening of the distribution.

Example 9.12 Consider the isothermal, free-radical batch polymerization of a monomer with $[M]_o = 1.6901$ gmol/L and $[I]_o = 1.6901 \times 10^{-3}$ gmol/L. At the reaction temperature, $k_p^2 / k_t = 26.1$ L/mol/s and $k_d = 4.369 \times 10^{-7}$/s^{-1}. This particular polymer terminates by disproportionation ($\xi = 2$). Neglect chain transfer. Assume perfect reactor stirring and $f = 1$. Because this reaction is carried out in a rather dilute solution, volume change is negligible. Calculate and plot $\bar{x}_n$, $\langle \bar{x}_n \rangle$, $\bar{x}_w$, $\langle \bar{x}_w \rangle$, and $\langle \bar{x}_w \rangle / \langle \bar{x}_n \rangle$ versus conversion for this system.

Hint: For this particular system, (k_p^2 / k_t) is very large and k_d is small (compare with Example 9.1). This has some important consequences. First, the chain lengths will be tremendous. (Although this example is based on a real system, in real life it is likely that chain lengths would be transfer limited, that is, the chains would transfer to *something* in the reaction mass before reaching the lengths calculated.) Second, [I] remains essentially constant throughout the course of the reaction. As a result, $\bar{x}_n$ is linear with X (combine Equations 9.27a and 9.14 to see why). This, in turn, allows easy analytical evaluation of Equations 9.66–9.69, but also renders Equation 9.65 useless.

Solution. Following the suggestion in the hint yields

$$\overline{x}_n = \frac{k_p[M]_0}{\xi(fk_dk_t)^{1/2}[I]^{1/2}}(1-X) = 158900(1-X)$$

Plugging this relation into Equation 9.66 gives

$$<\overline{x}_n> = \frac{158900X}{\int_0^X[dX/(1-X)]} = \frac{158900X}{-\ln(1-X)}$$

Now, for $\xi = 2$ and $\psi = 0$, Equation 9.64 or 9.49 gives $\overline{x}_w = 2\overline{x}_n$, so that

$$\overline{x}_w = 317800(1-X)$$

When this relation is plugged into Equation 9.68, we get

$$<\overline{x}_w> = \frac{1}{X}\int_0^X x_w dX = \frac{317800}{X}\int_0^X(1-X)dX = 317800\left(1-\frac{X}{2}\right)$$

Plots based on these equations are shown in Figure 9.6. Note that at high conversions, the polydispersity index begins to shoot up well above the minimum (instantaneous) value of 2.0.

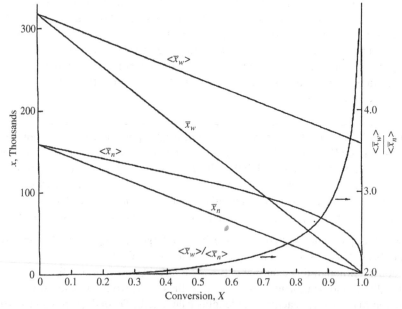

FIGURE 9.6 Instantaneous and cumulative number- and weight-average chain lengths and the cumulative polydispersity index for an isothermal, free-radical, batch polymerization (Example 9.12).

If it is necessary to maintain minimum polydispersity (albeit with shorter chains), the drift in instantaneous chain length can often be compensated for by varying the rate of addition of chain-transfer agent to the reactor. In Example 9.12, a high initial rate of a very active agent could be reduced as conversion increased, thereby counteracting the decrease in $\bar{x}_n$ (and increase in polydispersity) that would otherwise occur. Adjusting the rate of addition of chain-transfer agent to maintain $\bar{x}_n$ constant has been discussed for batch [3] and continuous [2] reactors.

Example 9.12 constitutes one of the simplest possible illustrations of what is sometimes termed "polymer reaction engineering." Even a minor complication, such as substitution of the parameters of Example 9.1 or the inclusion of chain transfer would necessitate a numerical solution. While the basic principles are there, additional detail is beyond the scope of this chapter, but you might wish to consider the application of these principles to a nonisothermal reactor, etc.

9.12 EMULSION POLYMERIZATION

The preceding discussion of free-radical addition polymerization has considered only homogeneous reactions. Considerable polymer is produced commercially by a complex *heterogeneous* free-radical addition process known as emulsion polymerization. This process was developed in the United States during World War II to manufacture synthetic rubber. A rational explanation of the mechanism of emulsion polymerization was proposed by Harkins [4] and quantified by Smith and Ewart [5] after the war, when information gathered at various locations could be freely exchanged. Perhaps the best way to introduce the subject is to list the feed sent to a typical reactor.

Typical Emulsion Polymerization Feed
100 parts (by weight) monomer (water insoluble)
180 parts water
2–5 parts fatty acid surfactant (emulsifying agent)
0.1–0.5 part *water-soluble* initiator
0–1 part chain-transfer agent (monomer soluble)

The surfactant is the critical ingredient that helps to form an emulsion where the monomer is dispersed in droplets. Surfactants are generally the sodium or potassium salts of organic acids or sulfates that have alkane-based R groups;

$$[R-\overset{\overset{O}{\|}}{C}-O]^- \, Na^+ \qquad\qquad [R-O-\overset{\overset{O}{\|}}{\underset{\underset{O}{\|}}{S}}-O]^- \, Na^+$$

When they are added to water in low concentrations, they ionize and float around freely much as sodium chloride ions would. The anions, however, consist of a highly polar hydrophilic (water-seeking) "head" (COO^- or SO_3^-) and an organic, hydrophobic (water-fearing) "tail" (R). As the surfactant concentration is increased, a value is suddenly reached

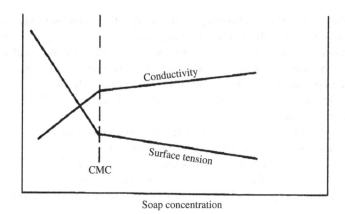

FIGURE 9.7   Variation of solution properties to define the CMC.

where the anions begin to agglomerate in *micelles* rather than float around individually. These micelles have dimensions on the order of 5–6 nm, far too small to be seen with a light microscope (although a successful emulsion of micelles will turn a solution cloudy, as a bottle of oil and vinegar salad dressing would look after shaking). The micelles consist of a tangle of the hydrophobic tails in the interior (getting as far away from the water as possible) with the hydrophilic heads on the outside. This process is easily observed by following the variation of a number of solution properties with surfactant concentration, for example, electrical conductivity or surface tension (see Figure 9.7). The break in the slope occurs when micelles start to form with higher concentrations of surfactant, and is known as the *critical micelle concentration* or CMC.

When an organic monomer is added to an aqueous micelle solution, it naturally prefers the organic environment within the micelles. Some of it congregates there, swelling the micelles until equilibrium is reached with the contraction force of surface tension. Most of the monomer, however, is distributed in the form of much larger (1 μm or 1000 nm) droplets stabilized by surfactant. This complex mixture is an *emulsion*. The cleaning action of surfactants is why they are widely used in soaps and detergents, where they emulsify oils and greases.

Despite the fact that most of the monomer is present in the droplets, the swollen micelles, because of their much smaller size, present a much larger surface area than the droplets. This is easily seen by assuming a micelle volume to drop volume ratio of 1/10 and using the ballpark figures given above. Since the surface/volume ratio of a sphere is 3/R,

$$\frac{S_{\text{micelle}}}{S_{\text{droplet}}} = \left(\frac{V_{\text{mic.}}}{V_{\text{drop.}}}\right)\left(\frac{R_{\text{drop.}}}{R_{\text{mic.}}}\right) \approx \left(\frac{1}{10}\right)\left(\frac{1000}{5}\right) = 20$$

Figure 9.8 illustrates the structures present during emulsion polymerization.

Free-radicals for classical emulsion polymerizations are generated *in the aqueous phase* by the decomposition of water-soluble initiators, usually potassium or ammonium persulfate:

$$S_2O_8^{2-} \quad \rightarrow 2\,^-SO_4\cdot$$

Persulfate   Sulfate ion radical

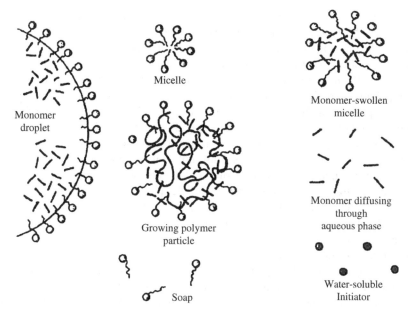

FIGURE 9.8 Structures in emulsion polymerization.

Redox systems, so called because they involve the alternate oxidation and reduction of a trace catalyst, are one of the alternatives for generating free-radicals in the initiation step. For example,

$$1. S_2O_8^{2-} + HSO_3^- \rightarrow SO_4^{2-} + {}^-SO_4 \cdot + HSO_3 \cdot$$

Persulfate Bisulfite

$$2. S_2O_8^{2-} + Fe^{2+} \rightarrow SO_4^{2-} + Fe^{3+} + {}^-SO_4 \cdot$$

$$3. HSO_3^- + Fe^{3+} \rightarrow HSO_3 \cdot + Fe^{2+}$$

$$\text{Net } S_2O_8^{2-} + HSO_3^- \rightarrow SO_4^{2-} + {}^-SO_4 \cdot + HSO_3 \cdot$$

The original wartime GR-S, poly(butadiene-co-styrene), polymerization was carried out at 50 °C using potassium persulfate initiator ("hot" rubber). The use of the more efficient redox systems allowed a reduction in polymerization temperature to 5 °C ("cold" rubber). The latter has superior properties because the lower polymerization temperature promotes *cis*-1-4-addition of the butadiene.

The radicals thus generated in the aqueous phase bounce around until they encounter some monomer. Since the surface area presented by the monomer-swollen micelles is so much greater than that of the droplets, the probability of a radical entering a monomer-swollen micelle rather than a droplet is large. As soon as the radical encounters the monomer within the micelle, it initiates polymerization. The conversion of monomer to polymer within the growing micelle lowers the monomer concentration therein, and monomer begins to diffuse from uninitiated micelles and monomer droplets to the growing, polymer-containing micelles. Those monomer-swollen micelles not struck by a radical during the early stages of conversion thus disappear, losing their monomer and surfactant to those that have been initiated. This first phase of the reaction was termed *Stage I* by Smith and Ewart. The reaction mass now consists of a stable number of growing polymer particles (originally micelles) and the monomer droplets, termed *Stage II*.

In Stage II, the monomer droplets simply act as reservoirs supplying monomer to the growing polymer particles by diffusion through the water. The monomer concentration in the growing particles maintains a nearly constant dynamic equilibrium value dictated by the tendency toward further dilution (increasing the entropy) and the opposing effect of surface tension attempting to minimize the surface area. Although most organic monomers are normally thought of as being water "insoluble," their concentrations in the aqueous phase, though small, are sufficient to permit a high enough diffusion flux to maintain the monomer concentration in the polymerizing particles [6]. Smith and Ewart then subdivided Stage II into three subcases where the ratio of growing free-radicals per particle is $\ll \frac{1}{2}$ (case 1), $= \frac{1}{2}$ (case 2), or $\gg \frac{1}{2}$ (case 3). The description that follows applies to their case 2 that often happens in practice.

A monomer-swollen micelle that has been struck by a radical contains *one* growing chain. With only one radical per particle, the chain cannot terminate by disproportionation or by combination, and it continues to grow until a second initiator radical enters the particle. Under conditions prevailing within the particle in case 2 (tiny particles and highly reactive radicals), the rate of termination is much greater than the rate of propagation, so the chain growth is terminated essentially immediately after the entrance of the second initiator radical [6]. The particle then remains dormant until a third initiator radical enters, initiating the growth of a second chain. This second chain grows until it is terminated by the entry of the fourth radical, and so on.

9.13 KINETICS OF EMULSION POLYMERIZATION IN STAGE II, CASE 2

Thus, in Stage II, the reaction mass consists of a stable number of monomer-swollen polymer particles that are the loci of all polymerizations. At any given time (for case 2), a particle contains either one growing chain or no growing chains (assumed to be equally probably). Statistically, then, if there are N particles per liter of reaction mass, there are $N/2$ growing chains per liter of reaction mass.

The polymerization rate is given, as before, by

$$r_p = k_p[M][P\cdot] \tag{9.7}$$

where k_p is the usual homogeneous propagation rate constant for polymerization within the particles and $[M]$ is the equilibrium monomer concentration within a particle. Now,

$$[P\cdot] = \frac{N}{2A} \text{ moles radical/liter of raction mass} \tag{9.70}$$

where A is Avogadro's number (6.02×10^{23} radicals/mol radicals). The rate of polymerization is then

$$r_p = k_p \frac{N}{2A}[M] \tag{9.71}$$

where typical units for the various terms are as follows:

$$k_p [=] \dfrac{\dfrac{\text{moles monomer}}{\text{liter of particles-second}}}{\dfrac{\text{moles monomer}}{\text{liter of particles}} \times \dfrac{\text{moles monomer}}{\text{liter of particles}}}$$

$$\dfrac{N}{2A} [=] \dfrac{\text{moles radicals}}{\text{liter of reaction mass}}$$

$$[M] [=] \dfrac{\text{moles monomer}}{\text{liter of particles}}$$

$$r_p [=] \dfrac{\text{moles of monomer}}{\text{liter of reaction mass-second}}$$

Equation 9.73 thus gives the rate of polymerization *per total volume of reaction mass*. Surprisingly, it predicts the rate to be independent of initiator concentration. Moreover, since both N and $[M]$ are constant in Stage II, a constant rate is predicted. This is borne out experimentally, as shown in Figure 9.9 [4]. Deviations from linearity are observed at low

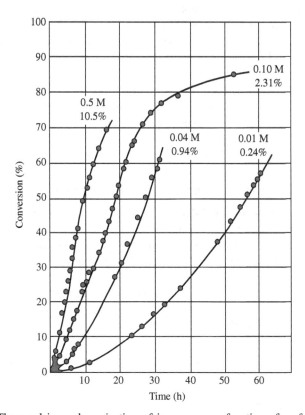

FIGURE 9.9 The emulsion polymerization of isoprene as a function of surfactant (potassium laureate) concentration—from low (0.01 molar) and slow to high (0.5 molar) and fast. Reprinted from Harkins [4]. Copyright 1947 by the American Chemical Society. Reprinted by permission of the copyright owner.

conversions as N is being stabilized in Stage I, and at high conversions in Stage III, when the monomer droplets are used up and are no longer able to supply the monomer necessary to maintain $[M]$ constant within the growing polymer particles. Thus, in Stage III, the rate drops off as the monomer is exhausted within the particles.

Note that the rate increases with surfactant (emulsifier) concentration! The more surfactant used, the more micelles are established in Stage I, and the higher N will be in Stage II. The predicted independence of rate on initiator concentration must be viewed with caution. It is valid as long as N is held constant, but in practice, N increases with $[I]_0$. The more initiator added at the start of a reaction, the greater the number of monomer-swollen micelles that start growing before N is stabilized. Methods are available for estimating N [6–8].

Example 9.13   The Putrid Paint Division of Crud Chemicals, Inc., has a latex available that is 10% (by weight) poly(vinyl acetate) and contains 1×10^{14} particles/cm^3. To obtain optimum characteristics as an interior wall paint, a larger particle size and higher concentration of polymer are needed. It is proposed to obtain these by adding an additional four parts (by weight) of monomer per part of polymer to the latex and polymerizing without further addition of emulsifier. The reaction will be carried to 85% conversion, and the unreacted monomer will be steam-stripped and recovered. (The conditions of this "seeded" polymerization are set up so that the entire reaction proceeds in Stage II.) Estimate the time required for the reaction and the rate of heat removal (in Btu/gal of original latex, per hour) necessary to maintain an essentially isothermal reaction at 60 °C. At what conversion would the monomer droplets disappear and the rate cease to be constant?

$$\text{Data} : k_p = 3700 \, \text{L/mol-s at } 60\,°\text{C}$$
$$\Delta H_p = -21 \, \text{kcal/mol monomer unit}$$
$$\rho(\text{polymer}) = 1.2 \, \text{g/cm}^3$$
$$\rho(\text{monomer}) = 0.8 \, \text{g/cm}^3$$
$$M(\text{monomer}) = M_r = 86 \, \text{g/mol}$$

The concentration of monomer in the growing polymer particles is 10% (by weight).

Solution. It must first be assumed that in the absence of additional emulsifier, no new micelles will be established, so the original latex particles act as the exclusive loci for further polymerization. Then, $[M]$ and N must be obtained for use in Equation 9.71.

By assuming additivity of monomer and polymer volumes in the latex particles, we get

$$[M] = 1.33 \, \text{mol monomer/L of particles}$$

Since $N = 1 \times 10^{17}$ particles/L of original latex,

$$\frac{N}{2A} = 8.31 \times 10^{-8} \, \text{mol free radicals/L of original latex}$$

and Equation 9.71 gives

$$r_p = 4.09 \times 10^{-4} \, \text{mole monomer/L of original latex-s}$$
$$= 1.47 \, \text{mole monomer/L of original latex-h}$$

By again assuming additivity of water and polymer volumes, the density of the original latex is found to be $1.02 \, \text{g/cm}^3$. One liter of the original latex therefore has a mass of 1020 g and contains 102 g of polymer, to which $102 \times 4 = 408 \, \text{g}$ or 4.74 mol of monomer are added. The reaction converts $4.74 \times 0.85 = 4.02$ mol of monomer, which takes

$$\frac{4.02 \text{ mol monomer}}{\text{liter of original latex}} \times \frac{\text{liter of original latex-h}}{1.48 \text{ mol monomer}} = 2.27 \, \text{h}$$

For an isothermal reaction, the rate of heat removal must be the same as the rate of heat generation by the exothermic reaction, and is the product of ΔH_p and r_p, or 31,000 cal/L of original latex/h or 467 Btu/gal original latex/h.

When the monomer droplets just disappear, the reaction enters Stage III, and all the monomer and polymer will be in the swollen polymer particles, a total of 102 g (original polymer) + 408 g (added monomer) = 510 g/L of original latex. At this point, the particles still contain 10% monomer, so there are $(0.1) \times 510 \, \text{g} = 51 \, \text{g}$ of unconverted monomer. Therefore,

$$X = 1 - \frac{M}{M_0} = 1 - \left(\frac{51}{408}\right) = 0.875 \text{ or } 87.5\%$$

Beyond this point, the rate will drop off as the monomer concentration in the particles falls.

Despite the secondary effect of $[I]_o$ on the rate, it has a strong influence on the average chain length. The greater the rate of radical generation, the greater will be the frequency of alternation between chain growth and termination in a particle, resulting in a lower chain length. If r_c represents the rate of radical capture/L of reaction mass (half of which produces dead chains), in the absence of chain transfer:

$$\bar{x}_n = \frac{k_p(N/2A)[M]}{(r_c/2)} = \frac{k_p N[M]}{Ar_c} \tag{9.72}$$

The rate of generation of radicals is based on the presence of initiator in the aqueous phase alone:

$$r_{\text{gen}}\left(\frac{\text{mol radicals}}{\text{liter of aq. phase-second}}\right) = 2k_d/\text{s}[I]\left(\frac{\text{mol initiator}}{\text{liter of aq. phase}}\right) \tag{9.73}$$

If a steady-state radical concentration is assumed,

$$r_c\left(\frac{\text{mol radicals}}{\text{liter of reac. mass-second}}\right) = r_{\text{gen}}\left(\frac{\text{mol radicals}}{\text{liter of aq. phase-second}}\right) \times \phi_a\left(\frac{\text{liter of aq. phase}}{\text{liter of reaction mass}}\right) \tag{9.74}$$

where ϕ_a is the volume fraction aqueous phase. Thus,

$$r_c = 2k_d[I]\phi_a = 2k_d[I'] \tag{9.75}$$

where $[I'] = [I] \ \phi_a$, the moles of initiator per total volume of the reaction mass and

$$\bar{x}_n = \frac{k_p N [M]}{2 A k_d [I']} \tag{9.76}$$

The chain length is, therefore, inversely proportional to the *first* power of initiator concentration (compare with the second-order behavior for homogeneous polymerizations in Eq. 9.27).

It must be emphasized that the preceding is a simplified view of an extremely complex process. Even within Stage II, Smith and Ewart pointed out that their case 2 was merely the middle of a spectrum. At one end of the spectrum, case 1, the average number of radicals per particle is much less than 1/2. This is believed to occur sometimes because of radical escape from the particles. At the other extreme, the average number of radicals per particle is much greater than 1/2. This comes about from large particles and/or small k_t, both of which slow termination. Under these conditions, each particle acts as a tiny homogeneous reactor, according to the kinetics and the mechanism previously developed for homogeneous free-radical addition. The transition from case 2 to case 3 kinetics is often observed in seeded emulsion polymerizations, where the particles are fairly large to begin with and grow from there as the reaction proceeds.

The Smith–Ewart theory was developed for monomers such as styrene, with very low water solubility. Monomers such as acrylonitrile, with appreciable water solubility (on the order of 10%), may undergo significant homogeneous initiation in the aqueous phase. In some emulsion systems, the particles flocculate (coalesce) during polymerization, not only making a kinetic description difficult, but also sometimes badly fouling reactors. Good reviews of this subject are available [8–11], as well as complete books [7,12–15].

9.14 SUMMARY

Addition polymerizations are industrially important, and the basics of these reactions have been introduced in a generalized sense that is applicable to most monomers with a polymerizable double bond (including vinyl and acrylic monomers). Free-radical polymerizations proceed by a three-step mechanism (initiation, propagation, and termination). Some of the important characteristics of addition polymerization are contrasted with step-growth polymerization in Table 9.1, as well as the cartoon shown in Figure 9.1. It is difficult to have a direct comparison of the two reaction mechanisms, since some terms (such as critical gelation point) are only relevant in one case. A great resource for the kinetics of free-radical and many other types of polymerization are covered in Reference 16. In the next chapters, specialized polymerizations as well as the design of polymerization processes used in the industry are described.

PROBLEMS

1 Pure styrene undergoes significant thermal polymerization without any added initiator above about 100 °C. A mechanism has been postulated to explain thermal polymerization which involves a reaction between two monomer molecules to form diradicals that initiate chain growth from each end.

$$\underset{\underset{\text{H}}{\overset{\text{H H}}{\underset{|}{\overset{|}{\text{C=C}}}}}}{} + \underset{\underset{\text{H}}{\overset{\text{H H}}{\underset{|}{\overset{|}{\text{C=C}}}}}}{} \xrightarrow{k_d} \underset{\underset{\text{H H}}{\overset{\text{H H H H}}{\underset{|}{\overset{|}{\text{·C-C-C-C·}}}}}}{}$$

(a) Assume that the above reaction is rate controlling and derive an expression for the rate of polymerization according to this mechanism.

(b) Styrene terminates by disproportionation at these temperatures. Obtain an expression for $\bar{x}_n$ according to this mechanism. Neglect chain transfer.

(c) Interpret your answer to part (b) in terms of the relation between $\bar{x}_n$, $<\bar{x}_n>$, and their variation with conversion in an isothermal batch reactor.

(d) What do you think the distribution of chain lengths would be for this mechanism?

2 In a free-radical polymerization, 1 mol of acrylic acid monomer $CH_2=C(H)(COOH)$ is placed in a reactor with 0.001 mol of hydrogen peroxide.

In this problem, ignore volume shrinkage during polymerization and assume no chain transfer (no loss of radicals to anything except another monomer once a chain reaction has begun).

(a) How much monomer remains at 75% conversion?

(b) If the initiator efficiency is 0.37, what is the average chain length in the reactor, including the remaining monomer ($\bar{x}_n$)?

(c) After the monomer is removed (by evaporation under vacuum), what is the average chain length of PAA?

(d) What is the number-average molecular weight of the PAA in part (c)?

(e) How would you determine the polydispersity of PAA in part (c)?

3 Consider a free-radical addition reaction in which there is negligible conventional termination. Instead, growing chains are killed by a degradative chain-transfer reaction:

$$P_x \cdot + D \xrightarrow{k_t} P_x + D'$$

where D' is a stable radical that does not reinitiate polymerization.

(a) Obtain the rate expression for this mechanism.

(b) Obtain the expression for number-average chain length according to this mechanism.

4 Crud Chemicals has been carrying out an isothermal batch polymerization of styrene exactly like the one treated in Examples 9.1 and 9.9. One day, however, they get a little impatient and decide to "kick" the reaction by adding a second charge of initiator, identical to the first, at 50% conversion ($t = 8.36\,\text{h}$) (this is actually a fairly common practice in industry). To help them understand the effects of this,

(a) calculate the ratio of $\bar{x}_n$ just after the kick to $\bar{x}_n$ just before the kick and

(b) resketch Figure 9.4 qualitatively for this process.

5 Consider the isothermal batch polymerization of styrene treated in Examples 9.1 and 9.9. Now uppose that a chain-transfer agent with $C = 1.0$ is added to the initial charge at a concentration of $[R:H]_o = 0.01\,\text{mol/L}$.

Recalculate $\bar{x}_n(t \rightarrow \infty)$ and $<\bar{x}_n>(t \rightarrow \infty)$ for this batch. You may neglect volume change.

6 A chain-transfer agent is included in a homogeneous, batch free-radical polymerization. What fraction of the initial amount of chain-transfer agent remains at 99% conversion for $C = 0.1$, 1.0, and 10?

7 The classical kinetic scheme for free-radical addition polymerization neglects one other possible form of termination, primary-radical termination (PRT), in which a growing chain is killed by reaction with a free-radical R· from initiator decomposition:

For the polymerization of styrene at 60 °C in Examples 9.1, 9.9, and 9.10, estimate:

(a) the concentration of primary radicals $[I\cdot]$ at $X = 0$ and

(b) the ratio of rates of regular termination (r_t) to primary-radical termination (r_{prt}) at $X = 0$.

Altough we do not know the rate constants k_a and k_{prt}, because of the similarity of the species involved, let us assume (at least as an order-of-magnitude approximation) that

$$k_a \approx k_p \quad \text{and} \quad k_{prt} \approx k_t$$

8 Equations 9.30 and 9.31 give the temperature dependence of rate- and number-average chain length for a homogeneous free-radical polymerization. Obtain the analogous expressions for classical Stage II, case 2 emulsion polymerization. Use the following ballpark figures to calculate $r_p(70\,°\text{C})/r_p(60\,°\text{C})$ and $\bar{x}_n(70\,°\text{C})/\bar{x}_n(60\,°\text{C})$ for the emulsion polymerization:

$$E_p \approx 6\,\text{kcal/mol} \quad E_d \approx 30\,\text{kcal/mol} \quad E_t \approx 2\,\text{kcal/mol}$$

9 For the experiment describing the temperature-dependence of chain length discussed in Example 9.3,

(a) determine the temperature required to run the free-radical polymerization to obtain a polymer with double the molecular weight of the polymer produced at 60 °C and

(b) for the reaction temperature in a, find the rate of this reaction compared to 60 °C.

10 An initial experiment at an eyeglass manufacturer has produced poly(methyl methacrylate), whose monomer is $CH_2=C(CH_3)(COOCH_3)$ (used in eyeglasses and hard contact lenses), with a number-average molecular weight of 143,000, by carrying out the reaction at 65 °C for 45 min. Since the product design group has determined that a molecular weight of 100,000 is satisfactory, how fast can you speed up the production process and still achieve the desired MW? The activation energies for each reaction step are: propagation $= 7\,\text{kcal/mol}$, initiator decomposition $= 27\,\text{kcal/mol}$, and termination $= 3\,\text{kcal/mol}$.

11 Calculate and plot as a function of conversion $\bar{x}_w$, $<\bar{x}_w>$, and the polydispersity index $<\bar{x}_w>/<\bar{x}_n>$ for the homogeneous, isothermal styrene polymerization of Examples 9.1 and 9.10.

12 Crud Chemicals is producing a polymer with minimum polydispersity by homogeneous, free-radical polymerization in a CSTR. The polymer terminates inherently by disproportionation. To maintain and control the desired average chain length, they continuously feed chain-transfer agent to the reactor through a control valve. Things are going along nicely, with the reactor in steady-state operation, when suddenly, the valve in the chain-transfer feed line sticks open, adding chain-transfer agent at a much higher rate than previously. They fiddle with the valve for a while but are unable to free it, so before the reactor has had a chance to reach a new steady state, they simply close a gate valve in the line, shutting off the flow of chain-transfer agent entirely. The reactor then proceeds to a new steady state in the absence of additional chain-transfer agent.

Sketch qualitatively $\bar{x}_n$ and $<\bar{x}_n>$ on the same coordinates versus time for the period described.

13 Modify Equation 9.76 to include chain transfer. Define any terms introduced.

14 A classical, Stage II, case 2 emulsion polymerization of styrene has the following parameters:

$$N = 3.2 \times 10^{17} \text{ particles/L of reaction mass}$$
$$[M] = 1.5 \text{ mol/L} \qquad [I] = 1.0 \times 10^{-3} \text{mol/L}$$
$$k_p = 176 \text{ L/mol-s}$$
$$k_d = 4.96 \times 10^{-6} /\text{s}$$
$$\phi_a = 2/3$$

Calculate:

(a) the rate of polymerization,

(b) $\bar{x}_n$, and

(c) the average lifetime of a growing chain.

15 Calculate and compare the average chain lifetime at the start of the reaction with the monomer half-life for the system in Example 9.10. Is the concept of an instantaneous quantity valid for this system? $k_t = 14.5 - 10^6 \text{ L/mol-s}$.

16 A batch, free-radical polymerization is carried out with $[M]_o = 1.0 \text{ mol/L}$ and $[I]_o = 2 \times 10^{-4} \text{ mol/L}$. The reaction reaches a limiting conversion of 90%. Assume $f = 1$.

(a) Calculate $<\bar{x}_n> (t \to \infty)$ assuming termination by disproportionation.

(b) Repeat part (a) with the addition of $6 \times 10^{-4} \text{ mol/L}$ of a mercaptan to the initial charge.

17 A batch reactor producing a free-radical addition polymer is sampled at various conversions. Analysis of these samples reveals that the weight-average chain length varies linearly with conversion from 10,000 at $X = 0$ to 5000 at $X = 1$. The polymer is known to terminate by combination and there is no significant chain transfer occurring. Obtain an expression for the polydispersity index of the product as a function of conversion.

REFERENCES

[1] Levenspiel, O., *Chemical Reaction Engineering*, 2nd ed., Wiley, New York, 1972.

[2] Kenat, T., R.I. Kermode, and S.L. Rosen, *J. Appl. Polym. Sci.* **13**, 1353 (1969).

[3] Hoffman, B.F., S. Schreiber, and G. Rosen, *Ind. Eng. Chem.* **56**(5), 51 (1964).

[4] Harkins, W.D., *J. Am. Chem. Soc.* **69**, 1428 (1947).

[5] Smith, W.V. and R.H. Ewart, *J. Chem. Phys.* **16**(6), 592 (1948).

[6] Flory, P.J., *Principles of Polymer Chemistry*, Cornell UP, Ithaca, NY, 1953, Chapter V-3.

[7] Bovey, F.A., *et al.*, *Emulsion Polymerization*, Interscience, New York, 1955.

[8] Gardon, J.L., Emulsion polymerization, Chapter 6 in *Polymerization Processes*, C. E. Schild-knechtand I. Skeist (eds), Wiley, New York, 1977.

[9] Ugelstad, J. and F.K. Hansen, *Rubber Chem. Technol.* **49**, 536 (1976).

[10] Blackley, D.C., *Macromol. Chem. (London)* **2**, 31 (1982).

[11] Poehlein, G.W., Chapter 6 in *Applied Polymer Science*, 2nd ed., R.W. Tessand G. W. Poehlein (eds), ACS Symposium Series 285, American Chemical Society, Washington, DC, 1985.

[12] Blackley, D.C., *Emulsion Polymerization*, Wiley, New York, 1975.

[13] Piirma, I. and J.L. Gardon (eds), *Emulsion Polymerization*, ACS Symposium Series 24, American Chemical Society, Washington, DC, 1976.

[14] Bassett, D.R.and A.E. Hamielec (eds), *Emulsion Polymers and Emulsion Polymerization*, ACS Symposium Series 165, American Chemical Society, Washington, DC, 1981.

[15] Piirma, I. (ed.), *Emulsion Polymerization*, Academic, New York, 1982.

[16] Odian, G., *Principles of Polymerization*, Wiley, New York, 1991.

CHAPTER 10

ADVANCED POLYMERIZATION METHODS

10.1 INTRODUCTION

The previous two chapters covered two of the primary methods for forming polymers: step-growth (or condensation) polymerization and free-radical addition (or chain) polymerization. Of course, there are other ways to make really big polymers, and some of these techniques have gained commercial importance due to their capabilities to form more stereoregular (e.g., isotactic or syndiotactic) polymers, to decrease a polymer's polydispersity, or to provide other methods to carefully control the polymer structure. This chapter will introduce the basic concepts for some of the more common advanced polymerization techniques: cationic, anionic, heterogeneous catalyzed (specifically the Ziegler–Natta catalyst system), metallocene-catalyzed, ring-opening metathesis, and atom transfer radical polymerizations.

10.2 CATIONIC POLYMERIZATION [1–4]

Strong Lewis acids, that is, electron acceptors, are often capable of initiating addition polymerization of monomers with electron-rich substituents adjacent to the double bond. Cationic catalysts are most commonly metal trihalides such as $AlCl_3$ or BF_3. These compounds, although electrically neutral, are short of two electrons for having a complete valence shell of eight electrons. Traces of a cocatalyst, usually water, are usually required to initiate polymerization, first by grabbing a pair of electrons from the cocatalyst:

$$\text{F:}\overset{\text{F}}{\underset{\text{F}}{\text{B}}} + \text{:}\overset{\text{H}}{\underset{\text{H}}{\text{O}}}\text{:} \longrightarrow \text{F:}\overset{\text{F}}{\underset{\text{F}}{\text{B}}}\text{:}\overset{\text{H}}{\underset{\text{H}}{\text{O}}}\text{:} \longrightarrow [\text{F:}\overset{\text{F}}{\underset{\text{F}}{\text{B}}}\text{:O:H}]^- \ [\text{H}]^+$$

Fundamental Principles of Polymeric Materials, Third Edition. Christopher S. Brazel and Stephen L. Rosen.
© 2012 John Wiley & Sons, Inc. Published 2012 by John Wiley & Sons, Inc.

The leftover proton (the cation) is thought to be the actual initiating species, abstracting a pair of electrons from the monomer and leaving a cationic chain end that reacts with additional monomer molecules (similar to the propagation step in free-radical polymerization).

$$[BF_3OH]^- \quad [H]^+ \ + \ \underset{\underset{CH_3}{|}}{\overset{\overset{H}{|}}{C}} :: \underset{\underset{CH_3}{|}}{\overset{\overset{CH_3}{|}}{C}} \longrightarrow H : \underset{\underset{CH_3}{|}}{\overset{\overset{H}{|}}{C}} : \underset{\underset{CH_3}{|}}{\overset{\overset{CH_3}{|}}{C}}]^+ \quad [BF_3OH]^- \quad \text{etc.}$$

Gegen or counter ion

An important point is that the *gegen* or *counter* ion (here, $[BF_3OH]^-$) is electrostatically held near the growing chain end and therefore can exert a steric influence on the addition of monomer units. Termination is thought to occur by a disproportionation-like reaction that regenerates the catalyst complex. This complex, therefore, is a true catalyst, unlike free-radical initiators:

$$\sim\!\!\underset{\underset{H}{|}}{\overset{\overset{H}{|}}{C}} - \underset{\underset{CH_3}{|}}{\overset{\overset{CH_3}{|}}{C}}]^+ \quad [BF_3OH]^- \longrightarrow \sim\!\!\underset{\underset{H}{|}}{\overset{\overset{H}{|}}{C}} - \underset{}{\overset{\overset{CH_3}{|}}{C}} = CH_2 \ + BF_3 \cdot H_2O$$

These reactions proceed very rapidly at low temperatures. For example, the polymerization of isobutylene illustrated above is carried out commercially at $-150\,°F$. The average chain length actually formed increases as the temperature is lowered.

Cationic initiation is successful only with monomers having electron-rich substituents adjacent to the double bond (like isobutylene), including

$$\underset{\underset{O\text{–}R}{|}}{\overset{\overset{H}{|}}{C}} \overset{H}{\underset{\underset{H}{|}}{=}} C$$

Alkyl vinyl ethers

α-Methyl styrene

None of these monomers can be polymerized to high molecular weight with free-radical initiators, therefore cationic polymerization plays an important role in commercial production of these polymers.

10.3 ANIONIC POLYMERIZATION [5–7]

Addition polymerization may also be initiated by anions. Anionic polymerization has achieved tremendous commercial importance in the past few decades because of its ability to control molecular structure during polymerization, allowing the synthesis of materials that were previously difficult or impossible to obtain. A variety of anionic initiators has been investigated but the organic alkali-metal salts are perhaps the most common, as illustrated below for the polymerization of styrene with *n*-butyllithium:

$$H\text{–}C\text{–}C\text{–}C\text{–}C :]^- \quad [Li]^+ \ + \ C = C \longrightarrow H\text{–}C\text{–}C\text{–}C\text{–}C : C : C :]^- \quad [Li]^+$$

n-Butyllithium (*n*-BuLi)

Styrene

The anionic $(-)$ chain end then propagates the chain by adding another monomer molecule. Again, the gegen ion can sterically influence the reaction.

Sodium and lithium *metals* were used to polymerize butadiene in Germany during World War II. After the war, in the United States, it was discovered that under appropriate conditions, dispersions of lithium could lead to largely *cis*-1,4 addition of butadiene and isoprene (the latter being the synthetic counterpart of natural rubber). In these processes, a metal atom first reacts with the monomer to form an anion radical:

$$Li\cdot \ + \ \underset{\substack{| \quad \quad | \\ H \quad \quad H}}{\overset{\substack{H\;H\;H\;H \\ |\;\,|\;\,|\;\,|}}{C::C:C::C}} \ \longrightarrow \ \underset{\substack{| \quad \quad | \\ H \quad \quad H}}{\overset{\substack{H\;H\;H\;H \\ |\;\,|\;\,|\;\,|}}{\cdot C:C::C:C:]^-}} \ \ [Li]^+$$

Anion radical

These anion radicals then react in either of two ways. One may react with another atom of lithium

$$Li\cdot \ + \ \underset{\substack{| \quad \quad | \\ H \quad \quad H}}{\overset{\substack{H\;H\;H\;H \\ |\;\,|\;\,|\;\,|}}{\cdot C:C::C:C]^-}} \ [Li]^+ \ \longrightarrow \ ^+[Li] \quad \underset{\substack{| \quad \quad | \\ H \quad \quad H}}{\overset{\substack{H\;H\;H\;H \\ |\;\,|\;\,|\;\,|}}{^-[:C:C::C:C:]^-}} \ [Li]^+$$

$$2 \ \underset{\substack{| \quad \quad | \\ H \quad \quad H}}{\overset{\substack{H\;H\;H\;H \\ |\;\,|\;\,|\;\,|}}{\cdot C:C::C:C]^-}} \ [Li]^+ \ \longrightarrow \ ^+[Li] \quad \underset{\substack{| \quad\quad | \quad | \quad\quad | \\ H \quad\quad H\,H \quad\quad H}}{\overset{\substack{H\;H\;\;H\;H\,H\;H\;\;H\;H \\ |\;\,|\;\;|\;\,|\;\,|\;\,|\;\;|\;\,|}}{^-[:C-C=C-C:C-C=C-C:]^-}} \ [Li]^+$$

and/or two may rapidly undergo radical recombination.

Either way, the result is a dianion that propagates a chain from each end. Other dianionic initiators have also been developed [5–8].

There are a couple of very interesting aspects to these reactions. First, the rates of initiation and propagation vary with the monomer, gegen ion, and solvent. In general, the reactions proceed more rapidly in polar solvents as the species are more highly ionized. (The polarity of the solvent also strongly influences the stereospecific nature of the polymer.) In many important cases, the rate of initiation is comparable to the rate of propagation, unlike free-radical addition, in which $r_i \ll r_p$. This means that the initiator (in this case, it is an initiator rather than a catalyst) promptly starts growing chains. Second, in the absence of impurities, *there is no termination step*. The chains continue to grow until the monomer supply is exhausted. The ionic chain end is perfectly stable and the growth of the chains will resume if more monomer is added. For this reason, these materials are aptly termed *"living" polymers* and if additional monomer (or a different monomer) is later added, the reaction continues. However, proton-donating impurities such as water or acids quickly kill (terminate) them:

$$\sim\!\!\underset{\substack{| \quad | \\ H \quad \bigcirc}}{\overset{\substack{H\;\;H \\ |\;\;|}}{C-C:]^-}} \ [Li]^+ \ + \ H_2O \ \longrightarrow \ \sim\!\!\underset{\substack{| \quad | \\ H \quad \bigcirc}}{\overset{\substack{H\;\;H \\ |\;\;|}}{C-C:H}} \ + \ LiOH$$

This unique mechanism has a number of practically important consequences.

10.3.1 Block Copolymerization

If a second monomer is introduced after the initial monomer, charge is exhausted and the living chains resume propagation with the second monomer, neatly giving a block copolymer. Monomers can be alternated as desired to give AB (diblock) or ABA (triblock) or even more complicated block structures, conceivably even including three or more different monomers.

10.3.2 Synthetic Flexibility

Anionic polymerization allows the synthesis of all sorts of interesting and useful molecules. For example, bubbling carbon dioxide through a batch of living chains followed by exposure to water produces a carboxyl-terminated polymer:

$$\text{~}\overset{\displaystyle H}{\underset{\displaystyle H}{C}}\text{:]}^- \ \text{[Li]}^+ \ + \ CO_2 \ \longrightarrow \ \text{~}\overset{\displaystyle H}{\underset{\displaystyle H}{C}}-\overset{\displaystyle O}{C}-O\text{:]}^- \ \text{[Li]}^+$$

$$\text{~}\overset{\displaystyle H}{\underset{\displaystyle H}{C}}-\overset{\displaystyle O}{C}-O\text{:]}^- \ \text{[Li]}^+ \ + \ H_2O \ \longrightarrow \ \text{~}\overset{\displaystyle H}{\underset{\displaystyle H}{C}}-\overset{\displaystyle O}{C}-OH \ + \ LiOH$$

Similarly, ethylene oxide gives hydroxyl-terminated chains:

$$\text{~}\overset{\displaystyle H}{\underset{\displaystyle H}{C}}\text{:]}^- \ \text{[Li]}^+ \ + \ H_2C\!-\!CH_2 \ \longrightarrow \ \text{~}\overset{\displaystyle H}{\underset{\displaystyle H}{C}}-\overset{\displaystyle H}{\underset{\displaystyle H}{C}}-\overset{\displaystyle H}{\underset{\displaystyle H}{C}}-O\text{:]}^- \ \text{[Li]}^+$$

$$\text{~}\overset{\displaystyle H}{\underset{\displaystyle H}{C}}-\overset{\displaystyle H}{\underset{\displaystyle H}{C}}-\overset{\displaystyle H}{\underset{\displaystyle H}{C}}-O\text{:]}^- \ \text{[Li]}^+ \ + \ H_2O \ \longrightarrow \ \text{~}\overset{\displaystyle H}{\underset{\displaystyle H}{C}}-\overset{\displaystyle H}{\underset{\displaystyle H}{C}}-\overset{\displaystyle H}{\underset{\displaystyle H}{C}}-OH \ + \ LiOH$$

Note that if these reagents are added to a lithium-initiated dianion, *both* ends of the chain will be capped with the functional group. Such chains are macro diacids or diols. Carboxyl- or hydroxyl-terminated chains may then participate in the usual step-growth polymerizations (see Chapter 8).

Once the monomer is exhausted in a reaction, if a tetrafunctional monomer such as divinyl benzene (DVB), $H_2C=CH\phi HC=CH_2$, is added to a batch of living chains, it couples to itself and to the living chains. If it is assumed that a DVB molecule can react with another DVB only at one end (there are probably good steric reasons why this should be so), the following type of *star polymer* (or *dendrimer*) structures result with linear branches radiating from a DVB core (Figure 10.1).

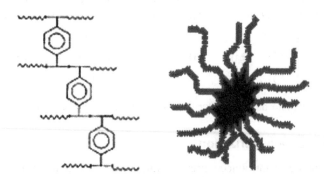

FIGURE 10.1 Schematic showing the attachment of several linear polymers to a core that contains several DVB units linked together. The cartoon on the right depicts how a multifunctional core can attach many polymer arms to make a star polymer.

It is conceivable that at this point, a second difunctional monomer could be added giving a star polymer with two different kinds of branches, but normally, the reactor would be opened and the reaction terminated. From the structure above, there are $(b - 1)$ moles of DVB per mole of star branches, where b is the average number of branches per star. Since the number of moles of star molecules is equal to the moles of living chains over the average number of branches, and with an initiator such as n-BuLi (n-butyl lithium), each molecule starts one living chain; to make a b-branch star polymer, one must add

$$N_{\text{DVB}} = \frac{(b-1)I_o}{b} \qquad (10.1)$$

moles of DVB, where I_o is the moles of initiator charged to the reactor.

Living chains can also be *linked* by, for example, dichloro compounds, instantly doubling their chain lengths:

$$
2 \overset{\displaystyle H}{\underset{\displaystyle H}{\sim\!\!\!\sim\text{C:}]^{-}}} \quad [\text{Li}]^{+} \;+\; \text{Cl--R--Cl} \longrightarrow \overset{\displaystyle H \quad H}{\underset{\displaystyle H \quad H}{\sim\!\!\!\sim\text{C:R:C}\sim\!\!\!\sim}} \;+\; 2\text{LiCl}
$$

With multifunctional linking agents, this technique can be used to form star polymers with various numbers of branches. It would be nice if all the functional groups on such compounds took part in the linking reaction, but sometimes steric factors lower the linking efficiency. Multifunctional initiators can also be used to grow star polymers. Various initiators, linking agents, and other reagents for anionic polymerization have been reviewed [5–8].

10.3.3 Monodisperse Polymers

Anionic polymerization is also quite useful at making polymers with narrow size distributions. Consider a batch reactor containing, say, a 20% solution of styrene in the solvent tetrahydrofuran (THF). A charge of n-BuLi is suddenly added. In this fairly polar solvent, the n-BuLi ionizes immediately and completely and with $r_i \approx r_p$ promptly starts chains growing—one chain for each molecule of n-BuLi. In the absence of terminating impurities, the number of growing chains remains constant and *they compete on an even basis for the available monomer*. They will therefore have essentially the same length, giving a nearly monodisperse polymer. Not only can anionic polymerization be used to synthesize essentially monodisperse homopolymers but the blocks in the block copolymers formed this way can also tailor to be monodisperse.

As with other modes of polymerization, the number-average chain length is given by the moles of monomer polymerized over the moles of chains present. With an initiator like n-BuLi, each molecule starts a single chain, therefore,

$$\bar{x}_n = \frac{M_0 - M}{I_0} = \frac{M_0 X}{I_0} \qquad (10.2)$$

(If the volume is constant, the moles may be bracketed to give concentrations.) Statistically, the distribution of chain lengths is obtained by answering the following question: given $M_o - M$ marbles and I_o buckets, what will be the distribution of marbles among the buckets

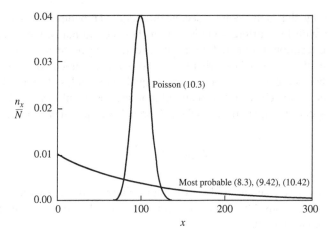

FIGURE 10.2 Comparison of Poisson and most-probable number-fraction distributions of chain lengths for $\bar{x}_n = 100$.

if the marbles are thrown completely randomly into the buckets? The result is the Poisson distribution from statistics:

$$\left(\frac{n_x}{N}\right) = \frac{(\bar{x}_n)^x \exp(-\bar{x}_n)}{x!} \approx \frac{\exp(x - \bar{x}_n)}{\sqrt{(2\pi x)}}\left(\frac{\bar{x}_n}{x}\right)^x \tag{10.3}$$

(As written above, the distribution does not include the initiator residue.) To a good approximation, the polydispersity index of this distribution is[1]

$$\frac{\bar{x}_w}{\bar{x}_n} \approx 1 + \frac{1}{\bar{x}_n} \tag{10.4}$$

from which it is seen that even at moderate $\bar{x}_n$ values, the polymer is essentially monodisperse. Figure 10.2 compares the Poisson distribution with the most probable distribution, Equations 8.3 and 9.42, for $\bar{x}_n$ values of 100. Because of their sharpness, the number and weight distributions are almost identical.

As in the case of step-growth polymerization, the average chain length increases continuously with conversion (Eq. 10.2). Unlike step-growth, the reaction mass consists only of monomer molecules and polymer chains of essentially a single length. Equations 10.2 and 10.3 characterize the polymer present.

Example 10.1 Starting with a batch reactor containing 1×10^{-3} moles of n-BuLi in dilute solutions, suggest *two* methods for making the block copolymer $[S]_{200}-[B]_{1000}-[S]_{200}$, where [S] represents a styrene repeat unit and [B] represents a butadiene repeat unit.

Solution. If we let all reactions go to completion, $X = 1$, then from Equation 10.2, $M_o = I_o x$ (since the blocks will be essentially monodisperse, the bar and n have been left off x).

---

[1] There are important practical limitations of not being able to add or mix reagents instantaneously and uniformly, and the almost inevitable presence of some terminating impurities. In practice, these always cause some additional broadening of the distribution.

(a) Add $200 \times 10^{-3} = 0.2$ mol of styrene to the reactor. When it has completely reacted, living polystyrene chains of $x = 200$ have been formed. Then add $1000 \times 10^{-3} = 1.0$ mol of butadiene, and when it has completely reacted, add another 0.2 mol styrene and react to completion before opening the reactor and terminating the chains.

(b) As before, start with 0.2 mol of styrene. When it has completely reacted, add 0.5 mol of butadiene and react to completion. Then, link the resulting 1×10^{-3} mol of $[S]_{200}-[B]_{500}^-$ living chains with 0.5×10^{-3} moles of a difunctional linking agent $Cl-R-Cl$ to give 0.5×10^{-3} mol of $[S]_{200}-[B]_{500}-R-[B]_{500}-[S]_{200}$. Normally, the single R in the middle is insignificant.

Commercially, procedure (b) is preferred. To obtain the desired rubbery properties in the central polybutadiene block, nonpolar hydrocarbon solvents are used to promote 1,4-addition. These solvents slow down the crossover reaction from butadiene back to styrene in procedure (a), leading to increased polydispersity in the second polystyrene block. The linking reaction in (b) is imperfect, and commercial materials contain some diblock chains along with the desired triblock. Also keep in mind that each time material is added to a reactor, the probability of getting some terminating impurities increases.

It will be left as an end-of-chapter exercise to show how the polymer in Example 10.1 could be made by starting with a dianionic initiator.

Example 10.2 Consider the anionic batch polymerization of 1.0 mol of styrene in tetrahydrofuran solution with 1.0×10^{-3} mol of n-BuLi initiator. Assume that $r_i \approx r_p$ and that mixing is perfect and instantaneous. At 50% conversion, 0.5×10^{-3} mol of water are added to the reaction mass and the reaction is allowed to continue.

(a) At 100% conversion, what chain lengths will be present in the reaction mass?

(b) Calculate $\bar{x}_n$ at 100% conversion.

Solution.

(a) The initiator starts 1×10^{-3} mol of chains growing. At 50% conversion, the water terminates half the chains. Thus, 1.0/4 mol of monomer is present in the 0.5×10^{-3} mol of terminated chains:

$$x(\text{terminated chains}) = \frac{1.0}{4 \times 0.5 \times 10^{-3}} = 500$$

The remaining 0.75 mol of monomer continues to grow in 0.5×10^{-3} mol of unterminated chains:

$$x(\text{unterminated chains}) = \frac{1.0(3/4)}{0.5 \times 10^{-3}} = 1500$$

(b) Since each initiator molecule starts a chain, at $X = 1$,

$$\bar{x}_n = \frac{M_0}{I_0} = \frac{1.0}{1 \times 10^{-3}} = 1000$$

or using Equation 5.2a

$$\bar{x}_n = \frac{\sum n_x x}{\sum n_x} = \frac{(0.5 \times 10^{-3})(500) + (0.5 \times 10^{-3})(1500)}{0.5 \times 10^{-3} + 0.5 \times 10^{-3}} = 1000$$

10.4   KINETICS OF ANIONIC POLYMERIZATION [5,6,9]

A general description of anionic polymerization kinetics is complicated by the associations that may occur, particularly in nonpolar (hydrocarbon) solvents. The rate of propagation is proportional to the product of the monomer concentration and the concentration of *active* living chains $[P_x^-]$:

$$r_p = -\left(\frac{1}{V}\right)\left(\frac{dM}{dt}\right) = k_p[M]\left[P_x^-\right] \qquad (10.5)$$

With negligible association (e.g., in THF solvent or hydrocarbons at n-BuLi concentrations less than 10^{-4} mol), each initiator molecule starts a growing chain, and in the absence of terminating impurities, the number of active living chains equals the number of initiator molecules added,

$$P_x^- = I_0 \quad \text{(constant)} \qquad (10.6)$$

$$r_p = -\left(\frac{1}{V}\right)\left(\frac{dM}{dt}\right) = k_p\left(\frac{M}{V}\right)\left(\frac{I_0}{V}\right) \qquad (10.7)$$

or making use of the definition of conversion X (Eq. 9.14) and assuming that the volume of the reaction mass is linear in the conversion (Eq. 9.17), we get

$$\frac{dX}{dt} = k_p[I]_0 \frac{1-X}{1+\varepsilon X} \qquad (10.8)$$

Unlike the case for free-radical addition, this can be readily integrated for a batch reactor ($[I]_o$ is constant) to give

$$\varepsilon X + (1+\varepsilon)\ln(1-X) = -k_p[I]_0 t \qquad (10.9)$$

but it is still not easier to get X as an explicit function of t.

Because anionic polymerizations are generally carried out in rather dilute solutions in inert solvents, volume changes with conversion tend to be much smaller than when undiluted monomer is polymerized. This often justifies the neglect of volume change ($\varepsilon = 0$) for which Equation 10.9 becomes

$$X = 1 - \exp(-k_p[I]_0 t) \quad (\varepsilon = 0) \qquad (10.10)$$

Example 10.3 For an isothermal reaction (constant k_p) subject to the assumptions in Example 10.2, obtain an expression that relates $\bar{x}_n$ to time. Neglect volume change.

Solution. Combining Equations 10.2 and 10.10 gives

$$x = \frac{M_0}{I_0}(1 - \exp\{-k_p[I]_0 t\})$$

Since the polymer will be essentially monodisperse, the bar and subscript have been left off x. To use this expression, it would be necessary to know V_o, the initial volume of the reaction mass, to get $[I]_o = I_o/V_o$.

In n-BuLi polymerizations at high concentrations in nonpolar solvents, the chain ends are present largely as inactive dimers, which dissociate *slightly* according to the equilibrium

$$(P_x^-)_2 \overset{K}{\rightleftharpoons} 2P_x^-$$
$$\text{Inactive dimer} \quad \text{Active chains}$$

where

$$K = \frac{[P_x^-]^2}{[(P_x^-)_2]} \ll 1 \tag{10.11}$$

The concentration of active chains is then

$$[P_x^-] = K^{1/2}[(P_x^-)]^{1/2} \tag{10.12}$$

Now it takes two initiator molecules to make one inactive dimer, therefore

$$\underbrace{[P_x^-]}_{\text{Negligible}} + 2[(P_x^-)_2] = [I]_0 \tag{10.13}$$

The rate of polymerization then becomes

$$r_p = -\left(\frac{1}{V}\right)\left(\frac{dM}{dt}\right) = k_p K^{1/2}\left(\frac{[I]_0}{2}\right)^{1/2}[M] \tag{10.14}$$

The low value of K, reflecting the presence of most chain, ends in the inactive associated state (dimers) giving rise to low rates of polymerization in nonpolar solvents. At very high concentrations, association may be even greater and the rate is essentially independent of $[I]_0$.

Example 10.4 Which type of isothermal reactor will produce the narrowest possible distribution of chain lengths in an anionic polymerization: continuous stirred tank (CSTR), batch (assume both are perfectly stirred), plug-flow tubular, or laminar-flow tubular?

Solution. As demonstrated in Example 10.3, the chain length depends on how long a chain is allowed to grow. In a batch reactor, and an *ideal* plug-flow reactor, all chains react for the same length of time; hence, the product will be essentially monodisperse. In CSTRs and laminar-flow tubular reactors (which have a parabolic flow profile), the residence time of chains in the reactor varies, causing a spread in the distribution. Keep in mind, however, that ideal plug flow is a practical impossibility, particularly with highly viscous polymer solutions. Compare these conclusions with those of Example 9.11.

10.5 GROUP-TRANSFER POLYMERIZATION

In the 1980s, the DuPont Company developed and patented [10] a new type of polymerization that mechanistically is similar to anionic polymerization. *Group-transfer polymerization* (GTP) has been defined as "polymerization of α,β-unsaturated esters, ketones, nitriles, or amides, initiated by silyl ketene acetals [11]." It has most commonly been used to polymerize acrylate and methacrylate monomers with the aid of anionic catalysts (they are true catalysts here), such as the bifluoride ion, $[FHF]^-$, or bioxyanions. GTP is illustrated below for the polymerization of methyl methacrylate (MMA) with silyl ketene acetal (SKA):

Initiation

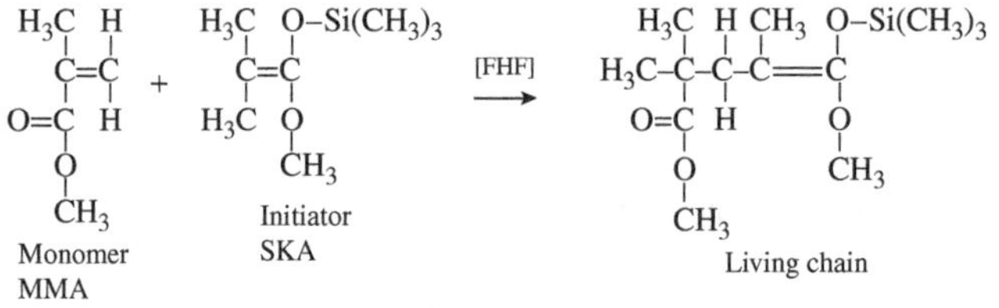

Propagation

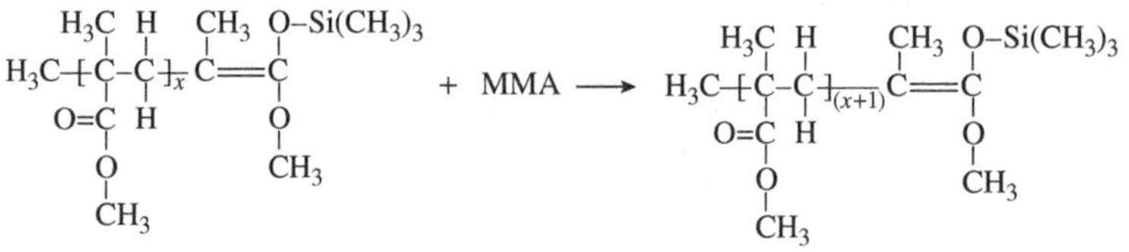

The initiating functionality is transferred to the growing end of the chain as each new monomer unit is added. They are living chains as in anionic addition and can likewise be used to produce monodisperse polymers, block copolymers, and, with the addition of appropriate reagents, chains with desired terminal groups. They can also control stereoregularity in the chain.

Unlike with anionic addition, chain transfer can occur:

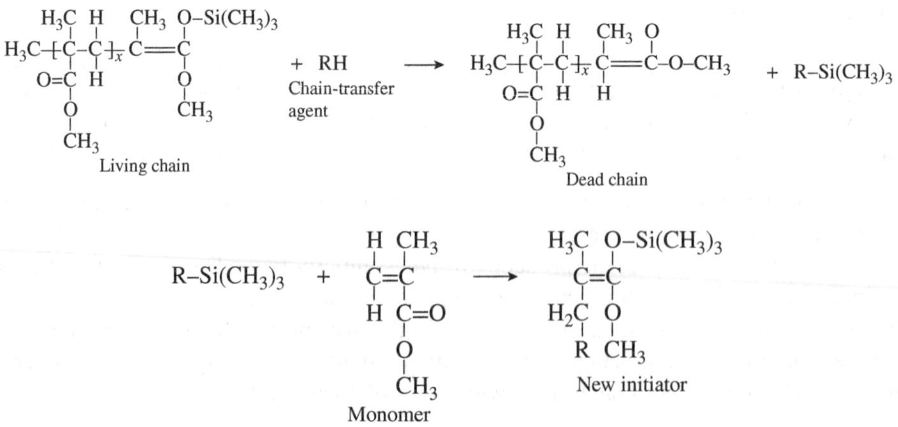

As in anionic addition polymerization, chain length can be reduced by using more initiators, but because these initiators are rather expensive, it is often preferable to use a chain-transfer agent instead. Also, at low monomer concentrations, termination can occur through cyclization of the chain end.

10.6   ATOM TRANSFER RADICAL POLYMERIZATION

A more recent development in living polymerizations that has shown great promise for tailoring polymer structure and molecular weight is atom transfer radical polymerization (ATRP). The discovery of ATRP in the 1990s [12] has led to the broadening of living polymerizations, which can include a wide range of monomer types (including vinyl, (meth)acrylates, styrenes, and epoxies). This is largely due to ATRP's robust method of synthesis that is less sensitive to the presence of impurities than anionic polymerization. The reaction proceeds through the use of a relatively inexpensive copper complex as the catalyst (Figure 10.3), which is normally bound to a ligand chemical, such as pyridine. The reaction produces high molecular weight, narrowly disperse polymers. In addition to the catalyst, ATRP requires an initiator, most often an alkyl halide (e.g., bromides and chlorides). As with free-radical polymerization, the molecular weight of the resulting polymer is inversely proportional to the amount of initiator used. One unique characteristic of the ATRP initiator is that multiple functionalities (i.e., multiple halide groups) can influence the structure of the polymer; for instance, an initiator with six Br atoms can grow up to six polymer chains from the same initiator molecule, creating a star polymer.

In ATRP, there are reactive and dormant polymer species in equilibrium during the polymerizations, which alternate between halide-capped polymers (dormant) and growing (reactive) polymers with a free radical on the end. The choice of catalyst controls this equilibrium which in turn influences the polymerization rate and the distribution of chain lengths. The mechanism offers flexibility to conduct reactions in bulk, solution, or emulsions/suspensions, just as free-radical polymerizations. Due to the capability to polymerize a large range of monomers with an inexpensive catalyst in a reactor, where purity is nearly as important as in anionic polymerizations, ATRP continues to grow in popularity. For further information, review articles written by the inventors are available [12,16].

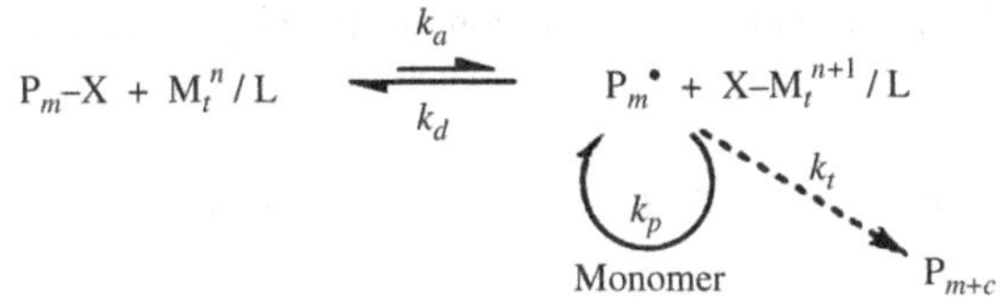

$$P_m-X + M_t^n/L \underset{k_d}{\overset{k_a}{\rightleftharpoons}} P_m^\bullet + X-M_t^{n+1}/L$$

$$\xrightarrow{k_t} P_{m+c}$$

$$k_p \quad \text{Monomer}$$

FIGURE 10.3   Metal (often copper) ligand (M/L) structure to support ATRP reactions. Here, P_m and P_{m+c} are polymers with length m and $m+c$, respectively, X is typically a halide atom, such as Cl or Br, and the k's are kinetic rate constants for activation (k_a), deactivation (k_{da}), and termination (k_t). Reprinted from [16] with permission from Elsevier.

10.7   HETEROGENEOUS STEREOSPECIFIC POLYMERIZATION [13–15]

Many monomers are gases at or near atmospheric temperature and pressure and can be polymerized using solid catalysts (hence, multiple phases in a reactor or heterogeneous reactions). In the early 1950s, Karl Ziegler in Germany observed that certain heterogeneous catalysts based on transition metals would polymerize ethylene to a linear, high-density material (HDPE, Chapter 5) at low pressure and temperature (compared to the existing free-radical process for low-density polyethylene). Giulio Natta in Italy showed that these catalysts would produce highly stereospecific poly α-olefins ($[H_2C-CHR]_x$, where $R = C_nH_m$) (notably polypropylene) and polydienes. Ziegler and Natta shared the 1963 Nobel Prize in chemistry for their work. Almost simultaneously, scientists at Amoco and Phillips in the United States developed heterogeneous catalysts based on Mo and Cr that also produced linear polyethylene under mild conditions.

One example of a *Ziegler–Natta catalyst* system is titanium tetrachloride, $TiCl_4$, and aluminum triethyl, $Al(C_2H_5)_3$. Vanadium and cobalt chlorides are also used as is $Al(C_2H_5)_2Cl$. When the substances are mixed in an inert solvent, a crystalline solid is obtained along with a highly colored supernatant (deep violet or brown). It is known that the reaction involves the reduction of the titanium to a lower valence state, probably $+2$, since $TiCl_3$ also forms an effective catalyst. The supernatant liquid alone will polymerize α olefins, but the resulting polymers show little stereospecificity. Commercial catalyst systems are based on the solid, either alone or on a support such as SiO_2 or $MgCl_2$. A "Phillips" catalyst typically consists of CrO_3 supported on silica or alumina.

Polymerization with Ziegler–Natta catalysts is thought to occur at active sites formed by interaction of the metal alkyl with a metal chloride on the surface of the metal chloride crystals. Monomer is chemisorbed at the site (thus accounting for its specific orientation when added to the chain), and propagation occurs by *insertion* of the chemisorbed monomer into the metal-chain bond at the active site.

$$\text{site–}P_x + M \rightarrow \text{site–}P_{x+1}$$

The chain thus grows out from the site like a hair from the scalp.

Hydrogen is used as a chain-transfer agent in these reactions:

$$\text{site} - P_x + H_2 \rightarrow \text{site} - H + H - P_x$$

Chain transfer with the metal alkyl component in Ziegler–Natta systems has been identified as

$$\text{site} - P_x + Al(R)_3 \rightarrow \text{site} - R + Al(R)_2 - P_x$$

and transfer to monomer may also occur.

Most commercial processes based on these catalysts are carried out with the monomers either in the gas phase or in the liquid phase at temperatures such that the polymer precipitates as it is formed. Solid, porous catalyst particles in the range of 10–100 μm are introduced to the reactor. As polymer begins to form, they rapidly break up into many

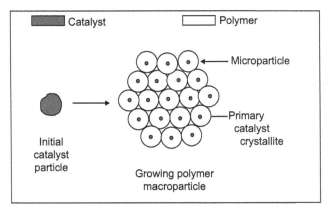

FIGURE 10.4   Mechanism of particle formation in Ziegler–Natta polymerization [17].

smaller (0.01–1 μm) fragments (primary crystallites) on which the polymer continues to grow. The reaction mass therefore consists of a suspension (not an emulsion—there is no surfactant) of macroparticles, which are in turn made up of an agglomeration of micro-particles, each of which surrounds a primary crystallite (Figure 10.4). According to Ray and coworkers [17], the macroparticle usually remains intact so that one is generated from each catalyst particle fed to the reactor. The primary catalyst crystallites are distributed more or less uniformly throughout the ultimate macroparticle.

If you are not yet convinced that these are complex systems, consider the following. The catalysts consist of two or more components, the nature and relative amounts of which influence the rate of polymerization. Each catalyst particle contains a multiplicity of surface sites on which polymer can grow and they may differ in activity (e.g., due to different crystal faces). Sites may be deactivated so that their number decreases with time. The monomer must be chemisorbed on the catalyst surface before it can be added to a chain. And before it even gets to the catalyst surface it must diffuse from the bulk fluid to the surface of the macroparticle through the interstices (pores) between the microparticles and finally through the layer of solid polymer that coats the catalyst fragment within each microparticle. Furthermore, the heat of polymerization is liberated at the surface of the catalyst fragments that must flow outward. This could give rise to significant temperature gradients, complicating the interpretation of experimental data and limiting the validity of isothermal rate expressions. It should not be surprising, therefore, that kinetic equations are significantly more complex to describe this reaction than those for free-radical and anionic addition polymerizations. However, the relevance of this catalyst system is of great industrial importance as tens of billions of pounds of polymer are produced with these catalyst systems each year. The following discussion illustrates some of the approaches that have been taken to describe the kinetics in systems with two-component Ziegler–Natta catalysts such as the $TiCl_4$–$Al(C_2H_5)_3$ system mentioned above.

The mechanism developed to describe the reaction is based on four assumptions [18]:

1. The propagation reaction occurs on active sites on the surface of the metal halide crystals.
2. Active sites are formed by the reaction of chemisorbed metal alkyls with metal halide on the crystal surface.

3. The propagation reaction occurs between a growing chain on the active site and an adjacent chemisorbed monomer molecule.

4. The metal alkyl competes with monomer for adsorption on the crystal surface.

Let C represent the number of catalytic sites on the catalyst surface, presumably associated with the Ti. Before a chain can grow on a site, however, the site must be *activated* by chemisorption of the $Al(C_2H_5)_3$ on an adjacent site. Therefore,

$$C_p^* = C\theta_A \tag{10.15}$$

where C_p^* is the number of growing chains stuck to the catalyst surface and θ_A is the fraction of the surface covered by chemisorbed alkyl.

The rate of polymerization is given by

$$r_p = k_s C_p^* \theta_M = k_s C \theta_M \theta_A \tag{10.16}$$

where k_s is the insertion rate constant, which characterizes the rate of insertion of chemisorbed monomer into the Ti-chain bond, and θ_M is the fraction of the surface covered by chemisorbed monomer.

Monomer and metal alkyl compete for chemisorption on the *uncovered* fraction of the active surface, $\theta_M = (1 - \theta_M - \theta_A)$

$$M + S \underset{k_{M-1}}{\overset{k_{M+1}}{\rightleftarrows}} M \cdot S \quad \text{(chemisorbed monomer)}$$

$$A + S \underset{k_{A-1}}{\overset{k_{A+1}}{\rightleftarrows}} A \cdot S \quad \text{(chemisorbed alkyl)}$$

where S represents a surface site:

$$\text{Rate of monomer adsorption} = k_{M+1}(1 - \theta_M - \theta_A)[M_s] \tag{10.17}$$

$$\text{Rate of monomer desorption} = k_{M-1}\theta_M \tag{10.18}$$

$$\text{Rate of alkyl adsorption} = k_{A+1}(1 - \theta_M - \theta_A)[A_s] \tag{10.19}$$

$$\text{Rate of alkyl desorption} = k_{A-1}\theta_A \tag{10.20}$$

$[M_s]$ and $[A_s]$ are the concentrations of monomer and alkyl *at the surface*. If we assume steady-state surface concentrations of both monomer and alkyl and equate the rates of adsorption and desorption of each we get

$$\theta_M = \frac{K_M[M_s]}{1 + K_M[M_s] + K_A[A_s]} \tag{10.21}$$

and

$$\theta_A = \frac{K_A[A_s]}{1 + K_M[M_s] + K_A[A_s]} \tag{10.22}$$

where $K_M = k_{M+1}/k_{M-1}$ and $K_A = k_{A+1}/k_{A-1}$ are the equilibrium constants for monomer and alkyl adsorption onto the catalyst. Equations 10.21 and 10.22 are Langmuir–Hinshelwood absorption isotherms, which when inserted into Equation 10.16 gives [19][2]:

$$r_p = k_s C \frac{K_A[A_s]K_M[M_s]}{(1 + K_A[A_s] + K_M[M_s])^2} \tag{10.23}$$

Equation 10.23 appears to give a reasonable description of experimental results. For example, the activity of a catalyst can be greatly enhanced kby ball milling and/or dispersing it on the surface of a support such as $MgCl_2$, both of which increase the exposed surface area, and therefore C. A maximum in rate is generally observed as the alkyl concentration is increased. Actually, almost all experiments show the rate to be the first order in monomer concentration, and Equation 10.23 is not linear in monomer. This simply means that $K_M[M_s] \ll 1$, which is generally agreed to be the case.

With a little creative algebra, Equation 10.23 can be put into a simpler and more convenient form:

$$r_p = k_s K_M C_p^* \theta_0 [M_s] = k_p C^* [M_s] \tag{10.24}$$

where $k_p = k_s K_M$ is a propagation rate constant and $C^* = C_p^* \theta_o$ is interpreted as the number of *active* growing chains, that is, those that possess an open adjacent site for the chemisorption of monomer or alkyl. Again, because

$$\theta_0 = (1 - \theta_M - \theta_A) = \frac{1}{1 + K_A[A_s] + K_M[M_s]} \tag{10.25}$$

is a function of monomer concentration, Equation 10.24 is not the first order in monomer unless $K_M[M_s] \ll 1$, which generally seems to be the case. Equation 10.24 is the form usually used by chemical engineers to describe Ziegler–Natta polymerizations.

Example 10.5 Estimate (a) the average lifetime of a growing chain and (b) the rate of polymerization for the polymerization of propylene in a heptane slurry (suspension) with a modern, high-activity Ziegler–Natta catalyst under conditions such that $k_p = 660$ L/mol-s, $C^* = 10^{-5}$ mol/g catalyst, $[M_s] = 4.0$ mol/L [19], and $\bar{x}_n = 5000$. (c) Also, estimate how long the reaction must be carried out under these conditions to achieve a catalyst yield of 10,000 g polymer/g catalyst (which is desirable to avoid a costly catalyst removal step). $[M_s]$ is kept constant by bubbling gas-phase monomer into the reactor as polymerization proceeds.

Solution.

(a) By analogy to Example 9.9, $\bar{t}$ is given by the average number of monomer units in a chain ($\bar{x}_n$) times the number of growing chains (C^*) divided by the rate at which

---

[2] It should be noted that when dealing with heterogeneous reactions such as these, rates are usually given per unit moles or mass of catalyst rather than per unit of reactor volume, as is the case for homogeneous reactions. In Equation 10.23, for example, the rate would typically be in mol monomer/mol Ti (or $TiCl_4$) s.

monomer is added to all chains (r_p):

$$\bar{t} = \frac{\bar{x}_n C^*}{r_p} = \frac{\bar{x}_n}{k_p[M_s]}$$

$$= \frac{5000 \text{ mol mon/mol chains}}{(660 \text{ L/mol chains} \cdot \text{s})(4.0 \text{ mol mon/L})} = 1.9 \text{s}$$

(b)
$$r_p = -\left(\frac{1}{m_{\text{cat}}}\right)\left(\frac{dM}{dt}\right) = k_p[M_s]C^*$$

$$= \left(660 \frac{\text{L}}{\text{mol chains} \cdot \text{s}}\right)\left(4.0 \frac{\text{mom mon}}{\text{L}}\right)\left(\frac{10^{-5} \text{ mol chains}}{\text{g cat}}\right)$$

$$= 0.0264 \text{ mol mon/g cat-s}$$

or, because the molecular weight of propylene is 42

$$r_p = 0.0264 \times 42 \text{ g/gmol} \times 3600 \text{ s/h} = 4000 \text{ g polymer/g cat-h}$$

(c) So, to get the specified catalyst yield, the reaction must be carried out for 10,000/4000 = 2.5 h. Note that the rate of polymerization is based on the mass of catalyst rather than the volume. The reaction time here is much greater than the average chain lifetime (like free-radical addition and unlike anionic addition), but the ratio is not as great as in a typical free-radical process.

Unfortunately, things are not always as simple as the last example in real life. For one thing, catalyst activity usually drops with time (due to fouling) even in the absence of polymerization. Catalyst half-lives typically are in the order of minutes to hours. This is explained by a progressive deactivation of the sites responsible for polymerization, C^*, and is described quantitatively with an nth-order deactivation law:

$$\frac{dC^*}{dt} = -k^*(C^*)^n \tag{10.26}$$

Currently, second-order deactivation $n=2$ seems to be the most popular, but $n=1$ and $n=3$ have also been used to fit data.

Example 10.6 Rework Example 10.5(c), but now assume that the catalyst deactivates by a second-order mechanism with a half-life of 1 h.

Solution. Integration of Equation 10.26 with $n=2$ and $C^* = C_o^*$ at $t=0$ gives

$$C^* = \frac{C_0^*}{1 + C_0^* k^* t}$$

because $C^*/C_o^* = 0.5$ at $t = t_{1/2}$, $t_{1/2} = 1/C_o^* k^*$. By inserting the above into Equation 10.24, we get

$$r_p = -\frac{1}{m_{cat}}\frac{dM}{dt} = \frac{k_p C_0^* [M_s]}{1 + C_0^* k^* t}$$

Separating variables (remember that $[M_s]$ is constant here) gives

$$-\frac{1}{m_{cat}}\int_{M_O}^{M} dM = k_p C_0^* [M_s] \int_0^t \frac{dt}{1 + C_0^* k^* t}$$

and integrating gives

$$\frac{M_0 - M}{m_{cat}} = \frac{k_p C_0^* [M_s]}{C_0^* k^*} \ln(1 + C_0^* k^* t) = k_p C_0^* [M_s] t_{1/2} \ln\left(1 + \frac{t}{t_{1/2}}\right)$$

Now $(M_o - M)/m_{cat} = 10{,}000$ g/g cat/42 g/mol $= 238$ mol/g cat. Using this value, $t_{1/2} = 1$ h, the parameters from Example 10.5 and solving for t gives $t = 11.3$ h. Compared with the result (2.5 h) of Example 10.5(c), this illustrates the practical consequences of a deactivating catalyst.

Another complication is the possibility of mass-transfer limitations on the reaction rate. And, because the polymer layer grows, mass transfer resistance will increase during a heterogeneous polymerization; it is quite possible that the polymerization will start out reaction-limited and change to mass transfer-limited as the conversion increases. To polymerize, monomer must first diffuse from the bulk fluid surrounding the macroparticles, where its concentration is $[M_b]$, to the catalytic surface within the microparticles, where its concentration is $[M_s]$. Where the inherent rate of reaction on the catalytic surface and/or the resistance to diffusion are high, the overall reaction rate can be determined by the physics of diffusion rather than the chemistry of the reaction.

Another way of putting this is that $[M_s] < [M_b]$. To use Equation 10.23 or 10.24, $[M_s]$ must be known. Relating $[M_s]$ to $[M_b]$ is a complex problem in mass transfer and is beyond the scope of this chapter. It is treated and explained in the literature [20,21]. Suffice it to say that to make calculations, you need to know the geometry of the particles, the mass-transfer coefficient at the surface of the macroparticles, the effective diffusivity for the monomer in the interstices (pores) between the microparticles, and the diffusivity of the monomer through the layer of polymer coating the primary catalyst particles (in addition to knowing the kinetic parameters of the reaction).

Calculations show that rates can be mass-transfer limited, depending on the activity of the catalyst and the physical properties of the system. One rather unusual result is that the resistance to mass transfer can decrease with time, causing an increase in rate (at constant $[M_b]$). This increase, when combined with a deactivating catalyst, gives rise to a maximum in the polymerization rate with time. The parameters in Example 10.5 represent a case that is not mass-transfer limited, therefore we can assume that $[M_s] \approx [M_b]$.

If all the polymer-producing sites on a catalyst surface had the same activity and were exposed to the same monomer concentration, and if chain transfer occurred randomly, we would expect the polymer to have a most-probable distribution of chain lengths (Eq. 9.47) with a polydispersity index of 2.0 [22]. Polymers produced with these heterogeneous catalysts typically have large polydispersity indices (from 3 to 20), however. Two reasons have been advanced for these broad distributions: (1) mass-transfer resistance, which causes with monomer concentration to vary with location and perhaps time in the particles, and (2) a range of site activities (k_p's) on the catalyst surface. Different types of sites have been identified on the surfaces [22].

Recent calculations show that while mass-transfer limitations can contribute somewhat to a broadening of the distribution, a range of site activities is needed to account for the observed polydispersities. Furthermore, the different sites must have a pretty wide range of activities [20]. To examine the effects of site heterogeneity, Equation 10.24 must be generalized to

$$r_p = [M_s] \sum_{i=1}^{N} k_{pi} C_i^*$$

(10.27)

where N is the number of different types of sites and

$$k_p C^* = \sum k_{pi} C_i^*$$

(10.28)

You can get large polydispersity indices with only two types of catalyst sites, provided that they have a large enough difference (at least an order of magnitude) in activities. However, this results in a bimodal (two-peaked) molecular weight distribution, which is not generally seen in practice. In some applications, narrow molecular weight distributions are desirable. Relatively small polydispersity indices in the range of three to four have been reported for systems where the catalyst is supported on $MgCl_2$ (which could be due to selective promotion of certain types of sites) or when additives are used (which may selectively poison certain types of sites) [21]. Single-site (e.g., EXXPOL metallocene) catalysts have also been developed to avoid the problem of high polydispersity [22,23].

The question of temperature gradients in the particles has also been addressed. The conclusion is that they are generally not significant, except in the very early stages of gas-phase polymerizations, where temperatures may get high enough to melt the polymer [24].

In practice, Ziegler–Natta catalyst systems are difficult to work with. Great care must be exercised in their preparation and use, since they are easily poisoned by water, among other things. They are pyrophoric (spontaneously burst into flame on contact with oxygen) and are used in close proximity to large amounts of flammable monomers and solvents. Therefore, they can present a significant safety hazard both in the laboratory and in the plant.

10.8 GRAFTED POLYMER SURFACES

Grafting a polymer onto a surface is a great way to change the surface properties of a material. This can be done to improve the compatibility between layers in a composite material, to reduce surface fouling, or to improve biocompatibility (among many potential applications). Because living polymerizations can be used for good control of polymer properties, they are often used in graft polymerizations.

The grafting method normally follows one of two patterns (Figure 10.5): grafting to or grafting from. In the first method, oligomers or polymers are pre made (and can easily be characterized for molecular weight and polydispersity) and attached to a surface through any number of chemistries (several condensation reactions can be used here). In grafting

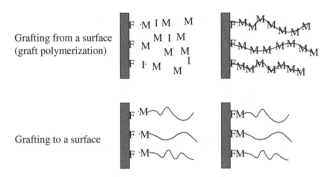

FIGURE 10.5 Depiction of methods to form a surface grafted with polymers. (a) Grafting from a surface by growing a polymer chain from surface functional groups (F). (b) Grafting to a surface by attaching a premade polymer to functional groups (F) on the surface.

from method, the surface must be activated (i.e., form a free-radical or growing living polymer complex) through which monomer can add. In this second technique, the grafts are more difficult to analyze, since the molecular weight and polydispersity cannot be determined without destroying the surface (or at least removing the grafts). Although this drawback may seem to favor grafting to techniques, a balancing disadvantage of the grafting to technique is that the large oligomers or polymers may have significant steric hindrances or diffusive resistance in reaching the surface for binding. This results in a surface that has a lower graft coverage (graft density). Grafting techniques are discussed further by Bhattarcharya and Misra [25].

10.9 SUMMARY

The various advanced polymerization methods described in this chapter highlight some of the more important mechanisms that can be used for polymerizations. Living polymerizations offer the ability for more precise control of polymer structures, including block, graft, and star polymers, whereas heterogeneous catalysis (such as the Ziegler–Natta polymerizations) offers the benefit of improved stereoregularity. It is beyond the scope of this textbook to go beyond an introduction of these techniques, but these polymerizations (and others such as reversible addition fragmentation chain transfer (RAFT) polymerizations and nitroxide-mediated polymerization (NMP)) are treated in greater depth elsewhere [25–27].

PROBLEMS

1 Given that a reactor contains 4.000 mol styrene in solution and supplies of n-BuLi and a tetrafunctional linking agent $R(Cl)_4$, list the amounts and order of addition of reagents necessary to produce a four-star polymer with branches of length $x = 1000$. Assume complete and instantaneous ionization, perfect linking, and no terminating impurities.

2 A reactor initially contains 1.00×10^{-3} mol of n-BuLi in a dry, inert solvent. 1.00 mol of styrene (S) is added to the reactor. At 50% conversion, 0.250×10^{-3} mol of a

difunctional linking agent, Cl−R−Cl, is added to the reactor. The reaction is then allowed to go to completion. The assumptions in Problem 10.1 apply.

(a) Describe quantitatively the reactor product (how many moles of what).

(b) Calculate $\bar{x}_n$ for the product.

3 Repeat Problem 10.2, but replace the linking agent with an additional 1.00×10^{-3} mol of n-BuLi.

4 Repeat Problem 10.2, but replace the linking agent with 1.00×10^{-3} mol of a dianionic initiator, $^+[Li]^-[:R:]^-[Li]^+$.

5 Consider anionic addition polymerization in which most of the chain ends are associated in inactive aggregates of n chain ends. The aggregates dissociate slightly to active chain ends according to the equilibrium relation

$$(P_x^-)_n \overset{K}{\rightleftharpoons} nP_x^-$$

where $K \ll 1$. Obtain an expression for the rate of polymerization.

6 Describe quantitatively how the polymer of Example 10.1 could be made with the dianionic initiator of Problem 10.4.

7 Describe how you would make a block copolymer of butadiene and nylon 6/6 (see Example 2.4E).

8 Obtain an expression for the number-average chain length at complete conversion in a group-transfer polymerization batch that contains I, M, and RH moles of initiator, monomer, and chain-transfer agent, respectively.

9 The following mechanism has been proposed to explain cationic polymerization (do not bet the farm on it):

$$\text{Ionization} \qquad GH \overset{K}{\rightleftharpoons} G^-H^+$$

$$\text{Initiation} \qquad G^-H^+ + M \overset{k_a}{\longrightarrow} P_1^+G^-$$

$$\text{Propagation} \qquad P_x^+G^- + M \overset{k_p}{\longrightarrow} P_{x+1}^+G^-$$

$$\text{Termination} \qquad P_x^+G^- \overset{k_t}{\longrightarrow} P_x + GH$$

Here, GH is the catalyst and G^- is the gegen ion. The ionization step is assumed to be essentially instantaneous, and always at equilibrium, with K being the equilibrium constant. Obtain expressions for the rate of polymerization and the number-average chain length according to this mechanism.

10 Consider an ideal anionic polymerization, the type that when carried out in a batch reactor would produce monodisperse polymer. Here, however, the reaction is to be carried out in a perfectly micromixed CSTR in which the steady-state monomer concentration is [M]. The average residence time $\tau = V/Q$, where V is the reactor

volume and Q is the volumetric flow rate. The residence–time distribution for such a reactor is

$$E(t) = \frac{1}{\tau}\exp\left(-\frac{t}{\tau}\right)$$

where $E(t)dt$ is the fraction of the material in the product stream that has resided in the reactor for time t. Assume that chains grow only while being in the reactor.

In terms of parameters defined above, obtain expressions for $\bar{x}_n$, (n/N), (w/W) (the number- and weight-fraction distributions of chain lengths), and the polydispersity index $\bar{x}_w/\bar{x}_n$. Does any of these look familiar?

11 Write the expressions giving conversion as a function of time for the reaction in Example 10.2(b).

12 Consider a Ziegler–Natta polymerization in which chains are terminated only by chain transfer with hydrogen. The rate of transfer is given by $r_{tr} = k_{tr}C^*[H_2]$.

    (a)   Obtain an expression for $\bar{x}_n$ in terms of concentrations and rate constants. Assume sites of uniform activity.

    (b)   Obtain the expression for the number-fraction distribution of chain lengths, (n/N) (x) for case (a). What is the polydispersity index for this distribution?

13 A gas-phase Ziegler–Natta polymerization is carried out in a constant-volume, isothermal batch reactor. Assume that surface monomer concentration is always proportional to bulk monomer concentration, $[M_s] = k[M_b]$. Obtain expressions for conversion as a function of time for two cases:

    (a)   With a constant-activity catalyst.

    (b)   With a catalyst that deactivates according to a second-order mechanism.

Define any parameters you need that are not defined in the chapter.

14 Define in terms of parameters the optimum alkyl concentration for maximum polymer production in Equation 10.23.

REFERENCES

[1] Plesch, P.H. (ed.), *The Chemistry of Cationic Polymerization*, MacMillan, New York, 1963.

[2] Russell, K.E. and G.J. Wilson. Cationic polymerizations, in *Polymerization Processes*, C. E. Schildknecht and I. Skeist (eds), Wiley, New York, 1977, Chapter 10.

[3] Percec, V., Recent developments in cationic polymerization, in *Applied Polymer Science*, 2nd ed., R.W Tess and G.W. Poehlein (eds), American Chemical Society, Washington, DC, 1985, Chapter 5.

[4] Odian, G., *Principles of Polymerization*, 4th ed., John Wiley and Sons, New York, 2004, Chapter 5-2.

[5] Morton, M., *Anionic Polymerizations: Principles and Practice*, Academic, New York, 1983.

[6] Swarc, M., *Living Polymerizations and Mechanisms of Anionic Polymerization*, Springer, Berlin, Germany, 1983.

[7] McGrath, J.E. (ed.), *Anionic Polymerization: Kinetics, Mechanisms and Synthesis*, American Chemical Society, Washington, DC, 1981.

[8] Franta, E. and P. Rempp, *Polym. Prepr.* **20**(1), 5 (1979).

[9] Odian, G., *op. cit.* Chapter 5-3.

[10] United States patents 4,414,372 and 4,417,034.

[11] Stinson, S.C., *Chem. Eng. News*, April 27, 1983, p. 43.

[12] Keii, T., *Kinetics of Ziegler-Natta Polymerization*, Chapman Hall, London, U.K., 1972.

[13] Chien, J.C.W. (ed.), *Coordination Polymerization*, Academic, New York, 1975.

[14] Karol, F.J., Coordinated anionic polymerization and polymerization mechanisms, in *Applied Polymer Science*, 2nd ed., R.W. Tess and G.W. Poehlein (eds), American Chemical Society, Washington, DC, 1985, Chapter 4.

[15] Matyjaszewski, K. and J. Xia, *Chem. Rev.* **101**(9), 2921 (2001).

[16] Coessens, V., T. Pintauer, and K. Matyjaszewski, *Prog. Polym. Sci.* **26**(3), 337 (2001).

[17] Floyd, S., T. Heiskanen, and W.H. Ray, *Chem. Eng. Progr.* **85**(11), 56 (1988).

[18] Tait, P.J.T., *Chem. Technol.* **5**, 688 (1975).

[19] Keii, T., et al., *Makromol. Chem.* **183**, 2285 (1982).

[20] Floyd, S., et al., *J. Appl. Polym. Sci.* **33**, 1021 (1987).

[21] Floyd, S., et al., *J. Appl. Polym. Sci.* **32**, 2935 (1986).

[22] Anon., *Mod. Plast.* **68**(7), **61** (1991); Leaversuch, R.D., *Mod. Plast.* **68**(10), 46 (1991); Anon., *Chem. Eng. News*, Dec. 23, 1991, p. 16.

[23] Anon., *Chem. Eng. Prog.* **87**(10), 21 (1991).

[24] Zucchini, U. and G. Cecchin, *Adv. Polym. Sci.* **51**, 101 (1983).

[25] Bhattarcharya, A. and B.N. Misra, *Prog. Polym. Sci.* **29**(8), 767 (2004).

[26] Matyjaszewski, K. *Controlled/Living Radical Polymerizations: From Synthesis to Materials*, ACS Symposium Series **944**, American Chemical Society, Washington, DC, 2006.

[27] Buchmeiser, M. *Metathesis Polymerization*, Advances in Polymer Science Series, Springer-Verlag, New York, 2009.

CHAPTER 11

COPOLYMERIZATION

11.1 INTRODUCTION

Copolymers can be easily made by either step-growth or free-radical addition polymerization. By linking together different monomers into the same polymer chain, several properties (hydrophilicity, rigidity, opacity, etc.) can be tuned. Although a mixture of many different monomers can be polymerized, the complexity and randomness of copolymers that have more than two or three different repeat units make them difficult to characterize and, generally, not useful in practice. However, in nature, proteins are synthesized by combination of 20 different amino acids, while synthetic copolymers tend to contain only two or three different repeat units. One important distinction is that biology has evolved an efficient way to code for the regular synthesis of proteins (polypeptides) by arranging the amino acid repeat units carefully. This specified order of amino acids allows proteins to fold into the three-dimensional shapes necessary for biological functions. Although some synthetic polypeptides can be made, this chapter focuses on more standard commercially important copolymers, their formation, and differences in characterization from homopolymers. One other caveat is that this chapter will focus on random copolymers, where monomers are mixed together prior to polymerization. More sophisticated techniques are required to make block and graft copolymers, since the monomers must be reacted in a specific order to form these more complex structures.

11.2 MECHANISM

We have seen in the previous chapters how the composition of a random copolymer can influence many of its important properties, including solubility, degree of crystallinity, T_g, and T_m. The control of copolymer composition is therefore of great practical importance.

Fundamental Principles of Polymeric Materials, Third Edition. Christopher S. Brazel and Stephen L. Rosen.
© 2012 John Wiley & Sons, Inc. Published 2012 by John Wiley & Sons, Inc.

From a practical standpoint, most copolymers are made by free-radical addition polymerization, although step-growth polymerization can also be used. For the sake of simplicity in our discussion, we will focus only on the free-radical mechanism. A quantitative treatment of random copolymerization is based on the assumption that the reactivity of a growing chain depends only on its active terminal unit. Therefore, when two monomers, M_1 and M_2, are copolymerized, there are four possible propagation reactions[1]:

$$\begin{array}{cc} \text{Reaction} & \text{Rate Equation} \\ P_1 \cdot + M_1 \xrightarrow{k_{11}} P_1 \cdot & k_{11}[P_1 \cdot][M_1] \end{array} \tag{11.1}$$

$$P_1 \cdot + M_2 \xrightarrow{k_{12}} P_2 \cdot \qquad k_{12}[P_1 \cdot][M_2] \tag{11.2}$$

$$P_2 \cdot + M_2 \xrightarrow{k_{22}} P_2 \cdot \qquad k_{22}[P_2 \cdot][M_2] \tag{11.3}$$

$$P_2 \cdot + M_1 \xrightarrow{k_{21}} P_1 \cdot \qquad k_{21}[P_2 \cdot][M_1] \tag{11.4}$$

The first subscript on the rate constants designates the nature of the chain end and the second identifies the monomer being added to the chain.

Application of the steady-state assumption to all polymer free-radical species ($P_1 \cdot$ and $P_2 \cdot$) requires that they be generated and consumed at equal rates. $P_1 \cdot$'s are generated in Reaction 11.4 and consumed in Reaction 11.2. Note that Reaction 11.1 just converts one $P_1 \cdot$ into another one, with not net change in their number. Therefore,

$$k_{12}[P_1 \cdot][M_2] = k_{21}[P_2 \cdot][M_1] \tag{11.5}$$

The rates of consumption of monomers M_1 and M_2 are

$$-\left(\frac{1}{V}\right)\left(\frac{dM_1}{dt}\right) = k_{11}[P_1 \cdot][M_1] + k_{21}[P_2 \cdot][M_1] \tag{11.6}$$

$$-\left(\frac{1}{V}\right)\left(\frac{dM_2}{dt}\right) = k_{12}[P_1 \cdot][M_2] + k_{22}[P_2 \cdot][M_2] \tag{11.7}$$

Dividing Equation 11.6 by Equation 11.7, eliminating the $[P \cdot]$'s with Equation 11.5, and recalling that $M_i = [M_i]V$ gives

$$\frac{dM_1}{dM_2} = \frac{M_1}{M_2}\left[\frac{r_1 M_1 + M_2}{M_1 + r_2 M_2}\right] \tag{11.8}$$

where r_1 and r_2 are the *reactivity ratios*.

---

[1] The notation has been changed a bit here to conform to the existing literature. The subscripts 1 and 2 designate the two monomers being copolymerized, not the number of repeating units. Thus, $P_1 \cdot$ represents a growing chain (of any length) with a terminal unit of monomer 1, that is, a chain to which a monomer M_1 was the last added. Although the mathematics of copolymerization looks the same for the various types of addition polymerization, there are some important mechanistic differences which will be discussed later.

$$r_2 = \frac{k_{22}}{k_{21}} = \text{relative preference of } P_2 \cdot \text{ for } M_2/M_1$$

$$r_1 = \frac{k_{11}}{k_{12}} = \text{relative preference of } P_1 \cdot \text{ for } M_1/M_2$$

The reactivity ratio is a measure of the likelihood of a polymer to react with a monomer of same type (Equations 11.1 or 11.3) as opposed to adding the opposite comonomer (Equations 11.2 or 11.4. Reactivity ratios are experimentally determined [1] or may be estimated [2]. In organic free-radical copolymerizations, for a given monomer pair, they are pretty much independent of initiator and solvent and are only weakly temperature dependent. In ionic copolymerizations, however, they depend strongly on the gegen ion and solvent.

This relation may be put in a more convenient form by defining the following:

f_1 = mole fraction of monomer 1 in the reaction mass *at any instant*

F_1 = mole fraction of monomer 1 in the copolymer formed *at that instant*

$$F_1 = (1 - F_2) = \frac{dM_1}{d(M_1 + M_2)} \tag{11.9}$$

$$f_1 = (1 - f_2) = \frac{M_1}{(M_1 + M_2)} \tag{11.10}$$

If monomer 1 is much more reactive than monomer 2, f_1 will decrease at the beginning of the polymerization, while F_1 will be higher than the feed composition of monomer 1. Combination of Equations 11.8–11.10 (if you enjoy algebra, you really ought to try this one) gives

$$F_1 = \frac{r_1 f_1^2 + f_1 f_2}{r_1 f_1^2 + 2 f_1 f_2 + r_2 f_2^2} = \frac{(r_1 - 1) f_1^2 + f_1}{(r_1 + r_2 - 2) f_1^2 + 2(1 - r_2) f_1 + r_2} \tag{11.11}$$

$$F_1 = F_1(r_1, r_2, f_1)$$

The quantity F_1, the instantaneous copolymer composition, is analogous to $\bar{x}_n$, the instantaneous number-average chain length in free-radical addition polymerization. Like $\bar{x}_n$, it depends on the conditions in the reactor at a particular instant. It, too, is really an average, since not all the copolymer formed at a particular instant has exactly the same composition. However, the instantaneous distribution of compositions is normally much narrower than the instantaneous distribution of chain lengths, and because the fact that it is an average is not normally of great practical importance and cannot be controlled anyhow, the overbar is left off of F_1.

11.3 SIGNIFICANCE OF REACTIVITY RATIOS

To gain an appreciation for the physical significance of Equation 11.11, let us look at some special cases of the reactivity ratios.

Case 1: $r_1 = r_2 = 0$. With both reactivity ratios zero, neither type of chain end can add its own monomer, so a *perfectly alternating* copolymer results. $F_1 = 0.5$ regardless of f_1, until one of the monomers is used up, the point at which polymerization stops.

Case 2: $r_1 = r_2 = \infty$. Here, P_1·'s can add only M_1 monomer, and P_2·'s only M_2, so the polymer formed will be a physical mixture of homopolymer 1 and homopolymer 2 chains.

Case 3: $r_1 = r_2 = 1$. Under these conditions, the growing chains find the monomers equally attractive, so the addition depends only on the ratio of the monomers in the vicinity of the chain ends, $F_1 = f_1$.

Case 4: $r_1 r_2 = 1$. This is the so-called ideal copolymerization, where each chain displays the same preference for one of the monomers over the other: $k_{11}/k_{12} = k_{21}/k_{22}$, so it does not matter what is at the end of the chain. In this case, Equation 11.11 reduces to

$$F_1 = \frac{r_1 f_1}{(r_1 - 1)f_1 + 1} \tag{11.12}$$

The reader familiar with distillation theory will note here the exact analogy between Equation 11.12 and the vapor–liquid equilibrium composition relation for ideal solutions with a constant relative volatility.

Case 5: $r_1 < 1$, $r_2 < 1$. This common situation corresponds to an azeotrope in vapor–liquid equilibrium. At the azeotrope

$$F_{1az} = f_{1az} = \frac{(1 - r_2)}{(2 - r_1 - r_2)} \tag{11.13}$$

Systems for which $r_1 > 1$, $r_2 > 1$ also form azeotropes, but have been rarely reported.

11.4 VARIATION OF COMPOSITION WITH CONVERSION

In general, $F_1 \neq f_1$, that is, the composition of the copolymer formed at any instant will differ from that of the monomer mixture from which it is being formed. Thus, as the reaction proceeds, the unreacted monomer mixture will be depleted in the more reactive monomer, and as the composition of the unreacted monomer changes, so will that of the polymer being formed, in accordance with Equation 11.11.

Example 11.1 Draw curves of instantaneous copolymer composition, F_1, versus monomer composition, f_1, for the following systems, and indicate the direction of composition drift as the reaction proceeds in a batch reactor.

(a) Butadiene (1), styrene (2), 60 °C; $r_1 = 1.39$, $r_2 = 0.78$.

(b) Vinyl acetate (1), styrene (2), 60 °C; $r_1 = 0.01$, $r_2 = 55$.

(c) Maleic anhydride (1), isopropenyl acetate (2), 60 °C; $r_1 = 0.002$, $r_2 = 0.032$.

Solution. Application of Equation 11.11 gives the plots in Figure 11.1. Note the system (a) approximates ideal copolymerization, Case 4 above. In system (b), styrene is the preferred monomer, regardless of the terminal radical; hence, the copolymer is largely styrene until styrene monomer is nearly used up. System (c) approximates Case 1 above.

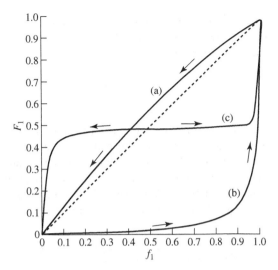

FIGURE 11.1 Instantaneous copolymer composition (F_1) versus monomer composition (f_1). (a) Butadiene (1), styrene (2) at 60 °C; $r_1 = 1.39$, $r_2 = 0.78$; (b) Vinyl acetate (1), styrene (2), 60 °C; $r_1 = 0.01$, $r_2 = 55$; (c) Maleic anhydride (1), isopropenyl acetate (2), 60 °C; $r_1 = 0.002$, $r_2 = 0.032$. The direction of composition drift in a catch reactor is indicated by arrows (Example 11.1). The dashed diagonal line represents Case 3, $F_1 = f_1$, $r_1 = r_2 = 1$.

The direction of composition drift with conversion is indicated by arrows. Note that system (c) forms an azeotrope at $F_1 = f_1 = 0.493$. To the left of the azeotrope, M_1 is the more reactive monomer $(F_1 > f_1)$, but to the right, M_2 is more reactive $(F_1 < f_1)$. Therefore, with an initial monomer composition $f_{1,o} < 0.493$, F_1 and f_1 will decrease with conversion, but if $f_{1,o} > 0.493$, they will increase. If the initial monomer charge is exactly the azeotropic composition, there will be no composition drift.

Consider a batch consisting of a total of M moles of monomer $(M = M_1 + M_2)$. At time t, the monomer has a composition f_1. In the time interval dt, dM moles of monomer polymerize to form copolymer with a composition F_1. Therefore, at time $t + dt$, there are $(M - dM)$ moles of monomer left whose composition has been changed to $(f_1 - df_1)$. Writing a material balance on monomer 1 gives that the initial amount of monomer either remains or is reacted (dM_1):

$$(M_1)|_t = (M_1)|_{t+dt} + dM_1 \tag{11.14a}$$

or

$$f_1 M = (M - dM)(f_1 - df_1) + F_1 dM \tag{11.14b}$$

Expanding and neglecting second-order differentials gives

$$\frac{dM}{M} = \frac{df_1}{F_1 - f_1} \tag{11.15}$$

At the start of the reaction, there are M_o moles of monomer present with a composition $f_{1,o}$, and at some later time, there are M moles of monomer left with a composition f_1. Integrating between these limits gives

$$\ln \frac{M}{M_0} = \int_{f_{10}}^{f_1} \frac{df_1}{F_1 - f_1} \tag{11.16}$$

This equation is the exact analog of the Rayleigh equation relating the amount and composition of the still-pot liquid in a batch distillation, which may be familiar to some readers. By choosing values for f_1 and calculating the corresponding F_1's using Equation 11.11, the integral may be evaluated to obtain a relation between the monomer composition and conversion ($X = 1 - (M/M_o)$). An analytic solution to Equations 11.11 and 11.16 has been obtained (3). For $r_1 \neq 1$, $r_2 \neq 1$ (if either is equal to 1, see the original reference):

$$\frac{M}{M_0} = \left[\frac{f_1}{f_{10}}\right]^{\alpha} \left[\frac{f_2}{f_{20}}\right]^{\beta} \left[\frac{f_{10} - \delta}{f_1 - \delta}\right]^{\gamma} = 1 - X \tag{11.17}$$

where

$$\alpha = \frac{r_2}{1 - r_2}$$

$$\beta = \frac{r_1}{1 - r_1}$$

$$\gamma = \frac{1 - r_1 r_2}{(1 - r_1)(1 - r_2)}$$

$$\delta = \frac{1 - r_2}{2 - r_1 - r_2}$$

Knowing the monomer composition f_1 as a function of conversion immediately gives the instantaneous copolymer composition F_1 as a function of conversion through Equation 11.11. This is important to know because if there is a large variation in the composition of the copolymer formed from the beginning of the reaction to high conversions, creating individual polymer chains may be blocky in some segments and more random in others. This can result in a wide variation in its properties.

Example 11.2 Discuss the possible influence of a wide variation in F_1 on the transparency of amorphous copolymers.

Solution. Amorphous, random copolymers of fairly uniform composition are normally transparent because on a macroscopic scale, they are homogeneous systems in which the discontinuities (groups of a few repeat units) are very much smaller than the wavelengths of visible light. Because of the general mutual insolubility of two different polymers, copolymers of widely differing F_1 values may actually form a heterogeneous system in which the discontinuities, globules of one phase in a matrix of another, will be larger than the wavelength of light. Because of the compositional difference of the phases, they almost certainly will differ in refractive index as well, leading to a system that scatters light and is therefore not transparent.

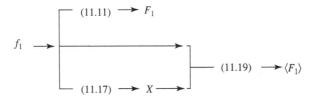

FIGURE 11.2 Schematic for calculation of copolymer compositions in a batch reaction.

In addition to the instantaneous copolymer composition F_1, another quantity of interest is $\langle F_1 \rangle$, the *cumulative composition* of the copolymer that has been formed up to a particular conversion. $\langle F_1 \rangle$ is exactly analogous to $\langle \bar{x}_n \rangle$, the cumulative number-average chain length from Chapter 9. For a batch reactor, it is obtained through a material balance:

moles M_1 charged = moles M_1 in copolymer + moles M_1 left in monomer

$$f_{10}M_0 = \langle F_1 \rangle (M_0 - M) + f_1 M \tag{11.18}$$

Rearranging gives

$$\langle F_1 \rangle = \frac{f_{10} - f_1(M/M_0)}{1 - (M/M_0)} = \frac{f_{10} - f_1(1 - X)}{X} \tag{11.19}$$

The distillation analog of this equation tells the well-educated bootlegger how much of his 20-proof sour mash he must distill over to have 120-proof white lightning in the jug under his condenser. Figure 11.2 illustrates the quantities defined in terms of the distillation analogy. The "well-stirred, adiabatic bathtub" analogy developed in Section 9.10 to visualize the relation between $\bar{x}_n$ and $\langle \bar{x}_n \rangle$ is equally applicable to the relation between F_1 and $\langle F_1 \rangle$. Here, F_1 is analogous to the temperature of the water entering the tub through the spigot and $\langle F_1 \rangle$ to the temperature of the water in the bathtub.

Calculation procedures for a batch reactor are summarized in Figure 11.2 for a system of known r_1, r_2, and $f_{1,o}$.

Example 11.3 For the styrene–butadiene system of Example 11.1, plot instantaneous copolymer composition F_1 and cumulative composition $\langle F_1 \rangle$ versus conversion for a batch reaction starting with a 50–50 (mol%) initial monomer charge.

Solution. As illustrated in the preceding diagram, calculations are facilitated by choosing f_1 as the independent variable. From Example 11.1, butadiene is the more reactive monomer, so as the reaction proceeds, the polymer will start out with a higher butadiene content and become more enriched in styrene (monomer 2), and f_1 will decrease from 0.5 to 0 as the butadiene is used up. This is best calculated and graphed using a spreadsheet. For a given value of f_1, the conversion X is calculated from Equation 11.17 with $\alpha = 3.55$, $\beta = -3.57$, $\gamma = 0.983$, and $\delta = -1.29$. F_1 is obtained through Equation 11.11 as in Example 11.1. $\langle F_1 \rangle$ is calculated from Equation 11.19 using the chosen f_1 and the values of X obtained from Equation 11.17.

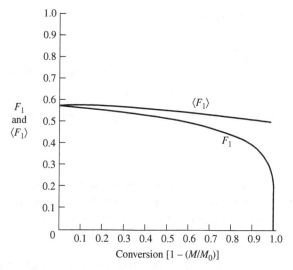

FIGURE 11.3   Instantaneous (F_1) and cumulative ($\langle F_1 \rangle$) copolymer composition versus conversion for butadiene(1)-styrene(2), $r_1 = 1.39$, $r_2 = 0.78$, $f_{1,o} = 0.50$ (Example 11.3).

The results are plotted in Figure 11.3. Note that in this particular case, $\langle F_1 \rangle$ does not vary much between 0% and 100% conversion, but F_1 changes considerably, particularly at high conversions (when only styrene is left in the reactor). Also note that $\langle F_1 \rangle$ at 100% conversion must always equal $f_{1,o}$. Because the entire initial monomer charge has been converted to polymer, their compositions must be the same.

The preceding mathematical developments apply equally to the copolymers formed by free-radical addition and anionic addition (though the numerical values of r_1 and r_2 would, in general, be different for the same monomer pair in each of the modes).

However, the nature of the molecules formed is quite different. Recall that in free-radical addition, the average chain lifetime is an infinitesimal fraction of the reaction time, while in anionic addition, the chains continue to grow throughout the reaction. Thus, in free-radical addition, F_1 is the composition of entire chains formed during a conversion increment dX. $\langle F_1 \rangle$ represents the average composition of the mixture of chains of different F_1 formed up to conversion X. So, free-radical addition copolymers will have a wide distribution of composition across all chains. In classic anionic addition, all the chains are essentially identical. As f_1 changes with conversion, we get a *composition gradient* in each chain as the reaction proceeds. F_1 is the composition of the portion of each chain formed over a conversion increment dX and $\langle F_1 \rangle$ is the average composition of each chain (as well as the entire copolymer) up to conversion X. Copolymerization with typical Ziegler–Natta catalysts should be more like that in free-radical systems.

Example 11.4   Suggest three techniques for producing copolymers of fairly uniform composition, that is, those in which F_1 does not vary much.

Solution. Three possibilities are sketched in Figure 11.4. With a semibatch reactor, the more reactive monomer is replenished as the reaction proceeds to maintain f_1 (and, therefore, F_1)

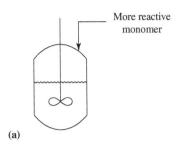

(a)

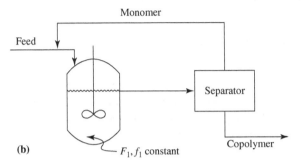

(b)

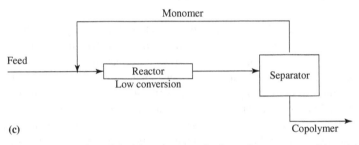

(c)

FIGURE 11.4   Techniques for minimizing the spread of copolymer composition: (a) semibatch reactor; (b) continuous stirred-tank reactor; (c) tubular reactor.

constant. A method for calculating the appropriate rate of addition has been described [4]. In a continuous stirred tank (CSTR, or back mix) reactor, both f_1 and F_1 are constant with time. In a continuous plug-flow reactor, the variation in F_1 can be kept small by limiting the conversion per pass in the reactor. Note that the last two techniques require facilities for separating unreacted monomer from the polymer, and in most cases, recycling it.

Example 11.5   Consider the semibatch reactor in Figure 11.4a.

Let $P(t)$ = the total moles of both monomers in the polymer formed (up to time t) and
   $A(t)$ = the moles of the more reactive monomer added up to time t.

In this setup, if the relation between $A(t)$ and $P(t)$ is arranged properly, not only will the copolymer be of uniform composition but also both monomers will be used up simultaneously. Obtain an expression that relates $A(t)$ to $P(t)$ and system constants.

Solution. First, make a total material balance on monomer: moles initially charged + moles added = moles unreacted + moles in copolymer:

$$M_0 + A(t) = M(t) + P(t)$$

Then, make a material balance on the more reactive monomer 1:

$$f_{10}M_0 + A(t) = M(t)f_1 + P(t)F_1$$

By eliminating $M(t)$ between the two, keeping in mind that if F_1 is to be constant, $\langle F_1 \rangle = F_1$ and $f_1 = f_{1,o}$ (constant), we find

$$A(t) = \frac{F_1 - f_{10}}{1 - f_{10}} P(t) = \text{constant} \times P(t)$$

Thus, the more reactive monomer is added in direct proportion to the amount of polymer formed (in terms of moles monomer polymerized). In practice, the real trick is keeping track of $P(t)$. One approach is to make a dynamic heat balance on the reactor. The liberated heat of polymerization is proportional to $P(t)$.

11.5 COPOLYMERIZATION KINETICS

The problem of free-radical copolymerization kinetics is not nearly in such good shape. In addition to the four propagation reactions, there are three possible termination reactions ($P_1 \cdot + P_2 \cdot, P_1 \cdot + P_1 \cdot, P_2 \cdot + P_2 \cdot$), each with its own rate constant. A general rate equation has been developed [5], but because of a lack of independent knowledge of the constants involved and the mathematical complexity, it has not been used much. An approximate integrated expression including first-order initiator decay has been presented [6]. It involves a single, average termination rate constant. This "constant" would not be expected to remain constant as composition varies. Nevertheless, there appears to be some experimental verification of the relation. In certain instances where the compositions are not varying too much (e.g., in the control of a continuously stirred tank reactor), simplification of Equations 11.6 and 11.7 to

$$-\left(\frac{1}{V}\right)\left(\frac{dM_1}{dt}\right) = K_1[M_1] \tag{11.20}$$

$$-\left(\frac{1}{V}\right)\left(\frac{dM_2}{dt}\right) = K_2[M_2] \tag{11.21}$$

where the "constants" $K_1 = (k_{11} [P_1 \cdot] + k_{21}[P_2 \cdot])$ and $K_2 = (k_{12} [P_1 \cdot] + k_{22}[P_2 \cdot])$ are determined experimentally, may prove satisfactory.

11.6 PENULTIMATE EFFECTS AND CHARGE-TRANSFER COMPLEXES

There are certain cases in which the equations developed in Section 11.2 do not adequately describe copolymerization. Two approaches have been taken to remedy these deficiencies: invoking penultimate effects and postulating the formation of charge-transfer complexes.

In the former, the next-to-last (or penultimate) monomer in a growing chain also exerts an influence on the addition of the next monomer molecule [7–10]. In the latter, a 1:1 complex forms reversibly between electron-donating and electron-accepting comonomers (introducing an equilibrium constant to the analysis). This complex may then polymerize (from either end—introducing four more reactivity ratios) with itself or with the uncomplexed monomers.

The idea of complex formation is particularly helpful in explaining free-radical copolymerizations in systems such as styrene–maleic anhydride. This system forms a 1:1 copolymer over most of the range of monomer compositions, and the addition of maleic anhydride greatly enhances the rate of polymerization over that of pure styrene, despite the fact that maleic anhydride will not homopolymerize at a noticeable rate. These observations are consistent with the formation of a strong, readily polymerized complex between the monomers. The general equations to describe such copolymerizations have been presented by Seiner and Litt [11] and applied in a number of special cases [12–15].

11.7 SUMMARY

The possibilities (and complexities) introduced by copolymerization are seemingly infinite. Monomers can be added at different ratios, but even so, the reactivity ratios must be considered to evaluate the arrangement of the different repeat units in the polymer. Also, the polymer chains that form early on in a reactor may have a different composition from those formed later. The kinetics also become more and more difficult to accurately describe. While copolymers are certainly interesting and valuable materials, this chapter has only broken the surface in the analysis of these reactions. Further complexities can be introduced—three or more different monomers, the addition of a crosslinking agent, and step-growth polymers all require detailed analysis that is beyond the scope of this book.

PROBLEMS

1 (a) Plot F_1 and $\langle F_1 \rangle$ versus conversion in a batch reactor for systems (b) and (c) of Example 11.1. Use $f_{1,o} = 0.50$.

   (b) If you were going to produce a copolymer of uniform composition $F_1 = \langle F_1 \rangle = 0.5$ in a semibatch reactor (Figure 11.4a), which monomer would have to be added to the reactor for systems (a), (b), and (c)?

2 Consider a copolymerization system in which $r_1 = r_2 = 0.5$. Sketch F_1 and $\langle F_1 \rangle$ versus conversion in a batch reactor for $f_{1,o}$'s of 0.25, 0.50, and 0.75.

3 Crud Chemicals wishes to make a copolymer for beverage bottles using styrene (St) for low cost and processability and acrylonitrile (AN) for its gas barrier (mainly to CO_2 and O_2) properties. They want a–uniform composition of 75 wt% acrylonitrile (component 1) and 25% styrene (component 2). This is 86%(1) and 14%(2) on a mole basis. With more AN, the copolymer becomes difficult to process (recall from Chapter 3 that homopolyacrylonitrile, though linear, is not thermoplastic). With less AN, the barrier properties suffer. For this system, $r_1 = 0.040$, $r_2 = 0.40$. Crud proposes to use the semibatch technique of Figure 11.4a.

For a product that is to contain 100 total moles of monomer at complete conversion, address the following.

(a) Which monomer must be added over the course of the reaction?

(b) How many moles of the monomer in (a) must be added over the course of the reaction?

(c) Calculate the initial reactor charge.

(d) Explain why this is a difficult copolymer to make.

(e) Suppose Crud were just to dump 86 mol of AN and 14 mol of St into the reactor and let it react to completion. Plot F_1 and $\langle F_1 \rangle$ versus conversion. Why is not this a good idea?

4 Most commercial copolymers of styrene and acrylonitrile, known as SAN, are about 75 wt% styrene and 25 wt% acrylonitrile. Compared to the polymer in Problem 11.3 (which is 75 wt% acrylonitrile, 25 wt% styrene), why is this copolymer much easier to make at approximately uniform composition?

5 We wish to make an acrylonitrile (1)–styrene (2) copolymer (see Problem 11.3) by the technique shown in Figure 11.4c. The range of instantaneous copolymer composition must be limited to $0.40 < F_1 < 0.60$.

(a) What is the feed composition to the reactor?

(b) What should the maximum conversion be?

(c) What will $\langle F_1 \rangle$ be?

6 Crud Chemicals wishes to make a pilot-plant batch of block copolymer $[M_2]_{500}$–$[M_1]_{500}$ by anionic polymerization. They start with 5.00 mol of M_2 and an appropriate quantity of n-BuLi in their reactor. They intend to let the M_2 react to completion and then add 5.00 mol of M_1. Unfortunately, they get impatient (as often happens at Crud), and dump in the 5.00 mol of M_1 after only 2.50 mol of M_2 have polymerized. For this system, $r_1 = 2.00$, $r_2 = 0.50$.

Plot F_1 and $\langle F_1 \rangle$ versus total moles monomer polymerized, from 0 to 10. Describe the molecules produced. Make the usual assumptions of complete ionization: no termination, instantaneous addition, and perfect mixing of reagents.

7 For a copolymerizing system of known r_1, r_2, and f_1, obtain an expression for the probability that a $P_1\cdot$ will add M_1 (rather than M).

8 Crud Chemicals is producing a copolymer of uniform composition in a semibatch reactor, as illustrated in Figure 11.4a. For their system, $r_1 = 2.0$, $r_2 = 0.5$, and $f_{1,o} = 0.5$. They adjust the rate of addition of the more reactive monomer according to the heat evolution in the reactor. Normally, the system works quite well. One day, however, the feed valve sticks full open at 50% conversion, suddenly dumping the remaining monomer into the reactor. Nevertheless, they let the reaction go to completion. Plot F_1 and $\langle F_1 \rangle$ versus conversion (of all monomer) for this screwed-up reaction.

9 Consider the use of a CSTR to produce a copolymer of uniform composition, Figure 11.4b. Here, $f_{1,o}$ is the composition of the gross feed (fresh feed + recycle) to the reactor and F_1 and f_1 are the steady-state compositions in the reactor and its effluent stream.

(a) Obtain the expression that relates $f_{1,o}$ to conversion X in such a reactor.

(b)   How is the recycle ratio (moles recycled per mole of fresh feed) related to X in such a process, assuming complete separation and recycling of monomer?

(c)   What must the mole fraction of monomer 1 be in the fresh feed to such a process, again assuming complete separation and recycling of monomer?

(d)   See Problem 11.3. Calculate the gross feed composition and the recycle ratio to make that copolymer in a CSTR operating at 50% conversion.

10   Your boss calls you in to tell you that you are going to be in charge of implementing a control system to produce a copolymer of uniform composition in a semibatch reactor (Figure 11.4a). The reactor is fitted with a jacket to which steam can be fed to heat the reactants to the desired reaction temperature (during which conversion is negligible). Once the reaction has begun, the steam is shut off and replaced by cooling water, which is throttled by a feedback control system to maintain the reactor temperature fairly constant. What information would you need to know, and how would you go about designing a control system to feed the more reactive monomer at the correct rate?

11   We have seen how the compositions in a copolymerizing system can be described in terms of two parameters, r_1 and r_2. What does it take to describe the terpolymerization of monomers M_1, M_2, and M_3? If you are feeling particularly masochistic, you might try to work out the terpolymerization equivalents of Equation 11.11.

12   Crud Chemicals is producing a copolymer of $F_1 = 0.5$ in a CSTR (Figure 11.4b). For their system, $r_1 = 2.0$, $r_2 = 0.5$. Things are going along fine until the feed pump for monomer 1 suddenly quits. It takes the maintenance people a couple of hours to get it back on line. Sketch qualitatively F_1 and $\langle F_1 \rangle$ versus time for the period from immediately before the breakdown to the restoration of steady-state operation (well after the repair).

REFERENCES

[1]   Greenly, R.Z., *Polymer Handbook*, 3rd ed., J. Brandrup and E.H. Immergut (eds), Wiley, New York, 1989, Chapter II, p. 153.

[2]   Alfrey, T., Jr., and C.C. Price, *J. Polym. Sci.* **2**, 101 (1947).

[3]   Meyer, V.E. and G.G. Lowry, *J. Polym. Sci.* **A3**, 2843 (1965).

[4]   Hanna, R.J., *Ind. Eng. Chem.* **49**, 208 (1957).

[5]   DeButts, E.H., *J. Am. Chem. Soc.* **72**, 411 (1950).

[6]   O'Driscoll, K.F. and R.S. Knorr, *Macromolecules* **1**, 367 (1968).

[7]   Merz, E.T., T. Alfrey, and G. Goldfinger, *J. Polym. Sci.* **1**, 75 (1946).

[8]   Barb, W.G., *J. Polym. Sci.* **11**, 117 (1953).

[9]   Ham, G.E., *J. Polym. Sci.* **54**, 1 (1961).

[10]   Ham, G.E., *J. Polym. Sci.* **61**, 9 (1962).

[11]   Seiner, J.A. and M. Litt, *Macromolecules* **4**, 308 (1971).

[12]   Litt, M. *Macromolecules* **4**, 312 (1971).

[13]   Litt, M. and J.A. Seiner, *Macromolecules* **4**, 314 (1971).

[14]   Litt, M. and J.A. Seiner, *Macromolecules* **4**, 316 (1971).

[15]   Spencer, H.G., *J. Polym. Sci., Polym. Chem. Ed.* **13**, 1253 (1975).

CHAPTER 12

POLYMERIZATION PRACTICE

12.1 INTRODUCTION

As we have seen in the past few chapters, there are a number of reaction mechanisms used to make polymers. Some reactor designs (semibatch and CSTRs) have been discussed as ways to form uniform or well-characterized polymers and copolymers. This chapter covers some of the important considerations with scaling up polymerizations, starting with bulk polymerization, and takes into account how transport processes (heat and mass transfer) may influence the design of polymerization reactors.

12.2 BULK POLYMERIZATION

The simplest and most direct method of converting monomer to polymer is known as *bulk* or *mass* polymerization. A typical feed for a free-radical bulk polymerization might consist of a liquid monomer, a monomer-soluble initiator, and perhaps a chain-transfer agent (no solvent!).

As simple as this seems, some serious difficulties can be encountered, particularly in free-radical bulk polymerizations. One of them is illustrated in Figure 12.1 [1], which indicates the course of polymerization for methyl methacrylate by either bulk polymerization or *solution* polymerization using various concentrations of benzene, an inert solvent. The reactions were carefully maintained at constant temperature. At low polymer concentrations, the conversion versus. time curves are described by Equation 9.19. As polymer concentrations increase, however, a distinct acceleration of the rate of polymerization is observed which does not conform to the classical kinetic scheme. This phenomenon is known variously as *autoacceleration*, the *gel effect*, or the *Tromsdorff effect*.

Fundamental Principles of Polymeric Materials, Third Edition. Christopher S. Brazel and Stephen L. Rosen.
© 2012 John Wiley & Sons, Inc. Published 2012 by John Wiley & Sons, Inc.

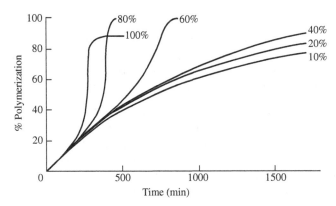

FIGURE 12.1 Polymerization of methyl methacrylate at 50 °C in the presence of benzoyl peroxide (a thermal initiator) in bulk (100%) or solution polymerization using various concentrations (10–80%) of monomer in benzene.

The reasons for this behavior lie in the difference between the propagation reaction (Eq. 9.3) and the termination steps (Eqs 9.4a and b), and the extremely high viscosities of the concentrated polymer solutions (10^4 poise is a ballpark figure–compare with water, which has a viscosity of 0.01 poise). The propagation reaction involves the approach of a small monomer molecule to a growing chain end, whereas termination requires that the ends of two growing chains get together. In bulk polymerizations, at high concentrations of polymer, it becomes exceedingly difficult for the growing chain ends to drag their chains through the entangled mass of dead polymer chains (eventually, the mass may *vitrify*, meaning that the chains are stuck in place and cannot move around enough to terminate). It is nowhere near as difficult for a monomer molecule to pass through the reaction mass, though. Thus, the rate of the termination reaction is limited not by the nature of the chemical reaction but by the rate at which the reactants can diffuse together, that is, it is *diffusion controlled.* This lowers the effective termination rate constant k_t, and since k_t appears in the denominator of Equation 9.13, the net effect is to increase the rate of polymerization. At very high polymer concentrations and below the temperature at which the chains become essentially immobile (or vitrified)—the T_g of the monomer-plasticized polymer—even the propagation step is diffusion limited, hence the bulk polymerization (100% monomer) curve levels off before reaching 100% conversion.

The difficulties of bulk polymerization are compounded by the inherent nature of the reaction mass. Vinyl monomers have rather large exothermic heats of polymerization, typically between –10 and –21 kcal/mol. Organic systems also have low heat capacities and thermal conductivities, about half those of aqueous solutions. Thus, the temperature can rise very quickly. To top it all off, the tremendous viscosities prevent effective convective (mixing) heat transfer. As a result, the overall heat-transfer coefficients are very low, making it difficult to remove the heat generated by the reaction. This raises the temperature, further increasing the rate of reaction (see Example 9.2) which in turn increases the rate of heat evolution, and can ultimately lead to disaster! To quote Schildknecht [2] on laboratory bulk polymerizations, "If a complete rapid polymerization of a reactive monomer in large bulk is attempted, it may lead to loss of the apparatus, the polymer or even the experimenter."

Example 12.1 The *maximum possible* temperature rise in a polymerizing batch may be calculated by assuming that no heat is transferred from the system (adiabatic assumption). Estimate the adiabatic temperature rise for the bulk polymerization of styrene, given that $\Delta H_p = -16.4$ kcal/mol and monomer molecular weight $= 104$.

Solution. The polymerization of 1 mol of styrene liberates 16,400 cal (assuming complete conversion). In the absence of heat transfer, all of this energy heats up the reaction mass. The heat capacities of organic compounds are often difficult to find; this is even more difficult here, as the properties of reaction mass change with conversion and probably also with temperature. To a reasonable approximation, however, the heat capacity of most liquid organic systems may be taken as 0.5 cal/g-°C. Thus,

$$\Delta T_{max} = 16400 \frac{cal}{mol} \times \frac{1\ mol}{104\ g} \times \frac{g°C}{0.5\ cal} \approx 315\ °C(!)$$

(The normal boiling point of styrene is 146 °C.)

Problems of overheating in bulk polymerizations can be circumvented in several ways:

1. By keeping at least one dimension of the reaction mass small, permitting heat to be conducted out. Poly(methyl methacrylate), PMMA, sheets are cast between glass plates as a maximum thickness of 5 in or so.

2. By maintaining low reaction rates through low temperatures, low initiator concentrations, and initiators that have relatively large energies of activation. The polymerization times for the PMMA sheets are on the order of 30–100 h and the temperatures are raised slowly as the monomer concentration drops. This approach has obvious economic disadvantages.

3. By starting with a *sirup* instead of the pure monomer. A sirup is a solution of the polymer in the monomer. It can be made in either of two ways: (a) by carrying the monomer to partial conversion in a kettle, or (b) by dissolving preformed polymer in monomer. Starting off with a sirup means that some of the conversion has already been accomplished, cutting heat generation and monomer concentration in the final polymerization. Since the density of a typical reaction mass increases on the order of 10–20% between 0% and 100% conversion in a bulk polymerization, the use of a sirup has the added advantage of cutting shrinkage when casting a polymer.

4. By carrying out the reaction continuously, with a lot of heat-transfer surface per unit conversion.

(Of course, a fifth option would be to use a solvent to aid in absorbing heat, but this is covered in a later section of this chapter.)

Batch bulk polymerization is often used to make objects with a desired shape by polymerizing in a mold. Examples are casting, potting, and encapsulation of electrical components and impregnation of reinforcing agents followed by polymerization. Continuous bulk polymerization is used for the production of thermoplastics by both step-growth and free-radical addition mechanisms. A continuous bulk process is outlined in Figure 12.2 [3]. Conversion is carried to about 40% in a stirred tank. The reaction mass then

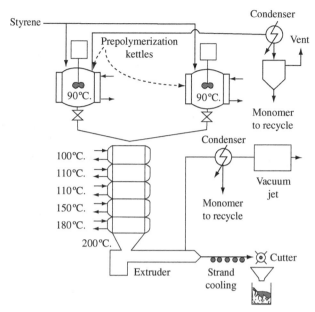

FIGURE 12.2   Continuous bulk polymerization of styrene. Reprinted by special permission from Reference 3. Copyright © 1962, by McGraw-Hill, Inc., New York, NY 10036.

passes down a tower with the temperature increasing to keep the viscosity at a manageable level and to drive up the conversion. The tower may be a simple gravity-flow device or it may contain slowly rotating spiral blades that scrape the walls, promoting heat transfer and conveying the reaction mass downward. The reaction mass is fed from the tower to a vented extruder at better than 95% conversion. Some additional conversion takes place in the extruder and a vacuum removes unreacted monomer, which is recycled. The extruded strands of molten polymer are then water cooled and chopped to form the roughly $^1/_8 \times {^1/_8} \times {^1/_8}$ in $(3 \times 3 \times 3\,\text{mm})$ pellets, which are sold to processors as "molding powder." Sheets of PMMA are also continuously cast from sirup between polished sheet-metal belts. Most bulk polymerizations are homogeneous. However, if the polymer is insoluble in its monomer and precipitates as the reaction proceeds, the process is sometimes known as *heterogeneous bulk* or *precipitation polymerization*. Two examples of such polymers are polyacrylonitrile and poly(vinyl chloride) (PVC). The latter is produced by a heterogeneous bulk process, which allows control of particle size and porosity for optimum plasticizer absorption [4]. Such heterogeneous polymerizations do not follow the kinetic scheme developed in Chapter 9 for homogeneous reactions.

Ethylene and a few other monomers are in the gas phase at atmospheric conditions. Thus, conventional low-density polyethylene (LDPE) is produced at pressures in the range of 15,000–50,000 psi (1000–3400 atm), and temperatures of about 150–300 °C [5]. These pressures are well above the critical pressure of ethylene (731 psi), so the monomer can be considered to be in the "fluid" phase. Depending on the temperature, pressure, and polymer molecular weight, one or two phases may be present as the polymer forms. Peroxide and azo free-radical initiators are used, as are small amounts of oxygen. Oxygen reacts with monomer to form peroxide initiators *in situ*.

Two types of rectors are used for high-pressure ethylene polymerization: autoclaves and tubular. The former are basically high-pressure CSTRs and the latter are continuous

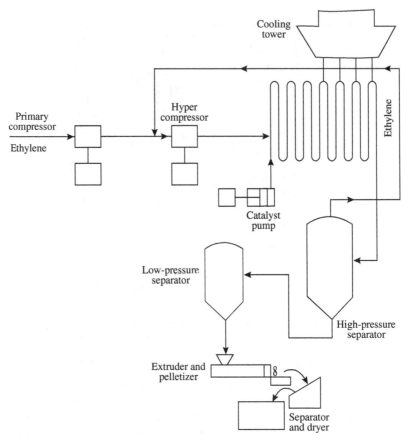

FIGURE 12.3 Tubular low-density polyethylene process. Reprinted with permission from *CHEMTECH* **13**(4), 223 (1983). Copyright 1983 American Chemical Society.

plug-flow reactors. Figure 12.3 illustrates a high-pressure tubular process[6]. The reactors are made of thick-walled pipe, typically from 1 in to 3 in in diameter and from 800 ft to 2500 ft long. They are also heat exchangers, with-heat transfer fluid circulating around the pipe, first to heat the ethylene to reaction temperature and then to remove the large heat of polymerization. Residence times are on the order of a minute, with single-pass conversions of 20% or less. Production rates are in the range of 8–125 million lb/yr per reactor [5].

Separation of polymer from unreacted monomer is basically a matter of reducing the pressure and allowing the monomer to flash off. Given the pressures involved, it is no surprise that the compressors represent a significant fraction of the capital and operating costs of these polymerization processes.

Bulk polymerization has several advantages:

1. Since only monomer, initiator, and perhaps chain-transfer agent are used, the purest possible polymer is obtained. This can be important in electrical and optical applications.
2. Objects may be conveniently cast to shape. If the polymer is the one that is crosslinked in the synthesis reaction, the only way of obtaining such objects short of machining from larger blocks.

3. Bulk polymerization provides the greatest possible polymer yield per reactor volume.

Among its disadvantages are the following:

1. It is often difficult to control.
2. To keep it under control, it may have to be run slowly, with the consequence of economic disadvantages.
3. As indicated by Equations 9.13 and 9.27, it may be difficult to get both high rates and high average chain lengths because of the opposing effects of [I].
4. It can be difficult to remove the last traces of unreacted monomer. After the reaction, the remaining monomer is left dispersed throughout the reaction mass and must diffuse to a surface to be evaporated. This is particularly important, for example, if the monomer is toxic and, particularly, when the polymer is intended for use in food-contact applications.

12.3 GAS-PHASE OLEFIN POLYMERIZATION

Gas-phase polyolefin processes based on Ziegler–Natta catalysts have undergone tremendous growth in the past decade. The popular UNIPOL® fluidized-bed process is shown in Figure 12.4 [6]. It typically operates at pressures in the range of 100–300 psi (7–20 atm) and at temperatures under 100 °C [6]. Monomer, catalyst particles, and hydrogen chain-transfer agent (see Section 10.6) are fed to the fluidized-bed reactor. The upward-circulating monomer suspends the growing polymer particles. The reactors have a characteristic bulge at the top which is a disengaging section. It lowers the velocity of the circulating monomer, keeping the growing polymer particles in the lower reaction zone. The monomer is circulated through a heat exchanger to remove the heat of polymerization. Polymer is discharged from the reactor in granular form, as powdered laundry detergent. It may be used as-is or extruded and pelletized to form the more conventional molding powder.

Impact-modified polypropylene is produced by using two reactors in series [7]. The first produces polypropylene particles, which are fed to the second reactor. A mixture of gaseous ethylene and propylene monomers (but no additional catalyst) is fed to the second reactor. The monomers diffuse to the still-active catalyst within the polypropylene particles from the first reactor, where they copolymerize to form a core of ethylene–propylene rubber within a polypropylene shell. This morphology greatly enhances the impact strength over that of polypropylene alone.

Compared to older processes, a number of significant advantages are claimed for the two-stage process above. With modern high-yield catalysts (10,000 g polymer/g catalyst), there is no need to remove the catalyst from the product; thus a difficult and costly step is eliminated. Modern catalysts also produce polypropylene with such a high isotactic content that removal of an atactic fraction, a common step in older processes, is not required. With low pressures, equipment and compression costs are low. No solvent is required. Separation of unreacted monomer is simple. With different catalysts and minor variations in process conditions, virtually the same equipment can be used to manufacture isotactic polypropylene (PP), high-density polyethylene (HDPE), and the so-called linear low-density polyethylene (LLDPE). Overall, major advantages in capital and operating costs, energy

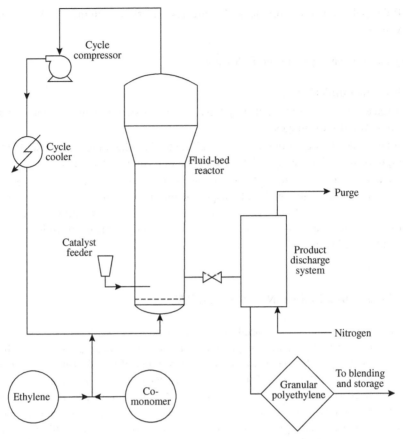

FIGURE 12.4 UNIPOL gas-phase polyolefin process. Reprinted with permission from *CHEMTECH* **13**(4), 225 (1983). Copyright 1983 American Chemical Society.

requirements, pollution, and superior product properties are cited by the licensors of this process.

12.4 SOLUTION POLYMERIZATION

As mentioned earlier, heat transfer problems can be mitigated by using a solvent for polymerization. A solvent can also overcome early polymer vitrification during the reaction. Thus, the addition of an inert solvent to a bulk polymerization mass can minimize many of the difficulties encountered in bulk systems (even though it also decreases the amount of product formed per reactor size). As shown in Figure 12.1, the solvent reduces the tendency toward autoacceleration in free-radical addition. The inert diluent adds its heat capacity without contributing to the evolution of heat and it cuts the viscosity of the reaction mass at any given conversion. In addition, the heat of polymerization may be conveniently and efficiently removed by refluxing the solvent. Thus, the danger of runaway reactions is minimized.

Example 12.2 Estimate the adiabatic temperature rise for the polymerization of a 20 wt% solution of styrene in an inert organic solvent.

Solution. In 100 g of the reaction mass, there are 20 g of styrene, so the energy liberated on its complete conversion to polymer is:

$$(20\,\text{g})\frac{(1\,\text{mol})}{104\,\text{g}}\frac{(16400\,\text{cal})}{\text{mol}} = 3150\,\text{cal}$$

The adiabatic temperature rise is then:

$$\Delta T_{\text{max}} = (3150\,\text{cal})\frac{(\text{g}^\circ\text{C})}{0.5\,\text{cal}}\frac{1}{100\,\text{g}} \approx 63\,^\circ\text{C}$$

The advantages of solution polymerization are:

1. Heat removal and control are easier than in bulk polymerization.
2. Since the reactions are more likely to follow known theoretical kinetic relations, the design of reactor systems is facilitated.
3. For some applications (e.g., lacquers and paints), the desired polymer solution is obtained directly from the reactor.

Among the disadvantages of solution polymerization are the following:

1. The use of a solvent lowers both rate and average chain length, which are proportional to [*M*].
2. Large amounts of expensive, flammable, and perhaps toxic solvent are generally required.
3. Separation of the polymer and recovery of the solvent require additional technology.
4. Removal of the last traces of the solvent and the monomer may be difficult, and these may have impact on the properties of the polymer (i.e., act as plasticizing agents).
5. Use of an inert solvent in the reaction mass lowers the yield per volume of the reactor.

Solution polymerizations are often used for the production of thermosetting condensation polymers, which are carried to a conversion short of the gel point in the reactor. The crosslinking is later completed in a mold. Such reactions may be carried out in a refluxing organic solvent. The water of condensation is carried overhead along with the solvent vapors. When the vapors are condensed, the water forms a second phase that is decanted before the solvent is returned to the reactor. Not only does this drive the reaction toward higher conversions but also the amount of water evolved provides a convenient measure of conversion (see Example 8.1).

Ionic polymerizations as well as many Ziegler–Natta polymerizations [5] are almost exclusively solution processes. They can be run under conditions such that the polymer product stays in solution, as in the production of stereospecific rubbers. The crystalline

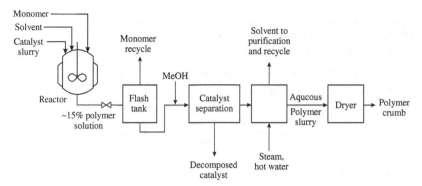

FIGURE 12.5 Ziegler–Natta solution polymerization process.

polymers polyethylene and isotactic polypropylene are commonly produced at temperatures sufficiently below T_m so that the polymer product is a solid that grows on the catalyst particles as in gas-phase polymerizations. Such processes are known as *slurry* polymerizations.

Figure 12.5 sketches a generic process utilizing a Ziegler–Natta catalyst system. Heat removal from the reactor(s) may be accomplished by refluxing the solvent, cooling jackets, external pumparound heat exchangers, or combinations of these. Catalyst deactivation with methanol or acid and separation by filtration or centrifugation are shown, although with modern high-yield catalysts these difficult and expensive steps can be eliminated.

In the Ziegler–Natta solution polymerization process, solvent and unreacted monomer are stripped with hot water and steam and recovered, leaving an aqueous polymer slurry, which is then dried to form a "crumb." With rubbers, the crumb is compacted and baled; with plastics, it is normally extruded and pelletized. Reactor designs for these processes are interesting and varied [8]. Most processes are continuous operations.

12.5 INTERFACIAL POLYCONDENSATION

Have you ever seen the example of forming nylon rope in a chemistry demonstration, where the polymer magically appears between two immiscible phases? This is an example of a variation of solution polymerization known as interfacial polycondensation. Besides being used for a "wow" experiment in demonstrations, it has been used in the laboratory for a long time [9], and is also applicable in industrial polymerizations. One monomer of a condensation pair is dissolved in one solvent and the other member of the pair in another solvent (*note*: this applies to AA and BB monomers, but does not work for AB-type monomers). The two solvents must be insoluble in each other. The polymer is soluble in neither and forms at the interface between them. One of the phases, generally, also contains an agent that reacts with the molecule of condensation to drive the reaction to completion.

An example of such a process is the preparation of nylon 6/10 from hexamethylene diamine and sebacoyl chloride (the acid chloride form of sebacic acid):

$$H_2N\{CH_2\}_6 NH_2 + Cl-\overset{O}{\overset{\|}{C}}\{CH_2\}_8\overset{O}{\overset{\|}{C}}-Cl \longrightarrow \{N\{CH_2\}_6\overset{H}{\overset{|}{N}}-\overset{H}{\overset{|}{N}}\overset{O}{\overset{\|}{C}}\{CH_2\}_8\overset{O}{\overset{\|}{C}}\}_x + HCl$$

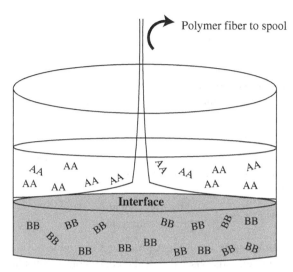

FIGURE 12.6   Interfacial polymerization of a nylon. The upper (organic) phase contains sebacoyl chloride (AA monomer) while the lower (aqueous) phase contains hexamethylene diamine (BB monomer). The reaction occurs where the two phases meet.

The acid chloride is dissolved in, for example, CCl_4, and the diamine in water, along with some NaOH to soak up the HCl that splits out during the reaction (Figure 12.6). In the classic "rope trick" demonstration, the aqueous layer is gently floated on top of the organic layer in a beaker. The reactants diffuse to the interface, where they react rapidly to form a polymer film. With care, the film can be withdrawn from the interface in the form of a continuous, hollow strand that traps considerable liquid. New polymer forms at the interface as the formed polymer is removed and replenished with fresh monomer on both sides. Commercially, it is probably easier to simply stir the phases together.

A major advantage of this technique is that these reactions usually proceed very rapidly at room temperature and atmospheric pressure, in contrast to the long times, high temperature and vacuum pressures usually associated with polycondensations. Another advantage of interfacial polymerizations for step-growth polymerizations is that the stoichiometry of an AA, BB reaction matches at the interface, therefore, we can avoid the problems noted in Chapter 8 with stoichiometric imbalance causing very low molecular weight polymers to form. These must be balanced against the cost of preparing the special monomers, such as the acid chloride above, and the need to separate and recycle solvents and unreacted monomers.

12.6 SUSPENSION POLYMERIZATION

Under the discussion of bulk polymerization in Section 12.2, it was mentioned that one of the ways of facilitating heat removal was to keep one dimension of the reaction mass small. This is carried to its logical extreme in *suspension* polymerization by suspending the monomer in the form of droplets 0.01–1 mm in diameter in an inert, nonsolvent liquid (almost always water). To achieve these suspended droplets, some physical mixing energy must be added (note that no surfactant is used here, so the droplets do not become an emulsion). However, similar to emulsion polymerization, suspended droplets behave as individual bulk reactors with dimensions small enough such that heat removal is not a problem. The heat can easily be soaked up and removed from the low-viscosity, inert suspension medium.

An important characteristic of these systems is that the suspensions are *thermodynamically unstable*, and must be maintained with agitation and suspending agents. A typical feed might consist of

$$\left.\begin{array}{l}\text{Monomer (water insoluble)}\\ \text{Initiator (monomer soluble)}\\ \text{Chain-transfer agent (monomer soluble)}\end{array}\right\}\ \text{Monomer phase}$$

Water

$$\text{Suspending agent}\ \left\{\begin{array}{l}\text{Protective colloid}\\ \text{Insoluble inorganic salt}\end{array}\right.$$

Two types of suspending agent are used. A protective colloid is a water-soluble polymer whose function is to increase the viscosity of the continuous water phase. This hydrodynamically hinders coalescence of the monomer droplets, but is inert with regard to the polymerization. A finely divided insoluble inorganic salt such as $MgCO_3$ may also be used. It collects at the droplet–water interface by surface tension and prevents coalescence of the droplets upon collision. Also, a pH buffer is sometimes used to help stability.

The monomer phase is suspended in the water at about a $1/2$ to $1/4$ monomer/water volume ratio. The reactor is purged with nitrogen and heated to start the reaction. Once underway, temperature control in the reactor is facilitated by the added heat capacity of the water and the low viscosity of the reaction mass, essentially that of the continuous (water) phase, allowing easy heat removal through a jacket heat exchanger.

The size of the product polymer beads depends on the strength of agitation, as well as the nature of the monomer and suspending system. Between 20% and 70% conversion, agitation is critical. Below this range, the organic phase is still fluid enough to redisperse and above it, the particles are rigid enough to prevent agglomeration; but if agitation stops or weakens between these limits, the sticky particles will coalesce or agglomerate in a large mass and finish polymerization that way. Again quoting Schildknecht [10], "After such uncontrollable polymerization is completed in an enormous lump, it may be necessary to resort to a compressed air drill or other mining tool to salvage the polymerization equipment."

Figure 12.7 [11] shows a typical suspension polymerization process. Since any flow system is bound to have some relatively stagnant corners, it has been impractical to run suspension polymerization continuously on a commercial scale. The stirred tank reactors are usually jacketed, stainless, or glass-lined steel kettles of up to 50,000-gal capacity. The polymer beads are filtered or centrifuged and water-washed to remove the protective colloid and/or rinsed with a dilute acid to decompose the $MgCO_3$. The beads are quite easy to handle when wet, but tend to pick up a static charge when dry, making themselves cling to each other and everything else. These beads can be molded directly, extruded and chopped to form molding powder, or used as-is, for example, as ion-exchange resins or the beads from which polystyrene foam cups and packing peanuts are made.

Ion-exchange resins are basically suspension beads of polystyrene crosslinked by copolymerization with a few percent divinyl benzene (a crosslinking agent), which are then chemically treated to provide the charged groups for functionality in separation and purification. To reduce mass-transfer resistance in the ion-exchange process, an inert solvent may be incorporated in the organic suspension phase. When polymerization is

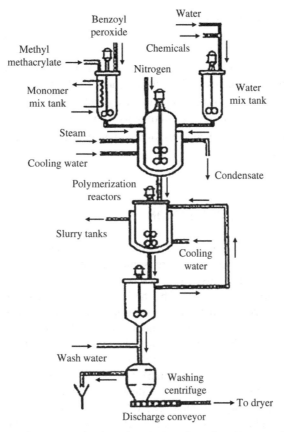

FIGURE 12.7 Suspension polymerization of methyl methacrylate. Excerpted by special permission from Reference 11. Copyright © 1966 by McGraw-Hill, Inc., New York, NY 10036.

complete, the solvent is removed, leaving a highly porous bead with a large internal surface area ("macroreticular"). The foam beads used in coffee cups are made from linear polystyrene containing an inert liquid blowing agent, usually pentane. Pentane may be added to the monomer prior to polymerization, but more commonly, it is added to the reactor after polymerization and is absorbed by the polystyrene beads. When exposed to steam in a mold, the beads soften and are foamed and expanded by the volatilized blowing agent to form the familiar cups and other foam items.

Commercially, suspension polymerization has been limited to the free-radical addition of water-insoluble liquid monomers. With a volatile monomer such as vinyl chloride, moderate pressures are required to maintain it in the liquid state. It is possible, however, to perform inverse suspension polymerizations with a hydrophilic monomer or an aqueous solution of a water-soluble monomer suspended in a hydrophobic continuous phase.

The major advantages of suspension polymerization then are as follows:

1. Easy heat removal and temperature control.
2. The polymer is obtained in a convenient, easily handled, and often directly useful form.

The disadvantages include the following:

1. Low yield per reactor volume.
2. A somewhat less pure polymer than from bulk polymerization, since there are bound to be remnants of the suspending agent(s) adsorbed on the particle surface.
3. The inability to run the process continuously; although if several batch reactors are alternated, the process may be continuous from that point on.
4. It cannot be used to make condensation polymers or used for ionic or Ziegler–Natta polymerizations.

12.7   EMULSION POLYMERIZATION

Emulsion polymerization was introduced while discussing some advanced polymerization techniques in Chapter 9, but there are some important characteristics of emulsion polymerization that merit discussion for industrial-scale processes. When the supply of natural rubber from the East was cut off by the Japanese in World War II, the United States was left without an essential material (tires for airplanes, especially at that time, needed replacement after every few landings). The success of the Rubber Reserve Program in developing a suitable synthetic substitute and the facilities to produce it in the necessary quantities is one of the all-time outstanding accomplishments of chemists and engineers, comparable to the Manhattan (atomic bomb) Project and the production of penicillin in its significance at the time. The styrene–butadiene copolymer rubber GR-S (government rubber-styrene) or SBR (styrene–butadiene rubber) as it is now called is still the most important synthetic rubber, and much of it is still produced, along with a variety of other polymers, by the emulsion polymerization process developed during the war.

The theory behind the emulsion reaction is discussed in Chapter 9. A generic commercial process is outlined in Figure 12.8. The reactors are usually stainless or glass-lined steel tanks, similar to those used for suspension polymerization. In contrast to suspension polymerization, however, a proper emulsion is thermodynamically stable (due

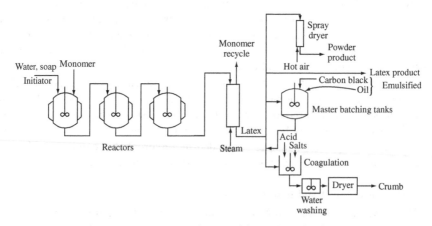

FIGURE 12.8   Schematic of an emulsion polymerization process.

to the surfactant or emulsifying agent), and therefore emulsion polymerizations can be run continuously. Newer processes often have several continuous stirred-tank reactors (CSTRs) in series.

The product of an emulsion polymerization is known as a *latex*—polymer particles on the order of 0.05–0.15 µm stabilized by the surfactant. These latexes are often important items of commerce in their own right (think of latex paints or white glue). This process can also be modified slightly for microencapsulation purposes, with the polymer forming a shell surrounding an inner fill material (these capsules are then used in controlled release formulations in a number of fields).

Where the polymer must be mixed with other materials, the process of *master batching* sometimes allows this to be done conveniently and uniformly. In rubber technology, carbon black and oil are emulsified and mixed with the rubber latex and then the mixture is coagulated together, giving a uniform and intimate dispersion of the additives in the rubber.

The increased recognition of possible adverse physiological effects of even small amounts of residual monomer makes emulsion polymerization attractive in certain applications. The extremely small size of the latex particles provides a very short diffusion path for the removal of small molecules from the polymer, for example, by steam stripping, permitting very low levels of residual monomer to be obtained.

For many applications, the solid polymer must be recovered from the latex. The simplest method is spray drying (a process that produces a fine mist from the solution while evaporating the solvent rapidly with hot air); but since no attempt is made to remove the surfactant, the product is an extremely "dirty" polymer. A latex may be "creamed" by adding a material such as acetone that is at least a partial solvent for the polymer. This makes the particles sticky and causes some agglomeration. The latex is then coagulated by adding an acid, usually sulfuric, which converts the surfactant from its sodium salt to its insoluble hydrogen form, and/or by adding an electrolyte salt, which disrupts the stabilizing double layer on the particles, causing them to agglomerate through electrostatic attraction. The former method leaves much insoluble material adsorbed on the particle surfaces, but in some applications, this may even be beneficial. For example, fatty acids from the surfactant act as lubricants in tire manufacturing. The coagulated polymer "crumb" is then washed, dried, and either baled or processed further.

As with suspension polymerization, commercial emulsion polymerization has pretty much been restricted to the free-radical solution addition of water-insoluble, liquid monomers (with volatile monomers such as butadiene and vinyl chloride, moderate pressures are required to keep them in the liquid phase). Inverse emulsion polymerizations, with a hydrophilic monomer phase dispersed in a continuous hydrophobic base are, however, possible.

The advantages of emulsion polymerization are as follows:

1. Ease of control. The viscosity of the reaction mass is much less than that of a true solution of comparable polymer concentration, the water also adds its heat capacity to improve temperature control, and the reaction mass may be refluxed.
2. It is possible to obtain both high rates of polymerization and high average chain lengths by using high surfactant and low initiator concentrations.
3. The latex product is often directly valuable or aids in obtaining uniform compounds through master batching.
4. The small size of the latex particles allows the attainment of low residual monomer levels.

The disadvantages include the following:

1. It is difficult to get a pure polymer product. The tremendous surface area of the tiny particles provides plenty of room for adsorbed impurities. This includes water attracted by residual surfactant, traces of which can cause problems in certain applications.
2. Considerable technology and expense is required to recover the solid polymer.
3. The water in the reaction mass lowers the yield per reactor volume.
4. It cannot be used to make condensation polymers or used for ionic or Ziegler–Natta polymerizations.

12.8 SUMMARY

Much as any industrial chemical plant must consider numerous factors beyond the chemistry of a reaction, industrial polymerization processes require adequate control of temperature (especially to remove large exothermic heats of reaction), provide mixing, and involve appropriate separation techniques. Bulk polymerizations offer pure polymers in a reaction vessel, allowing the greatest possible yield per reactor volume. However, disadvantages such as vitrification, the Tromsdorff effect, and extraction of unreacted monomer have led to other polymerization methods, including solution, suspension, and emulsion polymerization.

PROBLEMS

1 What difficulties would you foresee in carrying out anionic or Ziegler–Natta polymerizations by suspension or emulsion processes?

2 Suspension polymerization is routinely used to produce ion-exchange beads from copolymers of styrene and divinyl benzene. Why is it not used to make polymer for ion-exchange membranes of similar composition?

3 Emulsion polymerization is used extensively for the production of synthetic rubbers but suspension polymerization is not. Why?

4 (a) Autoacceleration is common in the free-radical polymerization of vinyl monomers. Why is it not observed in the anionic addition polymerization of the same monomers?

(b) What do you think would be a distinguishing molecular characteristic of polymer formed in a strongly autoaccelerated reaction, and why?

5 The semibatch method of producing copolymers of uniform composition (Figure 11.4a) can be used in conjunction with all modes of polymerization but suspension. Why?

6 Describe a suspension-polymerization formulation for the production of poly(acrylic acid). This monomer is similar to acetic acid in most properties.

7 The 100% monomer curve in Figure 12.1 levels off at about 90% conversion. What would you do to push the conversion to 99%?

8 Why do you suppose that such high pressures are needed for LDPE production? They are not needed for the other free-radical reactions we have seen.

9 To be useful in paint formulations, a polymer must form a continuous film on the surface to which it is applied.

(a) What key properties must the polymer phase in a latex-paint formulation have?

(b) One of the major advantages of latex paints over the older oil-based paints is easy cleanup with water. You may have noticed, however, that if you let your brushes sit around too long before rinsing them, you can no longer get the paint out with water. Why?

10 In the two-reactor, impact-modified polypropylene process described in Section 12.3, it was noted that no new catalyst was introduced to the second reactor. Suppose it was. How would that alter the product?

11 Crud Chemicals is preparing a batch of glyptal resin by refluxing 2 kmol of glycerin and 3 kmol of phthalic anhydride (see Example 8.2) in toluene. The water of condensation is decanted before returning the toluene to the reactor for the next batch. Calculate the maximum liters of water they can collect before shutting down the reaction.

REFERENCES

[1] Schulz and Harborth, G. *Makromol. Chem.* **1**, 106 (1947).

[2] Schildknecht, C.E., *Polymer Processes*, Interscience, New York, 1956, p. 38.

[3] Wohl, M.H., *Chem. Eng.* August 1, 1962, p. 60.

[4] Thomas, J.C., *Soc. Plast. Eng. J.* **23** (10), 61 (1967).

[5] Albright, L.F., *Processes for Major Addition-Type Plastics and Their Monomers*, McGraw-Hill, New York, 1974.

[6] Karol, F.J., *CHEMTECH* **12** (7), 222 (1983).

[7] Haggin, J., *Chem. Eng. News* March 31, 1986, p. 15; Bisio, A., *Chem. Eng. Progr.* **85** (5), 76 (1989).

[8] Gerrens, H., *CHEMTECH* **12** (7), 434 (1982).

[9] Morgan, P.W., *Soc. Plast. Eng. J.* **15** (6), 485 (1959).

[10] Schildknecht, C.E., *Polymer Processes*, Interscience, New York, 1956, p. 94.

[11] Guccione, E., *Chem. Eng.* June 6, 1966, p. 138.

PART III

POLYMER PROPERTIES

This part of the book moves from the chemical structure and reactions used to form polymers to their physical properties. Some of the material in this section may be familiar to one who has studied the mechanics of materials, but it is worth delving deeper into polymers and discover some of their unusual characteristics (viscoelasticity, for one). Topics that are addressed here include mechanical strength, flexibility, polymer responses to compression, stretching and shear forces, and the equipment used to test these properties. One chapter is devoted to mathematical models to describe viscoelastic behavior (using combinations of so-called springs and dashpots) to approximate the complex response of polymers to stresses that include both elastic stretch (which is particularly useful in the waistbands of underwear) and viscous deformation (which is very useful in automobile bumpers to dissipate the energy of a collision).

Fundamental Principles of Polymeric Materials, Third Edition. Christopher S. Brazel and Stephen L. Rosen
© 2012 John Wiley & Sons, Inc. Published 2012 by John Wiley & Sons, Inc.

CHAPTER 13

RUBBER ELASTICITY

13.1 INTRODUCTION

Natural and synthetic rubbers possess some interesting, unique, and useful mechanical properties. No other materials are capable of reversible extension of 600–700%. No other materials exhibit an increase in strength (elastic modulus) with increasing temperature. It was recognized long ago that vulcanization (crosslinking with sulfur) was necessary for rubber deformation to be completely reversible. We now know that this is a result of the crosslinks that prevent the bulk slippage of the polymer chains past one another, eliminating flow (irrecoverable deformation). More recently, this function of the covalent crosslinks has been achieved using physical (instead of chemical) means and using rigid domains (either glassy or crystalline) within some linear polymers, such as poly(vinyl alcohol). Thus, when a stress is applied to a sample of crosslinked rubber, equilibrium is fairly rapidly established. Once at equilibrium, the properties of the rubber can be described by thermodynamics.

13.2 THERMODYNAMICS OF ELASTICITY

Consider an element of material with dimensions $a \times b \times c$, as sketched in Figure 13.1.

Applying the first law of thermodynamics to this system yields:

$$dU = dQ - dW \tag{13.1}$$

where dU is the change in the system's *internal energy*, and dQ and dW are the heat and work exchanged between the system and its surroundings as the system undergoes a

Fundamental Principles of Polymeric Materials, Third Edition. Christopher S. Brazel and Stephen L. Rosen.
© 2012 John Wiley & Sons, Inc. Published 2012 by John Wiley & Sons, Inc.

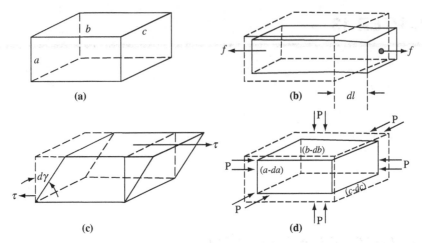

FIGURE 13.1   Types of mechanical deformation: (a) unstressed (b) uniaxial tension (c) pure shear, and (d) isotropic compression.

differential change. (We have adopted the convention here that work done by the system on the surroundings is positive.)

We will consider three types of mechanical work.

1. Work done by a uniaxial tensile force, f:

$$dW(\text{tensile}) = -fdl \tag{13.2}$$

   where dl is the differential change in the system's length arising from the application of the force f. This is the fundamental definition of work. The negative sign arises from the need to reconcile the mechanical convention of treating a tensile force (which does work on the system) as positive with the thermodynamic convention above.

2. Work done by a shear stress, τ:

$$dW(\text{shear}) = (\text{force})(\text{distance}) = -(\tau bc)(a\,d\gamma) = -\tau V d\gamma \tag{13.3}$$

   where γ is the shear strain (Figure 13.1c) and $V = abc =$ the system volume.

3. Work done by an isotopic (evenly applied on all sides) pressure in changing the volume:

$$dW(\text{pressure}) = P(cb)da + P(ac)db + P(ab)dc = PdV \tag{13.4}$$

   Note that no negative sign is needed here. A positive pressure causes a decrease in the volume (negative dV) and does work on the system.

If the deformation process is assumed to occur reversibly (in a thermodynamic sense), then:

$$dQ = TdS \tag{13.5}$$

where S is the system's entropy.

Combining the preceding five equations gives a general relation for the change of internal energy of an element of material undergoing a differential deformation:

$$dU = TdS - PdV + fdl + V\tau d\gamma \tag{13.6}$$

Now, let us consider three individual types of deformation.

1. *Uniaxial Tension at Constant Volume and Temperature:* Under these conditions, $dV = \tau = 0$. Dividing the remaining terms in Equation 13.6 by dl, restricting to constant T and V, and solving for f gives:

$$f = \left(\frac{\partial U}{\partial l}\right)_{T, V} - T\left(\frac{\partial S}{\partial l}\right)_{T, V} \tag{13.7}$$

2. *Pure Shear at Constant Volume and Temperature:* Here, $dV = f = 0$. Dividing the remaining terms in Equation 13.6 by $d\gamma$, restricting to constant T and V, and solving for τ gives:

$$\tau = \frac{1}{V}\left(\frac{\partial U}{\partial \gamma}\right)_{T, V} - \frac{T}{V}\left(\frac{\partial S}{\partial \gamma}\right)_{T, V} \tag{13.8}$$

3. *Only Isotropic Compression at Constant Temperature:*

$$P = -\left(\frac{\partial U}{\partial V}\right)_{r} + T\left(\frac{\partial S}{\partial V}\right)_{T} \tag{13.9}$$

It is very difficult to carry out tensile experiments at constant volume to obtain the partial derivatives in Equation 13.7. Most experimental tests are carried out at constant pressure (atmospheric), and in general, there is a change in volume with tensile straining. Fortunately, Poisson's ratio is approximately 0.5 for rubbers, so this change in volume is small, and also Equation 13.7 is approximately valid for tensile deformation at constant pressure. For precise work, the hydrostatic pressure must be varied to maintain V constant or theoretical corrections applied to the constant-pressure data to obtain the constant-volume coefficients [1,2]. In pure shear experiments, V should be constant and Equation 13.8 should be valid.

13.2.1 Types of Elasticity

Equations 13.7–13.9 reveal that there are *energy* (the first term on the right) and *entropy* (the second term on the right) contributions to the tensile force, shear stress, or isotropic pressure. In polymers, *energy elasticity* represents the storage of energy resulting from the rotation of bonds (Figure 6.4) and the straining of bond angles and lengths from their equilibrium values. Compare this behavior with other inelastic solids (metals, ceramics); because polymers deform, they are able to store energy temporarily due to any of the three cases above. Interestingly, energy elasticity is almost entirely intramolecular (within single polymer chains) in origin, that is, there is no change in the energy interaction between different polymer chains with deformation [2].

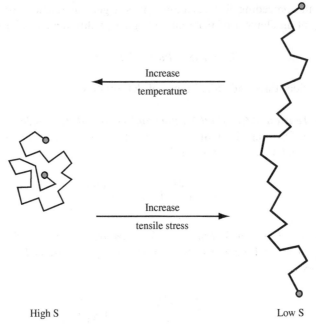

FIGURE 13.2   The effects of stress and temperature on chain configurations.

Entropy elasticity is caused due to decrease in entropy upon straining. This can be visualized by considering a single polymer molecule subjected to a tensile stress. In an unstressed state, the molecule is free to adopt an extremely large number of random, "balled-up" configurations (Figure 13.2a) switching from one to another through rotation about the bonds. Now imagine the molecule to be stretched out under the application of a tensile force (Figure 13.2b). It is obvious that there are far fewer configurational possibilities, and the more it is stretched, the fewer there are. Now, $S = k \ln \Omega$, where k is the Boltzmann's constant and Ω is the number of configurational possibilities, so *stretching decreases the entropy* (increases the molecular order). Raising the temperature has precisely the opposite effect. The added thermal energy of the chain segments increases the intensity of their lateral vibrations, favoring a return to the more random or higher entropy state. This tends to pull the extended chain ends together, giving rise to a retractive force.

It should be obvious from the discussion above that to exhibit significant entropy elasticity, the material must be above its glass transition temperature and cannot have appreciable amounts of crystallinity.

13.2.2   The "Ideal" Rubber

To demonstrate the "ideal" rubber, let us start with a gas subjected to an isotropic pressure, where the energy term in Equation 13.9 arises from the change in the intermolecular forces with volume and the entropy term arises from the increased space (and therefore greater "disorder") the molecules gain with increased volume. In an ideal gas, there are no intermolecular forces, $(\partial U/\partial V)_T = 0$. By analogy, in an ideal rubber, $(\partial U/\partial l)_{T,V} = (\partial U/\partial \gamma)_{T,V} = 0$ and

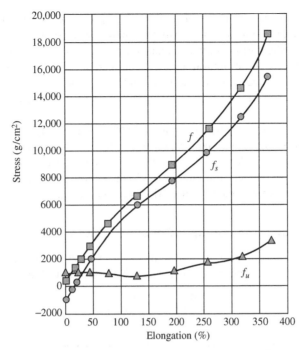

FIGURE 13.3   Energy (f_V) and entropy (f_S) contributions to tensile stress in natural rubber at 20 °C. Reprinted from Anthony *et al.* [3] Copyright 1942 by the American Chemical Society. Reprinted by permission of the copyright owner.

elasticity arises only from entropy effects. For many gases around room temperature and above, and around atmospheric pressure and below, $(\partial U/\partial V)_T < T(\partial S/\partial V)_T$, the ideal gas law is a good approximation. Similarly, as illustrated in Figure 13.3 [3], under some circumstances, $(\partial U/\partial l)_{T,V} < T(\partial S/\partial l)_{T,V}$ for rubbers and they behave just as ideal rubbers.

13.2.3   Effects of Temperature at Constant Force

Now let us consider what happens to the length of a piece of rubber when its temperature is changed while a weight is suspended from it, that is, when it is maintained at a constant tensile force, assuming constant volume (an approximation, in the usual constant-pressure experiment), $dU = TdS + f\,dl$. Solving for dl, dividing by dT, and restricting to constant f as well as V gives

$$\left(\frac{\partial l}{\partial T}\right)_{f,v} = \frac{1}{f}\left(\frac{\partial U}{\partial T}\right)_{f,v} - \frac{T}{f}\left(\frac{\partial S}{\partial T}\right)_{f,v} \tag{13.10}$$

As before, the first term on the right represents energy elasticity and the second represents entropy elasticity. Since internal energy increases with temperature, the partial derivative in the energy term is positive, as is f. The energy term, therefore, causes an increase in length with temperature (positive contribution to $(\partial l/\partial T)_{f,V}$). This is the normal thermal expansion observed in all materials (ceramics, metals, glasses, etc.), reflecting the increase in the average distance between atomic centers with temperature. All factors in the entropy term are positive, however, and since it is preceded by a negative sign, it gives rise to a *decrease* in length with increasing temperature. In rubbers, where the entropy effect overwhelms the normal thermal expansion, this is what is actually observed. In all other materials, where the structural units are confined to a single arrangement (e.g., the atoms in a crystal lattice cannot readily interchange), the entropy term is small. To visualize this, consider Figure 13.2 and the large amount of entropy gained by balling up a polymer chain.

The magnitude of the entropy contraction in rubbers is typically much greater than the thermal expansion of other materials. An ordinary rubber band will contract an inch or so when heated to 300 °F under stress, while the expansion of a piece of metal of similar length over a similar temperature range would not be noticeable without a microscope. According to a well-known metallurgist [4], "Polymers are all entropy."

13.2.4 Effects of Temperature at Constant Length

It is interesting to consider what happens to the force in a piece of rubber when it is heated while stretched to a constant length. The exact thermodynamic Maxwell relation

$$-\left(\frac{\partial S}{\partial l}\right)_{T,V} = \left(\frac{\partial f}{\partial T}\right)_{l,V} \tag{13.11a}$$

can be used to describe approximately the usual experiment conducted at constant pressure. A better approximation to the constant pressure experiment is

$$-\left(\frac{\partial S}{\partial l}\right)_{T,V} \approx \left(\frac{\partial f}{\partial T}\right)_{P,\alpha}$$

where $\alpha = l/l_o$ is the extension ratio, the ratio of stretched to unstretched length at a particular temperature. Combining Equations 13.11a and 13.7 gives

$$\left(\frac{\partial f}{\partial T}\right)_{l,V} = \frac{f}{T} - \frac{1}{T}\left(\frac{\partial U}{\partial l}\right)_{T,V} \tag{13.12}$$

Now, both f and T are positive, so the first term on the right causes the force to increase with temperature, a result of the greater thermal agitation (tendency toward higher entropy) of the extended chains. The partial derivative in the second term is usually (but not always!) positive, as energy is stored like a spring when it is extended. With the negative sign in front, this term predicts a relaxation of the tensile force with increasing temperature. Again, the second term reflects the ordinary thermal expansion obtained with all materials, but in rubbers, at reasonably large values of f, it is overshadowed by the first (entropy) term and the force increases with temperature. For an ideal rubber, $(\partial U/\partial l)_{T,V} = 0$, and integration of Equation 13.12 at constant volume yields

$$f = (\text{constant})T \quad (\text{ideal rubber}) \tag{13.13}$$

This is analogous to the linearity between P and T in an ideal gas at constant V. These observations are confirmed in Figure 13.4. When the polymer is not stretched much (e.g., the 3% elongation line), the polymer softens as the temperature is raised.

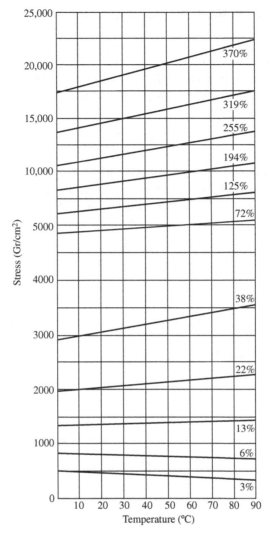

FIGURE 13.4 Force versus temperature in natural rubber maintained at constant extension (% relative to length at 20 °C) [3]. Copyright 1942 by the American Chemical Society. Reprinted by permission of the copyright owner.

This negative slope at low elongations arises from the predominance of thermal expansion when elongation, and hence f, is low. Note that there is an intermediate elongation, the *thermoelastic inversion* point (at approximately 13% elongation in Figure 13.4), at which force is essentially independent of temperature, where thermal expansion and entropy contraction balance. In the upper portions of the graph, the highly elongated samples require more stress to maintain greater elongation ratios, and the stress goes up appreciably with temperature, as the entropic terms dominate over thermal expansion, causing an even greater force of the polymer trying to coil back from its extended state.

13.3 STATISTICS OF IDEAL RUBBER ELASTICITY [1,2,5]

A typical rubber consists of long chains connected by short crosslinks every few hundred carbon atoms. The chain segments between crosslinks are known as *network chains*. (These segments are characterized by $\bar{M}_c$, the molecular weight between crosslinks.) The change in entropy upon stretching a sample containing N moles of network chains is

$$S - S_0 = NR\ln\frac{\Omega}{\Omega_0} \tag{13.14}$$

where the subscript 0 refers to the unstretched state, Ω is the number of configurations available to the N moles of network chains, and R is the gas constant. By statistically evaluating the Ω's, it is possible to show that for constant-volume stretching,

$$\begin{aligned} S - S_0 &= -\frac{1}{2}NR\left[\left(\frac{l}{l_0}\right)^2 + 2\left(\frac{l_0}{l}\right) - 3\right] \\ &= -\frac{1}{2}NR(\alpha^2 + 2\alpha^{-1} - 3) \end{aligned} \tag{13.15}$$

where $\alpha = l/l_o$ is the extension ratio. (Newer theories give somewhat different results [2], but this is adequate for our purposes.)

For an ideal rubber, in which the tensile force is given by

$$f = -T\left(\frac{\partial S}{\partial l}\right)_{T,\,V} \quad \text{(ideal rubber)} \tag{13.16}$$

differentiation of Equation 13.15 and insertion into Equation 13.16 yields

$$f = \frac{NRT}{l_0}(\alpha - \alpha^{-2}) \tag{13.17}$$

Also,

$$N = \frac{\text{mass}}{\bar{M}_c} = \frac{\rho V}{\bar{M}_c} = \frac{\rho l_0 A_0}{\bar{M}_c} = \frac{\rho l A}{\bar{M}_c} \tag{13.18}$$

where $\bar{M}_c$ is the number-average molecular weight of the chain segments between consecutive crosslinks, ρ is the density, and A is the cross-sectional area of the sample (since the volume change in stretching a piece of rubber is negligible, $A_o l_o = Al$). Therefore,

$$f = \frac{\rho A_0 RT}{\bar{M}_c}(\alpha - \alpha^{-2}) \tag{13.19}$$

The *engineering tensile stress*, σ_e, is defined as the tensile force divided by the initial cross-sectional area of the sample A_o and is, therefore,

$$\sigma_e \equiv \frac{f}{A_0} = \frac{\rho RT}{\bar{M}_c}(\alpha - \alpha^{-2}) \tag{13.20}$$

and the *true tensile stress* σ_t, the tensile force over the actual area A at length l, is

$$\sigma_t \equiv \frac{f}{A} = \frac{\rho RT}{\bar{M}_c}(\alpha^2 - \alpha^{-1}) \tag{13.21}$$

Since the tensile strain ε is

$$\varepsilon = \frac{l - l_0}{l_0} = \alpha - 1 \tag{13.22}$$

the slope of the true stress–strain curve (which gives the tangent Young's modulus) is

$$E = \left(\frac{\partial \sigma_t}{\partial \varepsilon}\right)_T = \left(\frac{\partial \sigma_t}{\partial l}\right)_T \left(\frac{\partial l}{\partial \varepsilon}\right)_T = \frac{\rho RT}{\bar{M}_c}(2\alpha + \alpha^{-2}) \tag{13.23}$$

and the initial modulus (as $\alpha \rightarrow 1$) for an ideal rubber becomes

$$E(\text{initial}) = \frac{3\rho RT}{\bar{M}_c} \tag{13.24}$$

The modulus is representative of the strength of a material (or how much resistance the material will give) when a force (axial tension, shear stress) is applied.

Equations 13.19–13.24 point out two important concepts: (1) the force (or modulus) in an ideal rubber sample held at a particular strain increases in proportion to the absolute temperature (in agreement with Eq. 13.13), and (2) the force is *inversely* proportional to the molecular weight of the chain segments between crosslinks. Thus, increased crosslinking, which reduces $\bar{M}_c$, is an effective means of stiffening a rubber. Equation 13.24 is often used to obtain $\bar{M}_c$ from mechanical tests and thereby evaluate the efficiency of various crosslinking procedures.

Even noncrosslinked polymers exhibit rubbery behavior above their T_g values for limited periods of time. This is due to mechanical (physical) entanglements acting as temporary crosslinks; $\bar{M}_c$ then represents the average length of the chain segments between those entanglements.

When compared with experimental data, Equation 13.19 does a reasonably good job in compression but begins to fail at extension ratios a greater than about 1.5, where the experimental force becomes greater than predicted. There are a number of reasons for this. First, Equation 13.15 is based on the assumption of a Gaussian distribution of network chains. This assumption fails at high elongations, and it is also in error if crosslinks are formed when chains are in a strained configuration. Second, it does not take into account the presence of chain end segments, which do not contribute to the support of stress. Third, some rubbers (natural, in particular) begin to crystallize as a result of chain orientation at high elongations. (Orientation reduces ΔS_m and raises T_m above the test temperature. See Example 6.6.) This causes the stress–strain curve to shoot up markedly. Some theoretical modifications to the theory are available that improve on the accuracy of these equations, for example, Mark and Erman [2] address the issues of non-Gaussian distribution of chain length and factor at the chain ends.

It is also important to keep in mind that in practice, rubbers are rarely used in the form of pure polymer. They are almost always reinforced with carbon black, and often contain other fillers, plasticizing, and extending oils, etc., all of which influence the stress–strain properties, and are not considered in the theories discussed here.

13.4 SUMMARY

Rubbery materials are a class of crosslinked polymers that are characterized by their response to mechanical stimuli (stretching, compression, etc.). This chapter explained the unusual behavior of polymers through the (often) competing combination of internal energy and entropic energy. The thermodynamic analysis used in this chapter served to develop equations to quantifiably predict the behavior of rubbers in stress environments and even use rubber elasticity experiments to estimate $\bar{M}_c$. This chapter served as an introduction to the thermodynamic basis for rubber elasticity; the practical implications of this behavior will be discussed in detail in the beginning of Chapter 15. While this chapter introduced the behavior of purely elastic materials, the next chapter covers the opposite extreme of viscous polymer melts (and solutions). These two concepts will be brought together in Chapter 15 on viscoelastic materials, which have features of both elastic solid rubbers and viscous liquid melts. Other texts are available for more in-depth study of rubber elasticity [1,2].

PROBLEMS

1 Recall the definition of enthalpy: $H \equiv U + PV$. Write the enthalpy analogs of Equations 13.7 and 13.8 for experiments carried out at constant pressure. Does it makes you wonder why we bother with internal energy at all?

2 Derive Equation 13.11a. *Hints*: Start with Equation 13.6, with $dV = d\gamma = 0$. Make use of the definition of Helmholtz free energy: $A \equiv U - TS$. Remember that the second mixed partial derivatives of thermodynamic state functions are independent of the order in which you take them.

3 The following data were obtained in a classroom demonstration in which a rubber band with inked-on gage marks was looped over the hook of a spring balance:

f, lb	l, cm
0	2.42
0.33	2.80
0.54	3.08
0.79	3.58
1.10	3.97
1.36	4.70
1.66	5.60
1.89	6.42
2.20	7.28
2.68	8.45
3.07	9.30

The initial cross section of the rubber band measured 1 mm × 6 mm. The density of the rubber was probably around 1 g/cm^3 and the temperature was 25 °C (remember, this is not rocket science). Assume that the band is made of natural rubber, $[C_5H_8]_x$. How well does Equation 13.19 fit these data? Estimate the average number of repeating units between crosslinks, $\bar{x}_c$ and $\bar{M}_c$.

4   Show that the fraction of the energy contribution (f_U) to the total force (f) in a uniaxially stretched rubber is given by

$$\frac{f_U}{f} = 1 - \frac{T}{f}\left(\frac{\partial f}{\partial T}\right)_{l,\,V} = -T\left(\frac{\partial \ln(f/T)}{\partial T}\right)_{l,\,V}$$

5   Use the equation of Problem 13.4 and the data in Figure 13.4 to calculate the fractional energy contribution to the force in the material.

6   Sketch qualitatively the expected variation in f_U/f with degree of crosslinking in a rubber.

7   It was mentioned that $(\partial U/\partial l)_{T,V}$ is not always positive (although it is still smaller than the entropy term). For example, it is negative for linear polyethylene, at least for low elongations. Why? *Hint*: Recall Figure 6.4.

8   This chapter has emphasized the differences between the behavior of polymers and nonpolymers in uniaxial tension. How would you expect them to be compare in isotropic compression? Why?

9   Repeat Problem 13.6 for rubbers in isotropic compression.

REFERENCES

[1] Flory, P.J., *Principles of Polymer Chemistry*, Cornell UP, Ithaca, NY, 1953, Chapter 11.

[2] Mark, J.E. and Erman, B. *Rubberlike Elasticity: A Molecular Primer*, Wiley, New York, 1988.

[3] Anthony, R.L., R.H. Caston, and E. Guth, *J. Phys. Chem.* **46**, 826 (1942).

[4] Paxton, H.W., personal communication.

[5] Tobolsky, A.V., *Properties and Structure of Polymers*, Wiley, New York, 1960, Chapter 2.

CHAPTER 14

INTRODUCTION TO VISCOUS FLOW AND THE RHEOLOGICAL BEHAVIOR OF POLYMERS

14.1 INTRODUCTION

In contrast to Chapter 13, where purely elastic materials were described, the opposite end of the spectrum is purely viscous materials, more commonly known as liquids, which for polymers means they are either molten or dissolved in a good solvent. *Rheology* is the science of the deformation and flow of materials, particularly of polymers that can melt and flow, as long as they are not crosslinked or degrade at temperatures below T_m. This area has been much-studied because polymers exhibit such interesting, unusual, and difficult-to-describe (at least from the standpoint of traditional materials) deformation behavior. The simple and traditional linear engineering models, Newton's law (for flow) and Hooke's law (for elasticity), often just are not reasonable approximations for the behavior of polymers. Not only are the elastic and viscous properties of polymer melts and solutions usually nonlinear but they also exhibit a combination of viscous and elastic response, the relative magnitudes of which depend on the temperature and the time scale of the experiment. Thus, we have a new term *viscoelasticity*. The viscoelastic response is dramatically illustrated by Silly Putty (a silicone polymer). When bounced (stressed rapidly), it is highly elastic, recovering most of the potential energy it had before being dropped. If stuck on the wall (stressed over a long time period), however, it will slowly flow down the wall, albeit with a high viscosity, and will show little tendency to recover any deformation. We limit ourselves here to one- and two-dimensional deformations. A detailed three-dimensional treatment of rheology is beyond the scope of this book. Several excellent resources are available [1–8].

Fundamental Principles of Polymeric Materials, Third Edition. Christopher S. Brazel and Stephen L. Rosen.
© 2012 John Wiley & Sons, Inc. Published 2012 by John Wiley & Sons, Inc.

14.2   BASIC DEFINITIONS

We begin our treatment of rheology with a discussion of *purely viscous flow*. For our purposes, this will be defined as a deformation process in which all the applied mechanical energy is nonrecoverably dissipated as heat in the material through molecular friction (i.e., no elastic response at all). This process is known as *viscous energy dissipation*. Purely viscous flow is in most cases a good approximation for dilute polymer solutions and often for concentrated solutions and melts, where the stress on the material is not changing too rapidly, that is, where steady-state flow is approached.

The *viscosity* of a material expresses its resistance to flow (compare a low viscosity fluid, such as water, with higher viscosity fluids, such as most oils, honey, or polymer melts). It is defined quantitatively in terms of two basic parameters: the *shear stress* τ and the *shear rate* $\dot{\gamma}$ (more correctly referred to as the rate of shear straining). Polymers can really complicate analysis of flow due to the high molecular weights and entanglements between (or even within) polymer chains. Shear stress and shear rate are defined in Figure 14.1 Consider a point in a laminar flow field (Figure 14.1). A rectangular coordinate system is established with the x axis (sometimes designated the one coordinate direction) in the direction of flow and the y axis (the two direction) perpendicular to surfaces of constant fluid velocity, that is, parallel to the *velocity gradient* (the velocity at position y is u, and due to the shear force F placed on the Cartesian element in Figure 14.1, the velocity at position $y + dy$ is $u + du$). This causes the material to deform. The z axis (three or neutral direction) is mutually perpendicular to the others. This is known as *simple shearing* flow. It is one example of a *viscometric flow*, a flow field in which the velocity and its gradient are everywhere perpendicular, with a third neutral direction mutually perpendicular to the others (see Bird *et al.* [2] for a more rigorous definition). Other examples of viscometric

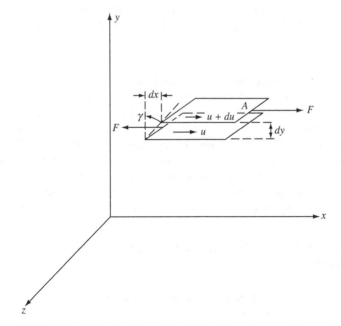

FIGURE 14.1   Cartesian element to define shear stress and shear rate.

flows are given later in this chapter and in Chapter 15. Viscometric flows can often be treated analytically. Fortunately, many laminar-flow situations encountered in practice either are viscometric flows or at least can be reasonably approximated by them.

As a fluid moves over a stationary surface, the friction at the surface slows down the flow so that the linear velocity of the fluid changes with distance from the surface. Thus, if we assume a fluid layer at a position y away from a surface moves with a velocity $u = dx/dt$ in the x direction, the layer at $y + dy$ has a velocity of $u + du$. The displacement gradient, dx/dy, is known as the shear strain, and is given by the symbol $\dot{\gamma}$:

$$\gamma = \frac{dx}{dy} = \text{shear strain} \quad \text{(dimensionless)} \tag{14.1}$$

The *time rate of change* of shear strain, $\dot{\gamma}$ (or $d\gamma/dt$, the dot is Newton's notation for the time derivative), is the so-called *shear rate*. Since the order in which the mixed second derivative is taken is immaterial (note that by sticking to two dimensions, we can write total rather than partial derivatives), then

$$\dot{\gamma} = \frac{d}{dt}(\gamma) = \frac{d}{dt}\left(\frac{dx}{dy}\right) = \frac{d}{dy}\left(\frac{dx}{dt}\right) = \frac{du}{dy}(\text{time}^{-1}) \tag{14.2}$$

Thus, in simple shearing flow, the shear rate is identical to the velocity gradient, du/dy. (This is not the case for all viscometric flows.)

The shear stress is the force (in the direction of flow) per unit area normal to the y axis:

$$\tau_{yx} = \frac{F(\text{in } x \text{ direction})}{A(\text{normal to } y \text{ direction})}\left(\frac{\text{force}}{\text{length}^2}\right) \tag{14.3}$$

The subscript yx will henceforth be dropped unless specifically needed.

Earlier in determining the viscosity-average molecular weights, we introduced the viscosity, η. Viscosity is actually defined as the ratio of shear stress to shear rate:

$$\eta \equiv \tau/\dot{\gamma} \tag{14.4}$$

Traditionally, most rheological work has been done in the CGS system, with force in dynes, mass in grams, length in centimeters, and time in seconds. In this system, the unit of viscosity is dyne-s/cm^2, or *poise* (P); centipoise is also commonly used since the viscosity of water is around 1 cp (for comparison, many polymer melts have viscosities of 10,000 P or higher). Since the early days, SI units have become *de rigueur*, with force in Newtons (N) and length in meters. SI unit for viscosities is in N·s/m^2 = Pa·s (1 Pa·s = 10 P = 1000 cp). Equations here will be written with either of these systems in mind. When using the English system of pound force, pound mass, feet, and seconds, each stress or pressure as written here must be multiplied by the dimensional constant $g_c = 32.2$ ft·lb$_m$/lb$_f$s^2, and viscosities are in lb$_m$/ft·s.

14.3 RELATIONS BETWEEN SHEAR FORCE AND SHEAR RATE: FLOW CURVES

When most materials are subjected to a constant shear rate, $\dot{\gamma}$ at a fixed temperature, a corresponding steady-state value of shear stress, τ, is soon established. A constant shear stress can also be applied, with the steady-state shear rate found based on the input τ. The

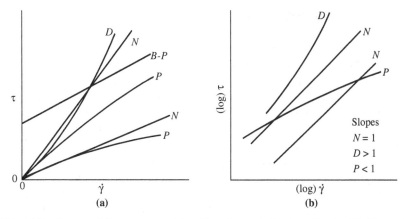

FIGURE 14.2 Types of flow curves: (a) arithmetic and (b) logarithmic. N, Newtonian; P, pseudoplastic; B-P, Bingham plastic (infinitely pseudoplastic); D, dilatant.

steady-state relation between shear stress and shear rate at constant temperature is known as a *flow curve*.

Newton's "law" of viscosity states that the shear stress is linearly proportional to the shear rate, with the proportionality constant being the viscosity, η:

$$\tau = \eta\dot{\gamma} \tag{14.5a}$$

Fluids that follow this hypothesis are termed Newtonian. The hypothesis holds quite well for many nonpolymer fluids, such as gases and water, and solvents, such as toluene. This type of flow behavior would be expected for small, relatively symmetrical molecules, where the structure and/or orientation do not change with the intensity of shearing (particularly those fluids that do not get entangled in one another, as polymers do).

An arithmetic flow curve (τ versus $\dot{\gamma}$) for a Newtonian fluid is a straight line through the origin with a slope η (Figure 14.2a). Because τ and $\dot{\gamma}$ often cover very wide ranges, it is usually preferable to plot them on log–log coordinates. Taking logarithms of both sides of Equation 14.5a yields

$$\log \tau = \log \eta + 1\log \dot{\gamma} \tag{14.5b}$$

Hence, a log–log plot of τ versus $\dot{\gamma}$, a logarithmic flow curve, will be a line of slope one for a Newtonian fluid, with the viscosity found by the $\log^{-1}$ of the y-axis intercept (Figure 14.2b).

Unfortunately, many fluids do not obey Newton's hypothesis. Both *dilatant* (shear-thickening) and *pseudoplastic* (shear-thinning) fluids have been observed (Figure 14.2). On log–log coordinates, dilatant flow curves have a slope greater than 1 and pseudoplastics have a slope less than 1. Dilatant behavior is somewhat uncommon but has been reported for certain slurries and implies an increased resistance to flow with intensified shearing. *Polymer melts and solutions are invariably pseudoplastic*, that is, their resistance to flow decreases with the intensity of shearing.

For non-Newtonian fluids, since τ is not directly proportional to $\dot{\gamma}$, the viscosity is not constant with the shear rate. Plots (or equations) giving η as a function of $\dot{\gamma}$ (or τ) are an equivalent method of representing a material's steady-state viscous shearing properties. Knowledge of the relation between any two of the three variables (τ, η, and $\dot{\gamma}$) completely defines the steady-state viscous shearing behavior, since they are related by Equation 14.4.

14.4 TIME-DEPENDENT FLOW BEHAVIOR

The types of non-Newtonian flow just described, although shear dependent, are time independent. As long as a constant shear rate is maintained, the same shear stress or viscosity will be observed at steady state. Some fluids, however, exhibit *reversible* time-dependent properties. When sheared at a constant rate or stress, the viscosity of a *thixotropic* fluid will decrease over a period of time (Figure 14.3), implying a progressive breakdown of structure. If the shearing is stopped for a while, the structure reforms, and the experiment may be duplicated. A classic example is ketchup (which contains natural polymers as thickening agents); it splashes all over after a period of vigorous tapping. Thixotropic behavior is important in the paint industry, where smooth, even application with brush or roller is required, but it is desirable for the paint on the surface to "set up" to avoid drips and runs after application. Thixotropic behavior is fairly common in polymer solutions and melts and is primarily due to chains disentangling and becoming aligned in the direction of flow over time, making it a bit easier to flow (lower viscosity). The opposite sort of behavior is manifested by *rheopectic* fluids, for example, certain drilling muds used by the petroleum industry. When subjected to continuously increasing and then decreasing shearing, time-dependent fluids give flow curves as in Figure 14.4.

Once in a while, polymer systems will falsely appear to be thixotropic or rheopectic. Careful checking (including before and after molecular weight determinations) invariably shows that the phenomenon is not reversible and is due to degradation or crosslinking of the polymer when in the viscometer for long periods of time, particularly at elevated temperatures. Other transient time-dependent effects in polymers are due to elasticity, and will be considered later, but for chemically stable polymer melts or solutions, the *steady-state* viscous properties are time independent. We treat only such systems from here on.

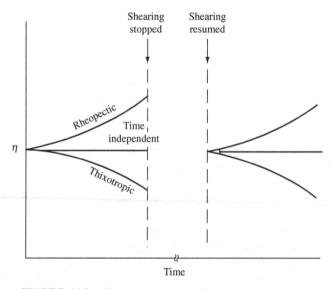

FIGURE 14.3   Viscosity behavior of time-dependent fluids.

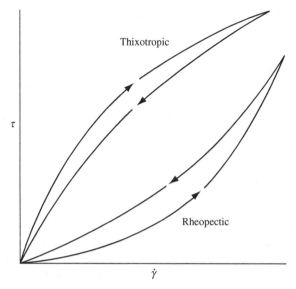

FIGURE 14.4   Flow curves for time-dependent fluids under continuously increasing and then decreasing shear.

14.5   POLYMER MELTS AND SOLUTIONS

When the flow properties of polymer melts and solutions can be measured over a wide enough range of shearing, the logarithmic flow curves appear as in Figure 14.5. It is generally observed that:

1. at low shear rates (or low stresses), a "lower Newtonian" region is reached with a so-called *zero-shear viscosity*, η_o,

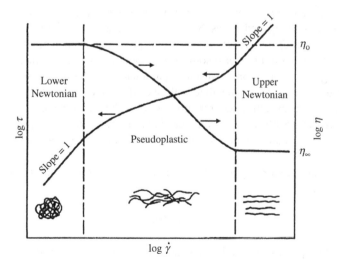

FIGURE 14.5   Generalized flow properties of polymer melts and solutions.

2. over several decades of intermediate shear stress, the material is pseudoplastic, and

3. at very high shear rates, an "upper Newtonian" region, with viscosity η_∞ is attained.

This behavior can be rationalized in terms of molecular structure. At low shear, the randomizing effect of the thermal motion of the chain segments overcomes any tendency toward molecular alignment in the shear field. The molecules are thus in their most random and highly entangled state, and have their greatest resistance to slippage (flow). As the shear is increased, the molecules will begin to untangle and align in the shear field, reducing their resistance to slippage past one another. Under severe shearing, they will be pretty much completely untangled and aligned, and will reach a state of minimum resistance to flow. This is illustrated schematically in Figure 14.5.

Intense shearing eventually leads to extensive breakage of main-chain bonds, that is, mechanical degradation. Furthermore, differentiation of Equation 13.3 with respect to time reveals that the rate of viscous energy dissipation per unit volume is equal to $\tau\dot{\gamma}$ It thus becomes exceedingly difficult to maintain the temperature constant under intense shearing, therefore good data that illustrate the upper-Newtonian region are relatively rare, particularly for polymer melts.

But what happens to the highly oriented molecules when the shear is removed? The randomizing effect of thermal energy tends to return them to their low-shear configurations, giving rise to an elastic retardation.

Some actual flow curves for polymer melts are shown in Figure 14.6 [9]. The data cover only a portion of the general range described above, because very few instruments are capable of obtaining data over the entire range. The polyisobutylene data cover the transition from lower Newtonian to pseudoplastic regions but the polyethylene data are confined to the pseudoplastic region. No trace of the upper Newtonian region is seen for either material at the shear rates investigated.

14.6 QUANTITATIVE REPRESENTATION OF FLOW BEHAVIOR

To handle non-Newtonian flow analytically, it is desirable to have a mathematical expression relating τ, η, and $\dot{\gamma}$, as Newton's law does for Newtonian fluids. A wide variety of such *constitutive relations* has been proposed, both theoretical and empirical [2,5,10]. All appear to fit at least some experimental data over a limited range of shear rates, but, in general, the more adjustable parameters in the equation, the better fit it provides. (There is an old saying that with six adjustable constants, you can draw an elephant, and with a seventh, make his trunk wave.) The mathematical complexity of the equations increases greatly with the number of parameters, soon outstripping the available data to establish the parameters and making the equations impractical for engineering calculations.

The traditional engineering model for purely viscous non-Newtonian flow is the so-called *power law*:

$$\tau = K(\dot{\gamma})^n \tag{14.6a}$$

This is a two-parameter model, the adjustable parameters being the consistency K and the flow index n. As written above, the dimensions of K depend on the

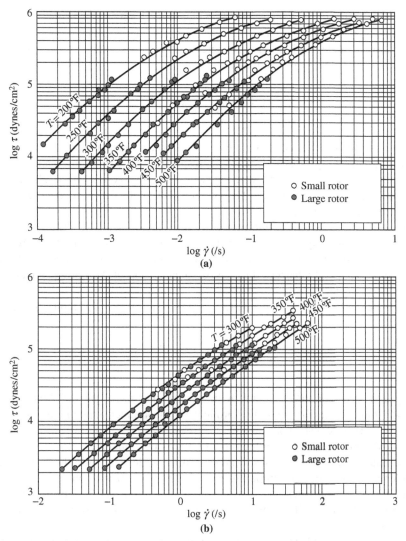

FIGURE 14.6   Flow curves for polymer melts [9]: (a) L-80 polyisobutylene; and (b) low-density polyethylene.

magnitude of n (dimensionless), so if you are a dimensional purist, the power law may be written as

$$\tau = K|\dot{\gamma}|^{n-1}\dot{\gamma} \qquad (14.6b)$$

This way, K has the usual viscosity units.

On $\log \tau$ versus $\log \dot{\gamma}$ coordinates, a power-law fluid is represented by a straight line with slope n. Thus, for $n = 1$, it reduces to Newton's law; for $n < 1$, the fluid is pseudoplastic, and for $n > 1$, the fluid is dilatant. The power law can reasonably approximate only portions of actual flow curves over one or two decades of shear rate (see Figure 14.6), but it does so with fair mathematical simplicity and has been adequate for many engineering purposes. Many useful relations have been obtained simply by replacing Newton's law with the power law in the usual fluid-dynamic equations.

The Carreau model [2] is a four-parameter model that can represent the features of the general flow curve in Figure 14.5. Here,

$$\frac{\eta - \eta_\infty}{\eta_0 - \eta_\infty} = [1 + (\lambda_c \dot{\gamma})^2]^{(n-1)/2} \tag{14.7}$$

The parameter λ_c is a time constant or characteristic time, and η_0 and η_∞ represent the viscosity, η, when $(\lambda_c \dot{\gamma})^2 \ll 1$ and $(\lambda_c \dot{\gamma})^2 \gg 1$, respectively. For mid-range values of $(\lambda_c \dot{\gamma})^2$, the equation generates a power-law region with a log–log slope of $n-1$. This model has been quite successful in fitting data for polymer melts and solutions over at least three or four decades of shear rate.

Not long ago, an equation with four constants would have been deemed excessively complex for engineering calculations, but the computer does not mind a bit. Actually, in many practical applications, the shear rates don not get high enough to approach the upper-Newtonian region, and a truncated (three-parameter) form of the Carreau equation, with $\eta_\infty = 0$, is adequate.

The (much) "modified Cross" model [11] has seen increasing application in recent years:

$$\eta(T, \dot{\gamma}) = \frac{\eta_0(T)}{1 + [C\eta_0(T)\dot{\gamma}]^{1-n}} \tag{14.8}$$

This is a three-parameter model, with the constant C having the dimensions of a reciprocal modulus (area/force). The product $C\eta_0$ has dimensions of time and may be thought of as a time constant or characteristic time. At low shear rates, $\eta \to \eta_0$, and at high shear rates, it gives a power-law region with a log–log slope of $n-1$.

The great virtue of this model is that unlike other models, it explicitly incorporates the dependence of viscosity on temperature as well as shear rate through the temperature dependence of the zero-shear viscosity (to be discussed shortly). Equation 14.8 is written to emphasize that point. This makes it particularly well suited for nonisothermal flow calculations.

As will be illustrated later, certain important calculations (e.g., the determination of velocity profiles and flow rates in tubes) are facilitated by flow equations that are explicit in τ, that is, have the form $\dot{\gamma} = \dot{\gamma}(\tau)$ or $\eta = \eta(\tau)$. The power law can be written in either form. Some of the equations not given here are in or can be put in one of these forms. Unfortunately, they tend not to fit data as well as the Carreau and modified Cross equations, neither of which can be made explicit in τ. This has led to the practice of fitting flow data with polynomials such as

$$\log \dot{\gamma} = a_0 + a_1 (\log \tau)^1 + a_2 (\log \tau)^2 + \cdots \tag{14.9a}$$

or

$$\log \eta = b_0 + b_1 (\log \tau)^1 + b_2 (\log \tau)^2 + \cdots \tag{14.9b}$$

where a_i and b_i are functions of temperature.

In principle, this is applicable and can give good fits, but in practice, two cautions are in order. First, it is easy to get carried away and include more terms in the equation than are justified by the data. This leads to the fitting of experimental scatter. Second, great care must be exercised when extrapolating the equations beyond the range of data used to determine them. Uncritical extrapolation can lead to artifacts such as shear stresses that *decrease* with increasing shear rate and negative viscosities, that violate both common sense and the second law of thermodynamics. These problems are less likely to arise when an established flow equation is used.

14.7   TEMPERATURE DEPENDENCE OF FLOW PROPERTIES

If you consider the number of processes that use molten polymers (e.g., injection-molded CDs), it is obvious that the temperature dependence of flow properties must be understood. The temperature dependence of the *zero-shear* viscosity can often be represented by the relation:

$$\eta_0(T) = Ae^{E/RT} \tag{14.10}$$

where E is the activation energy for viscous flow.

The η_o's from Ref. 12 are plotted according to Equation 14.10 in Figure 14.7. The fit is quite good. The slope is $E/2.303R$, from which $E = 11.4\,\text{kcal/mol}$. Thus, the combination of Equations 14.10 and 14.8 does a pretty good job of describing both the shear rate and the temperature dependence of the viscosity, at least for these data.

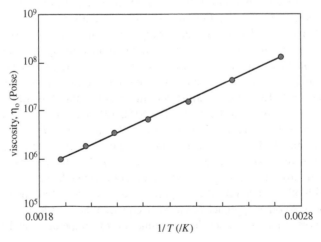

FIGURE 14.7 Temperature dependence of zero-shear viscosity. Zero-shear viscosities from modified Cross fits to the L-80 polyisobutylene data from Figure 14.6a [9] and plotted according to Equation 14.10.

Example 14.1 What must be the temperature to reduce the $100\,^\circ$C zero-shear viscosity of L-80 PIB by an order of magntiude?

Solution. The temperature dependence of zero-shear viscosity is given by Equation 14.11. Therefore, between temperatures T_1 and T_2,

$$\frac{\eta_{02}}{\eta_{01}} = \frac{\exp(E/RT_2)}{\exp(E/RT_1)} = \exp\left[\frac{E}{R}\left(\frac{1}{T_2} - \frac{1}{T_1}\right)\right]$$

Here, $T_1 = 100\,^\circ$C $= 373$ K and $\eta_{o2}/\eta_{o1} = 0.1$. With $R = 1.99$ cal/mol·K and the activation energy given above, solving for T_2 gives

$$T_2 = 439\,\text{K} = 166\,^\circ\text{C}$$

Temperature is therefore an effective means of controlling melt viscosity in processing operations, but two drawbacks must be kept in mind: (1) it takes time and costs money to put in and take out thermal energy, and (2) excessive temperature can lead to degradation of the polymer.

The famous Williams–Landell–Ferry (WLF) equation [13] is useful for describing the temperature dependence of several linear mechanical properties of polymers (see Chapter 16). For the zero-shear viscosity, it may be written as

$$\log_{10} \frac{\eta_0(T)}{\eta_0(T^*)} = \frac{-C_1(T - T^*)}{C_2 + (T - T^*)} \tag{14.11}$$

where T^* is a reference temperature. If the reference temperature is chosen as the glass transition temperature, $T^* = T_g$, the "universal" constants $C_1 = 17.44$ and $C_2 = 51.6$ (with T's in K) give a rough fit for a wide variety of polymers. The WLF equation is most useful in this form because T_g is extensively tabulated for many polymers [14]. Better fits can be obtained by using constants specific to the polymer, but these constants are less readily available. It has also been suggested that the fit can be improved by using $C_1 = 8.86$ and $C_2 = 101.6$, with T^* adjusted to fit specific data, if available. When this is done, T^* generally turns out to be $T_g + (50 \pm 5)\,^\circ$C [2]. Whichever constants are used, application of the WLF equation should be limited to the range $T_g < T < T_g + 100\,^\circ$C. The WLF Equation 14.11 can also be used in conjunction with Equation 14.8. However, Equation 14.10 often gives a better fit of η_o versus T data than Equation 14.11.

The modified Cross equation gives the temperature dependence of viscosity at finite shear rates. With other equations, the temperature dependence of all the constants would have to be known. Another approach has used an equation that is independent of any model for the flow curve. Since the viscosity is a function of temperature and shear stress or shear rate:

$$\eta = f(\tau, T) \quad \text{or} \quad \eta = f'(\dot{\gamma}, T)$$

By analogy to Equation 14.10, these functions are approximated by

$$\eta = B \exp\left(\frac{E_\tau}{RT}\right) \quad \text{(at constant } \tau) \tag{14.12a}$$

$$\eta = C \exp\left(\frac{E_{\dot\gamma}}{RT}\right) \quad \text{(at constant } \dot\gamma) \tag{14.12b}$$

where E_τ is the activation energy for flow at constant shear stress and $E\dot\gamma$ is the activation energy for flow at constant shear rate. Figure 14.8 shows plots of log $\dot\gamma$ versus $1/T$ at constant τ for L-80 polyisobutylene and low-density polyethylene (taking data for a constant τ from Figure 14.6) [9]. Since each of these data sets can fit to a straight line, the shear rate, $\dot\gamma$ follows Arrhenius behavior for PIB and PE. Since $\dot\gamma = \tau/\eta$, the slope of these lines is $E_\tau/2.303R$.

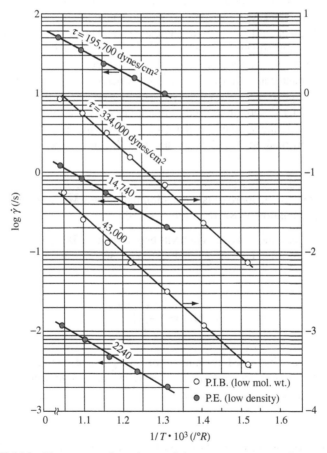

FIGURE 14.8   Temperature dependence of shear rate at constant shear stress [9].

It will be left as an end-of-chapter exercise to show that $E_\tau > E\dot{\gamma}$. In the limit $\tau \to 0$, $\dot{\gamma} \to 0$ for all Newtonian fluids, $E_\tau = E\dot{\gamma} = E$ (see note[1]).

14.8 INFLUENCE OF MOLECULAR WEIGHT ON FLOW PROPERTIES

It should be of no surprise that the larger the polymer, the more viscous the solution (after all, that is one way we used to estimate molecular weights in Chapter 5). It has long been known that the molecular weight has a strong influence on both the melt and the solution viscosities of polymers. Experiments show that the viscosity can be scaled to the weight-average molecular weights through

$$\eta_0 \propto \overline{M}_w^1 \quad \text{for } \overline{M}_w < \overline{M}_{wc} \tag{14.13a}$$

$$\eta_0 \propto \overline{M}_w^{3.4} \quad \text{for } \overline{M}_w > \overline{M}_{wc} \tag{14.13b}$$

where $\overline{M}_{wc}$ is a critical average molecular weight, thought to be the point at which molecular entanglements begin to dominate the rate of slippage of the molecules. It depends on the temperature and polymer type, but most commercial polymers are well above $\overline{M}_{wc}$.

Equation 14.13a quantitatively holds for just about all polymer melts. The addition of a low molecular weight solvent, of course, cuts down the entanglements and raises $\overline{M}_{wc}$. Equation 7.15 suggests that entanglements set in when the dimensionless product of intrinsic viscosity and concentration (the Berry number) exceeds one, $[\eta]c > 1$. Thus, if you know $[\eta]$ for your polymer in the solvent, you can get an idea of the concentration above which Equation 14.13b should be used (and the viscosity rises much more sharply!). In practice, even moderately concentrated (say 25% or more) polymer solutions have viscosities proportional to $\overline{M}_w^{3.4}$, provided that $\overline{M}_w$ is in the range of commercial importance.

Example 14.2 Obtain an approximate equation that relates $\overline{M}_{wc}$ to concentration and the MHS constants (Equation 5.18) for polymer solutions.

Solution. First, we assume that entanglements begin when the Berry number, $[\eta]c$, equals 1. Because all we are after here is a rough estimate, we can assume that $\overline{M}_w = \overline{M}_v$ (see Example 5.4). Thus,

$$\overline{M}_{wc} \approx \left(\frac{1}{Kc}\right)^{1/a}$$

As the shear rate is increased, the number of entanglements between chains is reduced, and as expected, the dependence of viscosity on molecular weight decreases (Figure 14.9).

---

[1] *Note:* The variable E in this chapter refers to the activation energy for flow (which is important when studying rheology). However, the modulus of elasticity, which is a measure of mechanical strength, is also represented by E, as first introduced in Chapter 13.

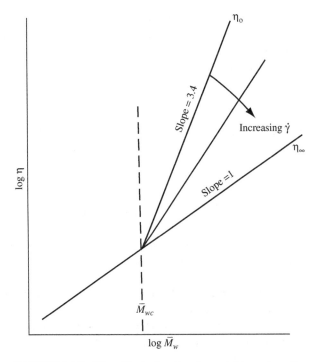

FIGURE 14.9   The effect of molecular weight on viscosity.

Example 14.3 By what percentage must $\overline{M}_w$ be changed to cut the zero-shear viscosity, η_o, in half?

Solution. From Equation 14.13*b*,

$$\frac{\eta_{02}}{\eta_{01}} = \left(\frac{\overline{M}_{w2}}{\overline{M}_{w1}}\right)^{3.4}$$

$$\frac{\overline{M}_{w2}}{\overline{M}_{w1}} = \left(\frac{\eta_{02}}{\eta_{01}}\right)^{1/3.4} = (1/2)^{1/3.4} = 0.816$$

$$\% \text{ change} = 100\left(\frac{\overline{M}_{w2} - \overline{M}_{w1}}{\overline{M}_{w1}}\right) = 100\left(\frac{\overline{M}_{w2}}{\overline{M}_{w1}} - 1\right) = -18.4\%$$

Thus, it takes only an 18% decrease in chain length to halve the zero-shear melt viscosity. Although things will not be quite so dramatic at high shear rates, this illustrates the importance of controlling molecular weight during synthesis to achieve a product with the desired processing properties.

14.9   THE EFFECTS OF PRESSURE ON VISCOSITY

For all fluids, viscosity increases with increasing pressure, as the free volume and, hence, the ease of molecular slippage is decreased. With liquids, including polymer melts, because of their relative incompressibility, the effect becomes noticeable only at fairly high pressures (hundreds of atmospheres). Nevertheless, such pressures are reached in certain processes (e.g., injection molding) and in certain types of capillary viscometers (later in this chapter). Available data have been fit by

$$\eta_0(P) = \alpha e^{\beta P} \tag{14.14}$$

where α and β are constants. Equation 14.14 may be combined with Equation 14.10 to give an equation that expresses both the temperature and pressure dependence of zero-shear viscosity:

$$\eta_0(T, P) = D \exp\left(\frac{E}{RT} + \beta P\right) \tag{14.15}$$

Presumably, Equation 14.15 could be used in conjunction with Equation 14.8 to express viscosity as a function of $\dot{\gamma}$, T, and P, but for pressure, there is as yet little experimental justification for this.

Carley [15] has critically reviewed work on polymer melts and concludes that pressure effects are of minor significance in most processing situations, provided the temperature is not too close to a transition. High pressures raise both T_g and T_m slightly, and, of course, the viscosity shoots up tremendously as either is approached.

14.10 VISCOUS ENERGY DISSIPATION

Regardless of the viscometric technique, determination of the true, isothermal flow properties at high shear rates can be complicated by high rates of viscous energy dissipation, which make them hard to maintain isothermal conditions. If we divide both sides of Equation 13.3 by $V\, dt$, we see that the *rate of viscous energy dissipation per unit volume* $\dot{E}$ is

$$\dot{E} = \tau \dot{\gamma} = \eta \dot{\gamma}^2 \tag{14.16}$$

Example 14.4 Obtain an expression for the adiabatic rate of temperature rise in a polymer sample subjected to a shear stress τ and shear rate $\dot{\gamma}$.

Solution. In the absence of heat transfer, the rate of temperature rise $\dot{T}$ is given by the rate of energy dissipation per unit volume divided by the volumetric heat capacity, ρc_p, as follows:

$$\dot{T} = \frac{\tau \dot{\gamma}}{\rho C_p} \tag{14.17}$$

Not only is temperature rise an important consideration in viscometry, where great care must be taken in the design of viscometers to permit adequate temperature regulation, but it must also be taken into account in the design of process equipment. In the steady-state operation of extruders, for example, virtually all of the energy required to melt and maintain the polymer in the molten state is supplied by the mechanical drive. Here, however, we will limit our considerations to isothermal flows.

14.11 POISEUILLE FLOW

Axial, laminar (Poiseuille) flow in a tube of cylindrical cross section is another example of a viscometric flow field. Here, geometry dictates the use of cylindrical coordinates. Fluid motion is in the x ("1") direction along the tube axis, the velocity gradient is directed everywhere in the outward radial r (or "2") direction, and the mutually perpendicular or neutral direction is the tangential θ (or "3") coordinate. Consider a cylindrical fluid element of radius r and length dx (Figure 14.10). Assume that the pressure in the fluid is a function of the distance along the tube (x coordinate) only. The net pressure force pushing the element in the x direction is the differential pressure drop across the element, $-dP$, times the area of the element's ends, πr^2. The motion of the element is resisted by a shear force at its surface, which is the product of the shear stress, τ_{rx}, and the surface area of the cylinder, $2\pi r\, dx$. In the steady-state flow, these forces must balance, so

$$\underset{\text{Surface shear force}}{2\pi r\, dx\, \tau_{rx}} = \underset{\text{Net pressure force}}{-\pi r^2 dP} \tag{14.18}$$

Thus, the dependence of shear stress on radius (denoted $\tau(r)$) is

$$\tau_{rx}(r) = -\left(\frac{r}{2}\right)\left(\frac{dP}{dx}\right) \tag{14.19}$$

For a tube with inner radius R (dropping the rx subscript on shear stress),

$$\tau_w = -\left(\frac{R}{2}\right)\left(\frac{dP}{dx}\right) \tag{14.20}$$

where τ_w is the shear stress at the tube wall ($r = R$) and

$$\tau(r) = \frac{r}{R}\tau_w \tag{14.21}$$

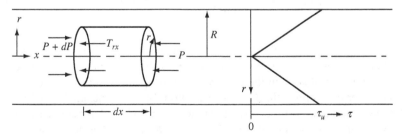

FIGURE 14.10   Force balance on an element in a cylindrical tube.

Therefore, the shear stress varies linearly with radius from zero at the tube center to a maximum of $\tau_w = -(R/2)\,(dP/dx)$ at the tube wall. *Note that this result does not depend in any way on the fluid properties.*

Since the axial fluid velocity u is a function of the radial position only (denoted $u(r)$), we may write

$$du(r) = \frac{du}{dr}\,dr \tag{14.22}$$

Integrating with the boundary conditions $u(R) = 0$ (i.e., the fluid sticks to the wall) and u at radius $r = u(r)$ gives

$$u(r) = \int_0^{u(r)} du(r) = \int_R^r \frac{du}{dr}\,dr \tag{14.23}$$

Realizing that the velocity gradient (du/dr) is the shear rate $\dot{\gamma}$ (just replace y with r in Eq. 14.2), we find that

$$u(r) = \int_R^r \dot{\gamma}[\tau(r)]\,dr = \int_R^r \frac{\tau(r)}{\eta[\tau(r)]}\,dr \tag{14.24}$$

where $\tau(r)$ is given by Equation 14.19. The functions $\dot{\gamma}(\tau)$ or $\eta(\tau)$ represent the material's steady-state viscous properties or flow curve, as discussed earlier. So, for a given pressure gradient, the velocity profile in the tube may always be calculated from the flow curve by

1. choosing an r and calculating τ from Equation 14.19,
2. obtaining $\dot{\gamma}$ (or η) at the τ above from the flow curve,
3. representing numerically or graphically the relation $\dot{\gamma}\,(r)$ (or $\tau(r)/\eta(r)$), and
4. integrating from R to r, which gives u at r.

If an analytic representation of the flow curve is available, Equation 14.24 may be integrated directly (provided the representation is simple enough). For example (verification will be left as an exercise for the reader, as they say in the math books), for a power-law fluid $\dot{\gamma}\,(\tau) = (\tau/K)^{1/n}$ and

$$u(r) = \frac{[(-dP/dx)/2K]^{1/n}}{(1/n) + 1}\left[R^{(1/n)+1} - r^{(1/n)+1}\right] \tag{14.25}$$

Equation 14.25 reduces to the familiar Newtonian parabolic profile for $n = 1$.

Note that the above calculations are facilitated if the flow curve can be written as an explicit function of τ, that is, in the form $\dot{\gamma} = \dot{\gamma}(\tau)$ or $\eta = \eta\,(\tau)$. Although the power law meets that criterion, neither the Carreau Equation 14.7 nor the modified Cross Equation 14.8 meets the same. For this reason, it is sometimes viewed as being easier to fit data with a polynomial such as Equations 14.9a or 14.9b, rather than to use an established constitutive equation to make these calculations (but note the cautions!).

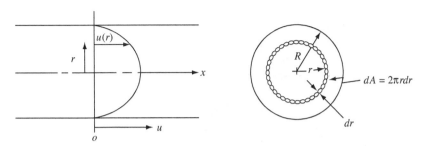

FIGURE 14.11   Determination of volumetric flow rate.

Looking at a differential ring of the tube cross section, with thickness dr, located at radius r, where the velocity is $u(r)$ (Figure 14.11), the differential volumetric flow rate is:

$$dQ = u(r)2\pi r \, dr \tag{14.26}$$

where $u(r)$ is the local velocity and $2\pi r \, dr$ is the area of the differential ring. Integrating over the tube cross section gives

$$Q = 2\pi \int_0^R u(r)r \, dr \tag{14.27}$$

Therefore, knowing the velocity profile allows calculation of the volumetric throughput Q. For a power-law fluid, insertion of Equation 14.25 into Equation 14.27 and turning the crank gives

$$Q = \left(\frac{-dP/dx}{2K}\right)^{1/n} \left[\frac{\pi}{(1/n)+3}\right] R^{(1/n)+3} \tag{14.28}$$

The average velocity V in the tube is defined by

$$V \equiv \frac{Q}{\pi R^2} \tag{14.29}$$

Velocity profiles for power-law fluids in a tube are plotted in a dimensionless form in Figure 14.12 for several values of n. The greater the degree of pseudoplasticity (i.e., the lower n), the flatter the profile becomes. For a Netwonian fluid, where $n = 1$, the profile is parabolic.

Further development of flow curves for polymer melts are discussed in more advanced rheology texts. Compared with many regular fluids, where simplifying assumptions can be used in the analysis of flow behavior, the significant of viscous dissipation for polymers must be considered as must entrance effects, where the polymer melt first enters a tube. However, successful models have been developed that even allow scale-up for commercially relevant processes.

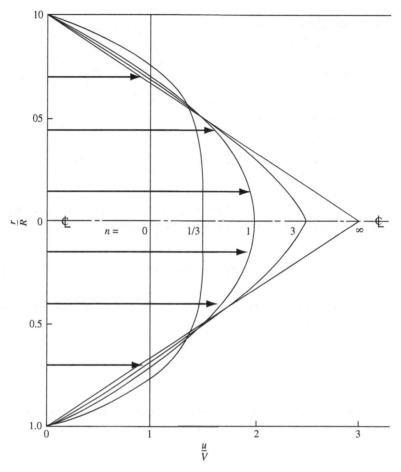

FIGURE 14.12 Velocity profiles for the laminar flow of power-law fluids in tubes.

14.12 TURBULENT FLOW

Due to the viscosity of polymer melts (and most solutions), laminar flow is all that you would expect to see (recall that viscosity is in the denominator of the Reynolds number (Re) and that large Re is required for turbulent flow). However, turbulent flow can be encountered in dilute polymer solutions. As with the turbulent flow of Newtonian fluids, pressure drops are conveniently handled in terms of the Fanning friction factor:

$$f \equiv \frac{Rg_c}{\rho V^2}\left(\frac{dP}{dx}\right) \tag{14.30}$$

(The dimensional constant $g_c = 32.2 \, \text{ft·lb}_m/\text{lb}_f\text{·s}^2$ is included here because these equations are still sometimes used with the English engineering system of units; in SI units,

$g_c = 1$ kg·m/N·s^2.) Since $\tau_w = (R/2)(dP/dx)$ (the minus sign has been dropped, it being understood that flow is in the direction of decreasing pressure), then

$$f = \frac{2\tau_w g_c}{\rho V^2} \tag{14.31}$$

Equations 14.30 and 14.31 apply to all fluids in both laminar and turbulent flow.

The next question is, "how do you define a Reynolds number for a fluid that has a variable viscosity?" Metzner and Reed [16] proposed that the known relation between the friction factor and the Reynolds number for the laminar flow of Newtonian fluids may be applied to the laminar flow of non-Newtonians as well:

$$f \equiv \frac{16}{\text{Re}} \quad \text{(laminar flow of all fluids)} \tag{14.32}$$

Since Equation 14.32 has been defined to apply to all fluids in laminar flow, it may now be used to obtain a *generalized Reynolds number*—applicable to all fluids in both laminar and turbulent flow—by combining it with Equation 14.31 as follows:

$$\text{Re} = \frac{8\rho V^2}{\tau_w g_c} \tag{14.33}$$

For a power-law fluid, $\tau g_c = K(\dot{\gamma})^n$, combined with Equations 14.20, 14.28, and 14.29 give

$$\text{Re} = \frac{8D^n V^{2-n} \rho}{K[6 + (2/n)]^n} \tag{14.34}$$

Equation 14.41 reduces to the familiar Newtonian relation, $\text{Re} = \rho\, V\, D/\eta$, with $\eta = K = K'$, when $n = 1$.

Equations 14.30–14.33 by definition should describe the behavior of all fluids in steady-state laminar viscous flow. They do so, provided that the true steady-state pressure gradient is used in Equation 14.30 [17]. Approximation of the steady-state gradient by $\Delta P/L$ can sometimes cause considerable error.

14.13 DRAG REDUCTION

Having defined the Reynolds number for non-Newtonian fluids, it would be nice to be able to say that one could then use the standard friction factor–Reynolds number correlations for solving turbulent flow calculations. Such is not always the case, however, because of a startling phenomenon known as *drag reduction* that is often observed in the turbulent flow of non-Newtonian fluids (Figure 14.13) [18] Small quantities of a polymeric solute can cut the friction factor significantly at higher Reynolds numbers, despite the fact that the solution viscosity is only slightly greater than that of the solvent. For example, 200 ppm of guar gum (a natural, water-soluble polymer) reduces the friction factor in Figure 14.13 by approximately a factor of five at $\text{Re} = 10^5$, while the viscosity of the solution is 24% greater than that of pure water.

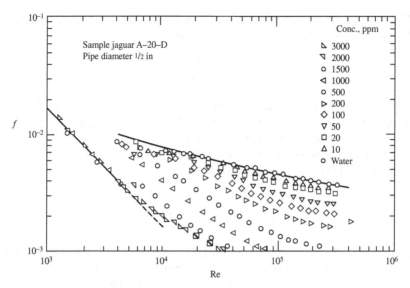

FIGURE 14.13   Turbulent drag reduction as evidenced by a drop in friction factors with the addition of polysaccharides to water [18]. Copyright 1972 by the American Chemical Society. Reprinted with permission of the copyright owner.

The drag reduction phenomenon has important practical applications. Pressure drops, and therefore pumping costs, can be reduced for a given flow rate or the capacity of a pumping system (e.g., the Trans-Alaska pipeline) can be increased by the addition of a drag-reducing solute. Fire departments add poly(ethylene oxide) to their pumpers to increase capacity. Experiments have also been conducted with the aim of increasing ship speed by squirting out a drag-reducing additive at the bow.

It has been suggested that drag reduction may be due to thickening of the laminar sublayer (boundary layer at the liquid/solid interface) caused by the elasticity of the polymer molecules (those polymer solutions that seem relatively inelastic are more likely to be amenable to the usual treatment). It has also been pointed out that most of the energy dissipation in turbulent flows occurs in small, high-frequency eddies. At these frequencies (the reciprocals of which are smaller than the material's relaxation time, giving high Deborah numbers, see Chapter 15), the material responds elastically (such as bouncing the Silly Putty rather than letting it flow down the wall), passing the stored elastic energy from eddy to eddy rather than dissipating it. Very high elongational viscosities of polymer solutions have also been implicated, even though the regular (shear) viscosities of the solutions are not much higher than those of the solvent.

Drag reduction is known to be most pronounced for solutions of high molecular weight, flexible-chain polymers, possibly because these form highly elastic solutions. Its magnitude is markedly reduced as the polymers are physically degraded by continued shear during flow. The effects of polymer molecular variables have already been discussed [19]. On the other hand, drag reduction has also been observed with slurries of rigid particles or fibers in Newtonian fluids. Perhaps, there is more than one cause.

As with Newtonian fluids, it is a pretty safe bet that flow is laminar for $Re < 2100$, but drag-reducing additives often seem to delay the laminar-turbulent transition to higher Re's. The friction factor–Reynolds number relation remains a function of pipe roughness.

Unlike the Newtonian case, though, the f–Re curve seems to depend on the pipe diameter when drag reduction is observed.

Unfortunately, the bottom line is that at present, in the absence of experimental data, there are no quantitative methods for predicting which polymeric solutes will produce drag reduction in a particular solvent, over what range of concentrations drag reduction will be observed, or what its magnitude will be. Fortunately, neglecting drag reductions gives conservative designs; any drag reduction that does occur will be a bonus in terms of increased flow rate and/or reduced pressure drop.

14.14 SUMMARY

This chapter has introduced the field of rheology, where the flow of polymer solutions and melts are studied. Polymers make for complex flow behavior, as their fluids are generally non-Newtonian and their viscosity changes with shear rate. The flow behavior can change with time and is dependent (at least to some degree) on temperature, pressure, and molecular weight. This makes the measurement of the viscosity of polymeric fluids particularly difficult. Methods used to measure viscosity and the rheological behavior of polymers are covered in Chapter 16. While this chapter serves as an introduction to rheology, there are numerous books and entire college courses on this subject.

PROBLEMS

1 The data below were taken from a prehistoric Ph.D. thesis (Rosen, S.L., Cornell University, 1964) recently unearthed by archeologists. They were obtained at 100 °C on a sample of poly(ethyl acrylate) with a viscosity-average molecular weight of 310,000 and a very broad molecular weight distribution.
   Compare the truncated ($\eta_\infty = 0$) form of the Carreau Equation 14.7 and the modified Cross Equation 14.8 for fitting these data.

τ, dyn/cm^2	$\dot{\gamma}$, s^{-1}	τ	$\dot{\gamma}$
3.15×10^3	0.00578	1.36×10^5	1.01
3.70	0.00696	1.67	1.60
4.53	0.00913	2.44	4.15
5.91	0.0112	2.76	5.08
7.34	0.0145	3.22	9.66
1.00×10^4	0.0213	3.59	12.76
1.47	0.0321	4.29	25.4
2.20	0.0548	5.03	50.8
2.72	0.0727	6.59	127
4.00	0.123	7.96	254
5.23	0.189	9.40	550
6.49	0.262	1.16×10^6	1500
7.73	0.358	1.33	3140
9.00	0.463	1.52	6610

2 Assume that the L-80 PIB in Figure 14.6a has $\overline{M}_w = 80{,}000$. Its $\eta_o = 6.67 \times 10^6$ P at 350 °F.

(a) If L-120 PIB has $\overline{M}_w = 120{,}000$, what will be its η_o at 350 °F? *Hint*: Use Equation 14.13b.

(b) In what proportions should L-80 and L-120 PIBs be mixed to produce a material with $\eta_o = 8.0 \times 10^6$ P at 350 °F? *Hint*: See Equation 5.7.

3 Write the power law in the form $\eta = \eta(\tau)$.

4 The following data were obtained by N.D. Sylvester (Ph.D. Thesis, Carnegie Institute of Technology, Pittsburgh, PA, 1968) at room temperature on a 1.10% aqueous solution of Separan® AP-30, a partially hydrolyzed polyacrylamide of molecular weight 2–3 million:

τ, dyn/cm^2	$\dot{\gamma}$, s^{-1}
365	415
434	748
471	929
536	1360
566	1630
616	2110
653	2360
695	2710
748	3400

Fit these data with one of the equations in the chapter. Obtain all the constants in the equation.

5 Some equations give the viscosity as an explicit function of shear stress, $\eta = \eta(\tau)$. Two equations of this form are

$$\eta = A + B\left[1 + \left(\frac{\tau}{C}\right)^2\right]^{-1} \qquad \eta = \frac{D}{1 + E(\tau)^m}$$
$$\text{(I)} \qquad\qquad\qquad \text{(II)}$$

where A, B, C, D, E, and m are constants.

(a) Which of the above models (if either) is capable of showing a zero-shear viscosity? What is the zero-shear viscosity in terms of the constants?

(b) Repeat (a) for an infinite shear viscosity.

(c) Which (if either) can exhibit a power-law region? What is the power-law constant n in this region in terms of the parameters in the equation?

6 Prove that for a pseudoplastic fluid, $E_\tau \geq E_{\dot{\gamma}}$, and that in the low-shear limit and for all Newtonian fluids, $E_\tau = E_{\dot{\gamma}}$ *Hint*: Consider viscosity to be a function of temperature and either shear rate or shear stress, and write an expression for its total differential.

7 What is the relation between E_τ and $E_{\dot\gamma}$ for a power-law fluid?

8 R.A. Stratton (*J. Colloid Interface Sci.* **22**, 517 (1966)) provides τ versus $\dot\gamma$ data for five essentially monodisperse polystyrenes of different molecular weight. Do these data better fit the Carreau or modified Cross equations? Stratton obtains η_o values by extrapolating plots of η versus τ to $\tau = 0$. Do these equations give substantially different values? Stratton's values at 183 °C are as follows:

$\overline{M}_w \times 10^{-5}$	$\eta_0 \times 10^{-4}, \mathrm{P}$
0.48	0.153
1.17	2.88
1.79	12.4
2.17	21.9
2.42	33.5

Do the above data conform to Equation 14.13ab?

9 Another four-parameter model for steady-state viscous flow is

$$\tau = \left(\frac{A + B}{1 + C\tau^m} \right) \dot\gamma$$

where A, B, C, and m are constants.

(a) In terms of model parameters, what is η_o?

(b) In terms of model parameters, what is η_∞?

(c) If the model is to represent a Newtonian fluid, what conclusion(s) can be drawn about the parameters?

10 Equations 14.10 and 14.11 are two different ways of representing the temperature dependence of zero-shear viscosity. For a material that follows the WLF Equation 14.11, obtain an expression for the temperature-dependent activation energy E in Equation 14.10.

11 A polymer is being processed at 500 K. Mechanically, things are fine, but discoloration of the parts indicates that some thermal degradation of the material is occurring during processing. A suggestion is made that the temperature be lowered to 450 K to minimize degradation. You point out that this could cause serious mechanical problems unless a switch was made to a different molecular weight polymer. Given that the polymer has a flow activation energy of 10 kcal/mol and the current material has $\overline{M}_w = 400,000$, what can you say about the $\overline{M}_w$ of a different grade of the same polymer that will allow adequate processing at 450 K?

12 Sometimes even liquids made up of small, symmetrical molecules appear to be pseudoplastic when measurements are made at very high shear rates in certain viscometers. Why?

13 Consider the sample of poly(methyl methacrylate) in Example 5.3. What is the minimum concentration of that polymer in acetone at 30 °C necessary to assure that Equation 14.13b is applicable?

14   A Bingham plastic (if such a thing really exists) has a flow curve $\tau - \tau_y = \eta' \, \dot{\gamma}$, where τ_y is a yield stress that must be exceeded to cause flow. Sketch the $\log \tau$ versus $\log \dot{\gamma}$ and $\log \eta$ versus $\log \dot{\gamma}$ for a Bingham plastic, showing limiting behavior at low and high $\dot{\gamma}$.

15   A power-law fluid is confined between two parallel, flat plates in simple shearing flow. The lower plate is fixed and the upper plate moves with a velocity V. The plates are separated by a distance δ. Calculate and sketch the velocity profiles for $n = 0.5$, $n = 1$, and $n = 1.5$.

16   Injection molding consists of forcing a heated (molten) thermoplastic into a cooled mold. It has been noticed that short shots (material freezing before the mold is filled) can sometimes be cured by forcing the material through a smaller gate (narrower diameter entrance to the mold). In fast, so-called "pin gates" (very small orifices at the mold entrance) seem very effective at this. Why?

17   A general scale-up equation for the laminar flow of a power-law fluid in cylindrical tubes may be written as

$$\left(\frac{\Delta P_2}{\Delta P_1}\right)^a \left(\frac{L_2}{L_1}\right)^b \left(\frac{D_2}{D_1}\right)^c \left(\frac{Q_2}{Q_1}\right)^d = 1$$

if we neglect entrance and exit effects. The subscripts 1 and 2 refer to two different flow situations with the same fluid. Given that exponent $a = 1$, determine exponents b, c, and d.

18   In their new plant, Crud Chemicals must transport 10 gal/min of a room-temperature polygunk solution 1000 ft with a maximum pressure drop of 500 psi. Laboratory data on the solution, obtained in a 0.10-in diameter, 10-in long tube, are summarized by

$$Q = 0.428 \, \Delta P + 0.00481 \, \Delta P^2$$

where ΔP is in psi and Q is in cm^3/min. This equation holds up to the maximum experimental ΔP of 40 psi. (*Note:* The fluid is pseudoplastic, but not power law.) Calculate the minimum pipe diameter needed to do the job and check to see if flow will indeed be laminar.

REFERENCES

[1] Barnes, H.A., J.F. Hutton, and K. Walters, *An Introduction to Rheology*, Elsevier, New York, 1989.

[2] Bird, R.B., R.C. Armstrong, and O.A. Hassager, Dynamics of Polymeric Liquids, Vol. **1**, *Fluid Mechanics*, Wiley, New York, 1977.

[3] Coleman, B.D., H. Markovitz, and W. Noll, *Viscometric Flows of Non-Newtonian Fluids*, Springer, New York, 1966.

[4] Christensen, R.M., *Theory of Viscoelasticity: An Introduction*, Academic, New York, 1971.

[5] Darby, R., *Viscoelastic Fluids*, Dekker, New York, 1976.

[6] Dealy, J.M. and K.F. Wissbrun, *Melt Rheology and Its Role in Plastics Processing*, Van Nostrand Reinhold, New York, 1960.

[7] Middleman, S., *The Flow of High Polymers*, Interscience, New York, 1968.

[8] Tanner, R.I., *Engineering Rheology*, Oxford University Press, New York, 1985.

[9] Best, D.M. and S.L. Rosen, *Polym. Eng. Sci.* **8** (2), 116 (1968).

[10] Bird, R.B., W.E. Stewart, and E.N. Lightfoot, *Transport Phenomena*, Wiley, New York, 1960, p. **94**, Example 3.5-1.

[11] Tseng, H.C., et al., *Preprints, Soc. Plast. Engrs. Annu. Tech. Conf.*, 1985, p. 716.

[12] Best, D.M., Thesis, Carnegie Institute of Technology, Pittsburg, PA, 1966.

[13] Tobolsky, A.V., *Properties and Structure of Polymers*, Wiley, New York, 1960, Chapter II.8.

[14] Brandrup, J. and E.H. Immergut (eds), *Polymer Handbook*, 3rd ed., Wiley-Interscience, New York, 1989.

[15] Carley, J.F., *Modern Plastics* **39** (4), 123 (1961).

[16] Metzner, A.B. and J.C. Reed, *Am. Inst. Chem. Eng. J.* **1**, 434 (1955).

[17] Patterson, G.K., J.L. Zakin, and J.M. Rodriguez, *Ind. Eng. Chem.* **61** (1), 22 (1969).

[18] Sellin, R.H.J. and R.T. Moses, *Drag Reduction in Fluid Flows, Techniques for Friction Control*, Ellis Horwood, Chichester (distributed by Wiley, NY), 1989.

[19] Zakin, J.L. and D.L. Hunston, *J. Macromol. Sci.-Phys.*, **B18** (4), 795 (1980).

CHAPTER 15

LINEAR VISCOELASTICITY

15.1 INTRODUCTION

In a traditional sense, engineers typically deal with elastic solids and viscous fluids as two separate and distinct classes of materials. Design procedures based on these concepts have worked pretty well because most traditional materials (water, air, steel, concrete), at least to a good approximation, fit into one of these categories. The realization has grown, however, that these categories represent only the extremes of a broad spectrum of material response. Polymer systems fall somewhere in between (thus, the use of the term "viscoelastic"), giving rise to some of the unusual properties of melts and solutions described previously. Other examples are important in the structural applications of polymers. In a common engineering stress–strain test, a sample is strained (stretched, bent, or compressed) at an approximately constant rate and the stress is measured as a function of strain. With traditional solids, the stress–strain curve is pretty much independent of the rate at which the material is strained. However, the stress–strain properties of many polymers are markedly *rate dependent*. Similarly, polymers often exhibit pronounced creep and stress relaxation behavior (to be defined shortly). While such behavior is exhibited by other materials (e.g., metals near their melting points), at normal temperatures it is negligible and is not usually included in design calculations. If the time-dependent behavior of polymers is ignored, the results can sometimes be disastrous.

15.2 MECHANICAL MODELS FOR LINEAR VISCOELASTIC RESPONSE

An as aid in visualizing viscoelastic response, we introduce two linear mechanical models to represent the extremes of the mechanical response spectrum. The spring in Figure 15.1a represents a *linear elastic* or Hookean *solid* that can be used to describe the response to a

Fundamental Principles of Polymeric Materials, Third Edition. Christopher S. Brazel and Stephen L. Rosen.
© 2012 John Wiley & Sons, Inc. Published 2012 by John Wiley & Sons, Inc.

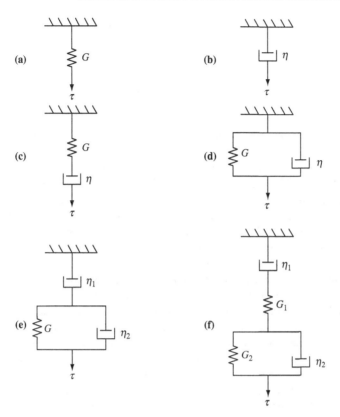

FIGURE 15.1 Linear viscoelastic models: (a) linear elastic; (b) linear viscous; (c) Maxwell element; (d) Voigt–Kelvin element; (e) three-parameter; (f) four-parameter.

shear force, whose constitutive equation (relation of stress to strain and time) is simply $\tau = G\gamma$, where G is a (constant) shear modulus of the material (or if subjected to a stretching force, $\sigma = E\varepsilon$, where E is the elastic modulus). Similarly, a *linear viscous* or Newtonian *fluid* is represented by a dashpot (a sort-of piston moving in a cylinder of Newtonian fluid) whose constitutive equation is $\tau = \eta\dot{\gamma}$, where η is a (constant) viscosity. In contrast to the spring, which snaps back to its original length after the strain is removed, a dashpot continues to expand as long as a force is applied, and when that force (strain) is removed, the dashpot remains at its stretched position. In the model, the strain is represented by the extension (stretching) of spring and/or dashpot.

The models developed here are visualized in tension, with tensile stress σ, tensile strain ε, and Young's modulus E. However, the same theory holds true for pure shear (viscometric) deformation, where a shear stress τ results in a shear strain γ with proportionality constant G (Hooke's modulus). η represents the Newtonian (shear) viscosity, while the elongational (Trouton) viscosity is given by η_e.

Some authorities strongly object the use of mechanical models to represent materials. They point out that real materials are not made of springs and dashpots. This is true, but they are not made of equations either, and it is much easier for most people to visualize the deformation of springs and dashpots than the solutions to equations.

A word is needed for the meaning of the term *linear*. For the present, a linear response will be defined as one in which the *ratio* of overall stress to overall strain, the overall modulus, $E(t)$ or $G(t)$, is a function of *time only*, not of the magnitudes of stress or strain:

$$E(t) \equiv \sigma/\varepsilon \left.\right\}$$
$$\left.\right\} = \text{function of time only for linear response}$$
$$G(t) \equiv \tau/\gamma \left.\right\}$$

(15.1a)

(15.1b)

The Hookean spring responds instantaneously to reach an equilibrium strain (ε or γ) upon application of a constant stress (σ_o or τ_o), and the strain remains constant as long as the stress is maintained constant. Sudden removal of the stress results in instantaneous recovery of the strain (Figure 15.2). Doubling the stress on the spring simply doubles the resulting strain, so the spring is linear, with $E(t) = E$ or $G(t) = G$, with E or G as constants, according to Equations 15.1a and 15.1b. (In assuming that the spring instantaneously reaches an equilibrium strain under the action of a suddenly applied constant stress, we have neglected inertial effects. Although it is not necessary to do so, including them would contribute little to the present discussion.)

If a constant stress (σ_o or τ_o) is suddenly applied to the dashpot, the strain increases with time according to $\sigma = (\varepsilon/\eta_e)t$ or $\gamma = (\tau_o/\eta)t$ (considering the strain to be zero when the stress is initially applied), Figure 15.3. Doubling the stress doubles the slope of the strain-time line, and at any time, the modulus $E(t) \equiv \sigma/\varepsilon = \eta_e/t = E$ or $G(t) \equiv \tau/\gamma = \eta/t = G$, where E or G are functions of time only. So the dashpot is also linear.

It may be shown that *any combination of linear elements must be linear*, so any models based on these linear elements, no matter how complex, can represent only linear responses. Just how realistic is a linear response? Its most conspicuous shortcoming is that it permits only Newtonian behavior (constant viscosity) in equilibrium viscous flow. For most polymers at strains greater than a few percent or so (or rates of strain greater than 0.1/s), a linear response is not a good quantitative description. Moreover, even within the

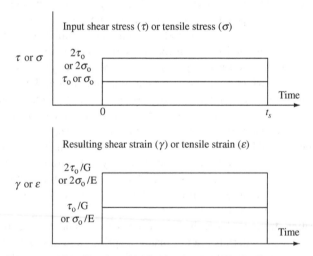

FIGURE 15.2 Response of spring (purely elastic component) to shear stress or tensile stress applied from time 0 to time t_s. The spring snaps back to its original position immediately after the stress is removed. Also note that the magnitude of the strain is directly proportional to the applied stress.

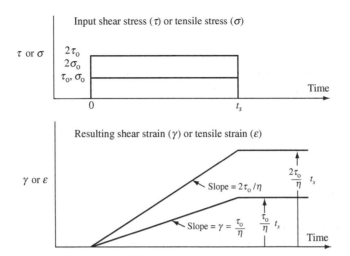

FIGURE 15.3   Response of a dashpot (purely viscous component) to shear stress or tensile stress applied from time 0 to time t_s. The dashpot is permanently deformed from its original position. Also note that the slope of the strain versus time curve is proportional to the applied stress. Not shown on the figure: for the tensile strain, the slope would be σ_o/η_e or $2\sigma_o/\eta_e$, respectively, for the two input stresses, with the permanent deformation being $\varepsilon = (\sigma_o/\eta_e)\, t_s$ for the input strain of σ_o or $(2\sigma_o/\eta_e)\, t_s$ for the input strain of $2\sigma_o$.

limit of linear viscoelasticity, a fairly large number of linear elements (springs and dashpots) are usually needed to provide an accurate quantitative description of the material's response. Hence, the quantitative applicability of simple linear models (those with a few springs and dashpots) is limited, but the models are extremely valuable in visualizing viscoelastic response and in understanding how and why variations in molecular structure influence that response.

15.2.1   The Maxwell Element

James Clerk Maxwell realized that neither a linear viscous element (dashpot) nor a linear elastic element (spring) was sufficient to describe his experiments on the deformation of asphalts, so he proposed a simple series combination of the two, the *Maxwell element* (Figure 15.1c). In a Maxwell element, the spring and dashpot support the same stress so

$$\sigma = \sigma_{\text{spring}} = \sigma_{\text{dashpot}} \tag{15.2a}$$

$$\tau = \tau_{\text{spring}} = \tau_{\text{dashpot}} \tag{15.2b}$$

Furthermore, the total strain (extension) of the element is additive (i.e., the sum of the strains in the spring and dashpot):

$$\varepsilon = \varepsilon_{\text{spring}} + \varepsilon_{\text{dashpot}} \tag{15.3a}$$

$$\gamma = \gamma_{\text{spring}} + \gamma_{\text{dashpot}} \tag{15.3b}$$

Differentiating Equations 15.3a and 15.3b with respect to time gives

$$\dot{\varepsilon} = \dot{\varepsilon}_{\text{spring}} + \dot{\varepsilon}_{\text{dashpot}} \tag{15.4a}$$

$$\dot{\gamma} = \dot{\gamma}_{\text{spring}} + \dot{\gamma}_{\text{dashpot}} \tag{15.4b}$$

Considering the shear strain in Equation 15.4b, and realizing that $\dot{\gamma}_{\text{dashpot}} = \tau/\eta$ and $\dot{\gamma}_{\text{spring}} = \dot{\tau}/G$, plugging in and rearranging gives the differential equation for the Maxwell element:

$$\tau = \eta\dot{\gamma} - \frac{\eta}{G}\dot{\tau} = \eta\dot{\gamma} - \lambda\dot{\tau} \tag{15.5}$$

The quantity $\lambda = \eta/G$ has the dimension of time and is known as a *relaxation time*. Its physical significance will be apparent shortly.

15.2.1.1 Creep Testing
Let us examine the response of the Maxwell element in two mechanical tests commonly applied to polymers. First consider a *creep* test, in which a constant shear stress is instantaneously (or at least very rapidly) applied to the material and the resulting strain is followed as a function of time. Deformation after removal of the stress is known as *creep recovery*.

As shown in Figure 15.4, the sudden application of stress to a Maxwell element causes an instantaneous stretching of the spring to an equilibrium value of τ_o/G (or σ_o/E if a tensile stress is applied), where τ_o is the constant applied shear stress (or σ_o is the constant applied extensional stress). The dashpot extends linearly with time with a slope of τ_o/η (or σ_o/η_e for tensile experiments), and will continue to do so as long as the stress is maintained. Thus, *the Maxwell element is a fluid*, because it will continue to deform as long as it is stressed. The creep response of a Maxwell element is therefore

$$\varepsilon(t) = \sigma_o/E + \sigma_o/\eta_e t \tag{15.6a}$$

$$\gamma(t) = \tau_o/G + \tau_o/\eta t \tag{15.6b}$$

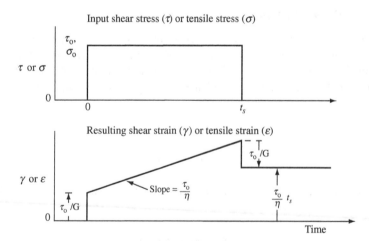

FIGURE 15.4 Creep response of a Maxwell element to an applied stress. The bottom figure shows the shear response, but a similar figure would also show the initial stretching of the spring due to a tensile stress (σ_o/E), and slope of the deformation due to the dashpot (σ_o/η_e), with the recoil of the spring being $-\sigma_o/E$ and the permanent deformation being $\sigma_o/\eta_e * t_s$.

or, in terms of compliance (for an applied tensile stress), where $J \equiv \varepsilon(t)/\sigma_o$, or a creep compliance (for an applied shear stress), where $J_c(t) \equiv \gamma(t)/\tau_o$,

$$J \equiv \varepsilon(t)/\sigma_o = 1/E + t/\eta \qquad (15.7a)$$

$$J_c(t) \equiv \gamma(t)/\tau_o = 1/G + t/\eta \qquad (15.7b)$$

The creep compliance $J_c(t)$, being independent of the applied stress τ_o (for a linear material), is a more general way to represent the creep response. When the stress is removed at time t_s, the spring immediately contracts by an amount equal to its original extension, a process known as *elastic recovery*. The dashpot, of course, does not recover, leaving a *permanent set* of $(\tau_o/\eta)t_s$, or $\sigma_o/\eta_e t_s$ (for a tensile stress), representing the amount the dashpot has extended during the application of stress. Although real materials never show sharp breaks in a creep test as does the Maxwell element, the Maxwell element does exhibit the phenomena of elastic strain, creep recovery, and permanent set, which are often observed with real materials.

Creep testing also demonstrates another term commonly used in material testing, that is, *deformation*. There are two types of deformation that may arise after a stress is applied to a material: *elastic deformation* and *plastic deformation*. Elastic deformation refers to materials whose shape changes temporarily, and once the force is removed, the material resumes its original shape; this is equivalent to the spring representation in viscoelasticity models. Plastic deformation refers to nonreversible changes in shape, represented by the dashpot here. It is interesting to note that even for metals, the term plastic deformation is used to indicate that a material has been deformed in a way that cannot be recovered.

15.2.1.2 Stress Relaxation

Another important test used to study viscoelastic responses is *stress relaxation*. A stress-elaxation test consists of suddenly applying a strain to the sample, and following the stress as a function of time as the strain is held constant. When the Maxwell element is strained instantaneously, only the spring can respond initially (for an infinite rate of strain, the resisting force in the dashpot is infinite) to a shear stress of $G\gamma_o$, where γ_o is the constant applied shear strain (or similarly for a tensile stress of $E\varepsilon_o$, where ε_o is the constant applied tensile strain). The extended spring then begins to contract, but the contraction is resisted by the dashpot. The more the spring retracts, the smaller is its restoring force, and, correspondingly, the rate of retraction drops. Solution of the differential equation for shear relaxation with $\dot{\gamma} = d\gamma/dt = 0$ and the initial condition $\tau = G\gamma_o$ at $t = 0$ shows that the stress undergoes a first-order exponential decay:

$$\tau(t) = G\gamma_o e^{-t/\lambda} \qquad (15.8a)$$

or, in terms of a *relaxation modulus*, $G_r(t) \equiv \tau(t)/\gamma_o$,

$$G_r(t) \equiv \frac{\tau(t)}{\gamma_o} = G e^{-t/\lambda} \qquad (15.8b)$$

Again, the relaxation modulus, $G_r(t)$, is a more general means of representing a stress-relaxation response because it is independent of the applied strain for linear materials.

From Equations 15.8, we see that the relaxation time λ is the time constant for the exponential decay, that is, the time required for the stress to decay to a factor of $1/e$ or 37%

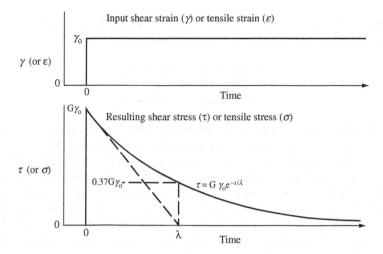

FIGURE 15.5   Shear stress relaxation of a Maxwell element to a constant applied shear strain.

of its initial value. The stress asymptotically drops to zero as the spring approaches complete retraction (Figure 15.5).

Stress-relaxation data for linear polymers actually look like the curve for the Maxwell element. Unfortunately, they cannot often be fitted quantitatively with a single value of G and a single value of λ, that is, the decay is not really first order.

Example 15.1   Examine the response of a Maxwell element in an engineering stress–strain test, a test in which the rate of tensile strain is maintained (approximately) constant at $\dot{\varepsilon}_o$.

Solution. Using Equation 15.5a gives

$$\sigma = \eta_e \dot{\varepsilon}_0 - \frac{\eta_e}{E}\dot{\sigma} \quad \text{or} \quad \sigma + \left(\frac{\eta_e}{E}\right)\left(\frac{d\sigma}{dt}\right) = \eta_e \dot{\varepsilon}_0 = \text{constant}$$

The solution to this differential equation with the initial condition $\sigma = 0$ at $t = 0$ is

$$\sigma = \eta_e \dot{\varepsilon}_0 [1 - e^{-(E/\eta_e)t}]$$

Since $d\varepsilon/dt = \dot{\varepsilon}_o$ is constant and we can assume that $\varepsilon = 0$ when $t = 0$

$$\varepsilon = \dot{\varepsilon}_0 t$$

The stress–strain curve is then

$$\sigma = \eta_e \dot{\varepsilon}_0 [1 - e^{-(E/\eta_e \dot{\varepsilon}_0)\varepsilon}]$$

This response is sketched in Figure 15.6. Note that at any given strain, σ increases with the rate of strain $\dot{\varepsilon}_o$, that is, the material appears "stiffer" (has a higher modulus). As crude as this model might prove to be in fitting experimental data, it does account, at least qualitatively, for some of the observed properties of linear polymers in engineering stress–strain tests.

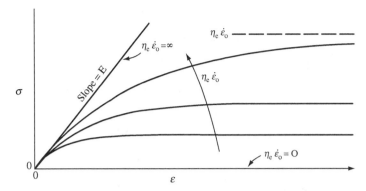

FIGURE 15.6   Response of a Maxwell element under constant rate of tensile strain (Example 15.1).

15.2.2   The Voigt–Kelvin Element

If a series combination of a spring and dashpot has its drawbacks, the next logical thing to try is a parallel combination, a Voigt or Voigt–Kelvin element (Figure 15.1d). Here, it is assumed that the crossbars supporting the spring and the dashpot always remain parallel, so that the strain is the same in the spring and in the dashpot at all times:

$$\gamma = \gamma_{\text{spring}} = \gamma_{\text{dashpot}} \tag{15.9}$$

The stress supported by the element is then the sum of the stresses in the spring and in the dashpot:

$$\tau = \tau_{\text{spring}} + \tau_{\text{dashpot}} \tag{15.10}$$

Combination of Equations 15.9 and 15.10 with the equations for the deformation of the spring and the dashpot gives the differential equation for the Voigt–Kelvin element:

$$\tau = \eta\,\dot{\gamma} + G\gamma \tag{15.11}$$

When the stress is suddenly applied in a creep test, only the dashpot offers an initial resistance to deformation, so the initial slope of the strain versus time curve is τ_o/η. As the element is extended, the spring provides an increasingly greater resistance to further extension, so the rate of creep decreases. Eventually, the system comes to equilibrium with the spring alone supporting the stress (with the rate of strain zero, the resistance of the dashpot is zero). The equilibrium strain is simply τ_o/G. Quantitatively, the response is an exponential rise,

$$\gamma(t) = \frac{\tau_0}{G}\left(1 - e^{-t/\lambda}\right) \tag{15.12a}$$

or, in terms of the creep compliance:

$$J_c(t) = \frac{1}{G}\left(1 - e^{-t/\lambda}\right) \tag{15.12b}$$

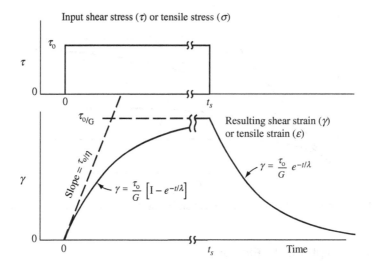

FIGURE 15.7   Creep response of a Voigt–Kelvin element to an applied shear stress.

If the stress is removed *after equilibrium has been reached*, the strain decays exponentially,

$$\gamma(t) = \frac{\tau_0}{G} e^{-t/\lambda} \tag{15.13}$$

Note that the Voigt–Kelvin element does not continue to deform as long as stress is applied, and it does not exhibit any permanent set (see Figure 15.7). It therefore represents a viscoelastic *solid*, and gives a fair qualitative picture of the creep response of some crosslinked polymers.

The Voigt–Kelvin element is not suited for representing stress relaxation. The instantaneous application of strain would be met by an infinite resistance in the dashpot, and so would require the application of an infinite stress, which is obviously unrealistic.

15.2.3   The Three-Parameter Model

The next step in the development of linear viscoelastic models is the so-called three-parameter model (Figure 15.1e). By adding a dashpot in series with the Voigt–Kelvin element, we get a liquid. The differential equation for this model may be written in operator form as

$$\left(1 + \lambda_1 \frac{d}{dt}\right)\tau = \eta_1 \left(1 + \lambda_2 \frac{d}{dt}\right)\dot{\gamma} \tag{15.14}$$

where $\lambda_1 = (\eta_1 + \eta_2)/G$ and $\lambda_2 = \eta_2/G$. Further, the form of Equation 15.13 suggests modification by adding higher order derivatives and more constants:

$$\left(1 + \lambda_1 \frac{d}{dt} + \xi_1 \frac{d^2}{dt^2} + \cdots\right)\tau = \eta_1 \left(1 + \lambda_2 \frac{d}{dt} + \xi_2 \frac{d^2}{dt^2} + \cdots\right)\dot{\gamma} \tag{15.15}$$

This, of course, will fit data to any desired degree of accuracy if enough terms are used.

15.3 THE FOUR-PARAMETER MODEL AND MOLECULAR RESPONSE

The four-parameter model (Figure 15.1f) is a series combination of a Maxwell element with a Voigt–Kelvin element. Its differential equation is

$$\ddot{\tau} + \left(\frac{G_1}{\eta_2} + \frac{G_1}{\eta_1} + \frac{G_2}{\eta_2}\right)\dot{\tau} + \frac{G_1 G_2}{\eta_1 \eta_2}\tau = G_1 \ddot{\gamma} + \frac{G_1 G_2}{\eta_2}\dot{\gamma} \tag{15.16}$$

Its creep response is the sum of the creep responses of the Maxwell and Voigt–Kelvin elements:

$$\gamma(t) = \frac{\tau_0}{G_1} + \frac{\tau_0}{\eta_1}t + \frac{\tau_0}{G_2}\left[1 - e^{-(G_2/\eta_2)t}\right] \tag{15.17a}$$

or, in terms of creep compliance:

$$J_e(t) = \frac{1}{G_1} + \frac{1}{\eta_1}t + \frac{1}{G_2}\left[1 - e^{-(G_2/\eta_2)t}\right] \tag{15.17b}$$

This is summarized in Figure 15.8.

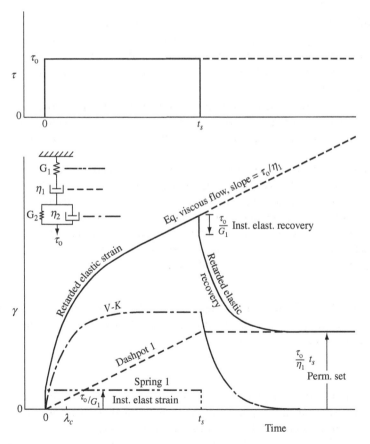

FIGURE 15.8 Creep response of a four-parameter model for an input shear stress. Note that the model also works for tensile stress, σ, with its corresponding tensile strain, ε.

The four-parameter model provides at least a qualitative representation of all the phenomena generally observed in the creep of viscoelastic materials: instantaneous elastic strain, retarded elastic strain, steady-state viscous flow, instantaneous elastic recovery, retarded elastic recovery, and permanent set. It also describes at least qualitatively the behavior of viscoelastic materials in other types of deformation. Of equal importance is the fact that the model parameters can be identified with the various molecular response mechanisms in polymers, and can, therefore, be used to predict the influences that changes in molecular structure will have on mechanical response. The following analogies may be drawn.

1. *Dashpot 1* (Figure 15.1f) represents molecular slip, the translational motion of molecules. This slip of polymer molecules past one another is responsible for flow. The value of η_1 alone (molecular friction in slip) governs the equilibrium flow of the material.

2. *Spring 1* represents the elastic straining of bond angles and lengths. All bonds in polymer chains have equilibrium angles and lengths. The value of G_1 characterizes the resistance to deformation from these equilibrium values. Since these deformations involve interatomic bonding, they occur essentially instantaneously from a macroscopic point of view. This type of elasticity is thermodynamically known as *energy elasticity*.

3. *Dashpot 2* represents the resistance of the polymer chains to uncoiling and coiling, caused by temporary mechanical entanglements of the chains and molecular friction during these processes. Since coiling and uncoiling require cooperative motion of many chain segments, they cannot occur instantaneously, and hence account for retarded elasticity.

4. *Spring 2* represents the restoring force brought about by the thermal agitation of the chain segments, which tends to return chains oriented by a stress to their most random or highest entropy configuration. This is, therefore, known as *entropy elasticity*.

The magnitude of the timescale shown in Figure 15.8 will of course depend on the values of the model parameters. The two viscosities, in particular, depend strongly on temperature. Well below T_g, for example, where η_1 and η_2 are very large, t_s needs to be on the order of days or weeks to observe appreciable retarded elasticity and flow. Well above T_g, t_s might be only seconds or less to permit the deformation shown. An important thing to keep in mind is that designs based on short-term property measurements will be inadequate if the object supports a stress for longer periods of time.

Example 15.2   Using the four-parameter model as a basis, qualitatively sketch the effects of (a) increasing molecular weight and (b) increasing degrees of crosslinking on the creep response of a linear, amorphous polymer.

Solution.

(a)   As discussed in Chapter 14, the equilibrium zero-shear (linear) viscosity of polymers, represented by η_1 in the model, increases with the 3.4 power of $\bar{M}_w$. Thus, the slope in the steady-state flow region τ_o/η_1 is greatly decreased as the molecular weight increases, and the permanent set $(\tau_o/\eta_1)t_s$ is reduced correspondingly (Figure 15.9).

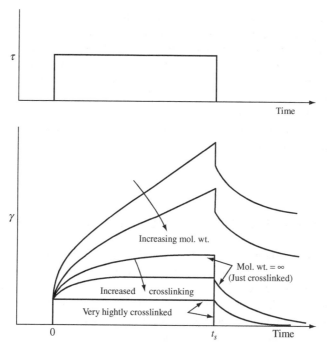

FIGURE 15.9 The effects of molecular weight and crosslinking on the creep response to application of a shear stress on an amorphous polymer. A similar response would be observed for elongational strain, ε, upon application of a fixed tensile stress, σ.

(b) Light crosslinking represents the limit of case (a) above, when the molecular weight reaches infinity, since all the chains are hooked together by crosslinks. Under these conditions, they cannot slip past one another, so η_1 becomes infinite, as indicated by the horizontal line in Figure 15.9, indicating that the dashpot allows no further movement after the initial spring response. If the crosslinking is light (crosslinks few and far between), as in a rubber band, coiling and uncoiling will not be appreciably hindered. Note that crosslinking converts the material from a fluid to a solid (it eventually reaches an equilibrium strain under the application of a constant stress) and it eliminates permanent set. The equilibrium modulus of such a lightly cross-linked rubber will be on the order of 0.1 to 1 MPa, the characteristic "rubbery" modulus. Further crosslinking begins to hinder the ability of the chains to uncoil and raises the restoring force (increases η_2 and G_2). At high degrees of crosslinking, as in hard rubber (ebonite), the only response mechanism left is straining bond angles and lengths, giving rise to an almost perfectly elastic material with a modulus on the order of 1 to 10 GPa, the characteristic modulus for a glassy polymer (below T_g).

The four-parameter model nicely accounts for the interesting examples of viscoelastic response mentioned earlier. For example, dashpot 1 allows viscous flow, while the elastic restoring forces of springs 1 and 2 provide the "rubber band" elasticity responsible for the Weissenberg effect, which is when a polymer solution is drawn up around a mixing rod rather than being flung outward. In engineering stress–strain tests, the moduli of polymers are observed to increase with the applied rate of strain. At high rates of strain, spring 1 provides the major response mechanism. As the rate of strain is lowered, dashpot 1 and the Voigt–Kelvin element contribute more and more to the overall deformation, giving a greater strain at any stress, that is, a lower modulus. When Silly Putty is bounced (stress applied rapidly for a short period of time), spring 1 again provides the major response mechanism. There is not time for appreciable flow according to dashpots 1 and 2, so not much of the initial potential energy is converted to heat through the molecular friction involved in slippage and uncoiling, and the material behaves in an almost perfectly elastic fashion. When it is stuck on the wall, the stress (in this case due to its own weight) is applied for a long period of time, and it flows downward as a result of the molecular slip represented by dashpot 1.

15.4   VISCOUS OR ELASTIC RESPONSE? THE DEBORAH NUMBER [1]

As discussed above, whether a viscoelastic fluid behaves as an elastic solid or a viscous liquid depends on the relation between the timescale of the deformation to which it is subjected and the time required for the material's time-dependent mechanisms to respond. Strictly speaking, the concept of a single relaxation time applies only to a first-order response, and thus is not applicable to real materials, in general. Nevertheless, a characteristic relaxation time λ_c for any material can always be defined as, for example, the time required for the material to reach $1 - 1/e$ or 63.2% of its ultimate retarded elastic response to a step change. A precise value is rarely necessary. The characteristic time is simply a means of characterizing the rate of a material's time-dependent elastic response, short λ_c's indicating rapid response and large λ_c's indicating sluggish response. The ratio of this characteristic material time to the timescale of the deformation is the *Deborah number*:

$$\text{De} \equiv \frac{\lambda_c}{t_s} \tag{15.18}$$

Response will appear elastic at high Deborah numbers (De $\gg$ 1) and viscous at low Deborah numbers (De $\rightarrow$ 0).

Consider, for example, the creep response of the four-parameter model (Figure 15.8). For this model, a logical choice for λ_c would be the time constant for its Voigt–Kelvin component, η_2/G_2. For De $\gg$ 1 ($t_s \ll \lambda_c$), the Voigt–Kelvin element and dashpot 1 will be essentially immobile, and the response will be due almost entirely to spring 1, that is, almost purely elastic. For De $\rightarrow$ 0 ($t_s \gg \lambda_c$), the instantaneous and retarded elastic response mechanisms have already reached equilibrium, so the only remaining response will be the purely viscous flow of dashpot 1, and the deformation due to viscous flow will completely overshadow that due to the elastic response mechanisms (imagine the creep curve of Figure 15.8 extended a meter or so beyond the page). Under conditions where De $\rightarrow$ 0, materials can be treated by the techniques outlined in Chapter 14 for purely viscous fluids.

Modifications of the devices described in Chapter 14 can also be used to obtain information on the material's elastic response. For example, if the stress is suddenly removed from a rotational viscometer, the creep recovery or elastic recoil of the material can be followed. This provides a value of λ_c for the material.

Example 15.3 Thermocouples and Pitot tubes (used to measure flowrate based on differential pressures) inserted in a flowing stream of a viscoelastic fluid often give erroneous results. Explain.

Solution. When a viscoelastic fluid in steady-state flow (De → 0) encounters a probe, it must make a sudden (De ≫ 0) jog to get around it. The retarded elastic response mechanisms simply cannot respond fast enough in the immediate vicinity of the probe, which for all practical purposes behaves as if it were covered with a solid plug. What is measured, therefore, is not characteristic of the fluid in an unobstructed stream.

Example 15.4 (This is believed due to Professor A.B. Metzner and has probably been verified by TV's "Mythbusters" team.) A paper cup containing water is placed on a tree stump. A 0.22-caliber bullet fired at the cup passes cleanly through, leaving the cup sitting on the stump. The water is replaced by a dilute polymer solution in a second cup. This time, the bullet knocks the cup 25 ft beyond the stump. Explain.

Solution. The characteristic relaxation time for a low molecular weight fluid such as water is extremely short, much shorter than the time it takes the bullet to pass through the cup (t_s). This, then, is a low-De situation. The water behaves as a viscous fluid. The bullet transfers a little momentum to it through viscous friction, but not enough to dislodge the cup. Adding a polymeric solute raises the characteristic relaxation time many orders of magnitude, to the point where this becomes a high-De experiment (the polymer chains cannot respond fast enough to get out of the way of the bullet). The bullet, in effect, slams into a solid and transfers much of its momentum to the fluid-cup system, carrying it beyond the stump.

15.5 QUANTITATIVE APPROACHES TO MODEL VISCOELASTICITY [2–5]

Although the four-parameter model is useful from a conceptual standpoint, it does not often provide an accurate fit of experimental data and therefore cannot be used to make quantitative predictions of material response. To do so, and to infer some detailed information about molecular response, more general models have been developed. The *generalized Maxwell model* (Figure 15.10) is used to describe stress-relaxation experiments. The stress relaxation of an individual Maxwell element is given by

$$\tau_i(t) = \gamma_0 G_i e^{-t/\lambda_i} \tag{15.19}$$

where $\lambda_i = \eta_i/G_i$. The relaxation of the generalized model, in which the individual elements are all subjected to the same constant strain γ_o is then:

$$\tau(t) = \sum_{i=1}^{n} \tau_i(t) = \gamma_0 \sum_{i=1}^{n} G_i e^{-t/\lambda_i} \tag{15.20a}$$

Expressed in terms of the time-dependent relaxation modulus $G_r(t)$, the response is

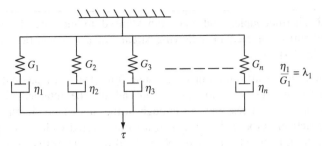

FIGURE 15.10   Generalized Maxwell model for material response to shear. Note that the model can also be used to show the response to extensional stress, σ.

$$G_r(t) \equiv \frac{\tau(t)}{\gamma_0} = \sum_{i=1}^{n} G_i e^{-t/\lambda_i} \qquad (15.20b)$$

Now, if n is large, the summation in Equation 15.20 may be approximated by the integral over a *continuous distribution of relaxation times* $G(\lambda)$:

$$G_r(t) = \int_0^{\infty} G(\lambda) e^{-t/\lambda} d\lambda \qquad (15.21)$$

Note that while the G_i's have modulus units (e.g., dyn/cm^2 or MPa), $G(\lambda)$ is in modulus/time units. Also note that if the generalized Maxwell model is to represent a viscoelastic solid such as a crosslinked polymer, at least one of the viscosities has to be infinite.

For creep tests, a generalized Voigt–Kelvin model is used (Figure 15.11). The creep response of an individual Voigt–Kelvin element is given by

$$\gamma_i(t) = \tau_0 J_i(1 - e^{-r/\lambda_i}) \qquad (15.22)$$

where $J_i = 1/G_i$ is the individual spring compliance. The response of the array, in which each element is subjected to the same constant applied stress τ_o, is then

$$\gamma(t) = \tau_0 \sum_{i=1}^{n} J_i(1 - e^{-t/\lambda_i}) \qquad (15.23a)$$

or, in terms of the overall creep compliance, $J_c(t)$:

$$J_c(t) \equiv \frac{\gamma(t)}{\tau_0} = \sum_{i=1}^{n} J_i(1 - e^{-t/\lambda_i}) \qquad (15.23b)$$

Again, for large n, the discrete summation above may be approximated by

$$J_c(t) = \int_0^{\infty} J(\lambda)(1 - e^{-t/\lambda}) d\lambda \qquad (15.24)$$

where $J(\lambda)$ is the *continuous distribution of retardation times* (1/modulus·time).

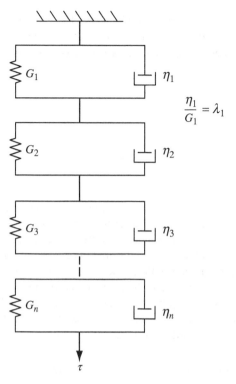

FIGURE 15.11   Generalized Voigt–Kelvin model for an input shear stress, τ. This model could also be used for modeling tensile stress, σ, with the corresponding moduli E_i.

If the generalized Voigt–Kelvin model is to represent a viscoelastic liquid such as a linear polymer, the modulus of one of the springs must be zero (infinite compliance), leaving a simple dashpot in series with all the other Voigt–Kelvin elements. Sometimes, the steady-flow response of this lone dashpot, $\gamma_{\text{dashpot}} = (\tau_o/\eta_o)t$, is subtracted from the overall response, leaving the compliances to represent only the *elastic* contributions to the overall response:

$$\gamma(t) = \frac{\tau_0}{\eta_0}t + \tau_0 \sum_{i=1}^{n} J_i^{\dagger}(1 - e^{-t/\lambda_i}) \tag{15.25a}$$

$$J_e(t) = \frac{t}{\eta_0} + \sum_{i=1}^{n} J_i^{\dagger}(1 - e^{-t/\lambda_i}) \tag{15.25b}$$

$$J_c^{\dagger}(t) = J_c(t) - \frac{t}{\eta_0} = \sum_{i=1}^{n} J_i^{\dagger}(1 - e^{-t/\lambda_i}) \tag{15.26}$$

$$J_c^{\dagger}(t) = J_c(t) - \frac{t}{\eta_0} = \int_{0}^{\infty} J^{\dagger}(\lambda)(1 - e^{-t/\lambda}) \tag{15.27}$$

Here, η_o is the steady-state (Newtonian) viscosity. The daggers indicate that the steady-state viscous flow has been removed and is treated separately.

Application of the discrete Equations 15.20 and 15.23 often involves a fairly large n to describe data accurately, thus requiring an impractically large number of parameters λ_i and $G_i = 1/J_i$. It has been suggested, however, that the individual parameters are related by [6]

$$\lambda_i = \frac{\lambda_0}{i^\alpha} \qquad (15.28)$$

$$G_i = \frac{1}{J_i} = \frac{\eta_0}{\displaystyle\sum_{i=1}^{n} \lambda_i} \qquad (15.29)$$

Equations 15.27 and 15.28 require that the G_i all be the same and that $\eta_o = \Sigma\eta_i$ (with $\eta_i = \lambda_i G_i$). They reduce the number of necessary parameters to three: η_o, the steady-state zero-shear viscosity; λ_o, a maximum relaxation time; and α, an empirical constant. The Rouse theory [7] for dilute polymer solutions predicts $\alpha = 2$, but for concentrated solutions and melts, better fits are obtained with α's between 2 and 4 [6].

Often, enough discrete parameters to provide reasonable response models can be extracted from experimental creep or stress-relaxation data using Tobolsky's "Procedure X" [4]. This procedure will be illustrated for stress-relaxation data in the form of the relaxation modulus $G_r(t)$. According to Equation 15.20b,

$$G_r(t) = G_1 e^{-t/\lambda_i} + G_2 e^{-t/\lambda_2} + G_3 e^{-t/\lambda_3} + \cdots \qquad (15.20b)$$

The procedure is based on two assumptions. First, the G_i's do not differ much in magnitude. This often turns out to be the case. As noted in Equation 15.29, theory suggests that the G_i should be identical. Second, there are a few discrete λ_i, with $\lambda_1 > \lambda_2 > \lambda_3 > \ldots$, and they differ enough so that at long times, the second-, third-, and higher-order terms approach zero, leaving only the first term to determine $G_r(t)$. These assumptions are easily tested. If a plot of $\ln G_r(t)$ versus t becomes linear at large t, the assumptions are valid. If that turns out to be the case, the slope of the linear region at long times is $-1/\lambda_1$ and its $t = 0$ intercept is $\ln G_1$. The known response of the first Maxwell element can then be subtracted from the overall response:

$$G_r(t) - G_1 e^{-t/\lambda_1} = G_2 e^{-t/\lambda_2} + G_3 e^{-t/\lambda_3} + \cdots \qquad (15.30)$$

and $\ln[G_r(t) - G_1 e^{-t/\lambda_1}]$ is plotted versus t. Again, if a linear region is reached at long times, the slope of that region is $-1/\lambda_2$ and its intercept is $\ln G_2$, and so on.

In principle, this procedure can be repeated indefinitely. In practice, the precision and timescale of typical single-temperature experimental data rarely justify going beyond $i = 3$. Even so, the resulting three-element generalized Maxwell model can often give a good fit to the data used to establish it. More importantly, the model can then be used to predict material response in other types of deformation, at least over similar timescales. The time–temperature superposition principle, which is discussed below, can extend timescales to the point where the parameters may be established to $i = 6$ or 7 or so.

Using the method developed here, a creep response can be predicted from stress-relaxation measurements, and vice versa. This interconvertability also applies to a variety of linear mechanical responses in addition to the two types discussed here, as is illustrated in the next section. The interconversion procedures have been discussed in detail by Schwarzl [8,9]. Furthermore, the shape of the distributions $G(\lambda)$ or $J(\lambda)$ provides the polymer scientist with information on molecular response mechanisms within the polymer. For example, peaks in a certain region of λ might imply motion of side chains on the molecules. This type of information can lead to the "design" of polymers with the type of side chains needed to provide particular mechanical properties.

15.6 THE BOLTZMANN SUPERPOSITION PRINCIPLE

Suppose a material initially free of stress is subjected to a test in which a strain $\gamma(t_o)$ is suddenly imposed at $t = 0$ and maintained constant for a while. This is classical stress relaxation, and the stress will decay according to the material's time-dependent relaxation modulus $G_r(t)$, that is, $\tau(t) = G_r(t)\gamma(t_o)$. Now, however, at time t_1, the strain is suddenly changed to a new level $\gamma(t_1)$, held there for a while, then at t_2 changed to $\gamma(t_2)$, and so on, as sketched in Figure 15.12a. What happens to the stress as a result of this strain history? Well, way back in 1876, Boltzmann suggested that the stresses resulting from each individual strain *increment* should be linearly additive, that is,

$$\tau(t) = \sum_{i=0}^{n} \Delta\tau_i = \sum_{i=0}^{n} G_r(t - t_i)\Delta\gamma(t_i) \quad (\text{for } t > t_i) \tag{15.31}$$

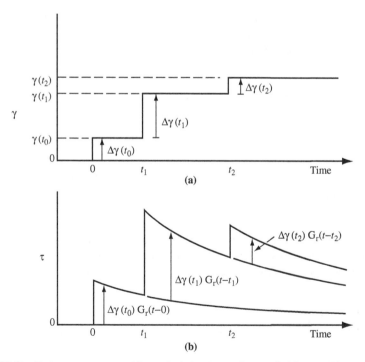

FIGURE 15.12 Boltzmann superposition principle: (a) applied strain history; (b) resulting stress history. The experiment can also be reversed, with an applied stress history causing a strain history. The shear stresses and strains shown can also be replaced with tensile stresses and strains.

where

$$\Delta\gamma(t_i) = \gamma(t_i) - \gamma(t_{i-1})$$ (15.32a)

and

$$\Delta\tau_i = G_r(t - t_i)\Delta\gamma(t_i)$$ (15.32b)

Here $\Delta\tau_i$ is the stress increment that results from the strain increment $\Delta\gamma_i$. The argument $t - t_i$ is the time after the application of a particular strain increment $\Delta\gamma_i$. This behavior is sketched in Figure 15.12b.

According to Boltzmann, the stress in the material at any time t depends on its entire past strain history, although since $G_r(t)$ is a decreasing function of time, the further back a $\Delta\gamma(t_i)$ has occurred, the smaller will be its influence in the present. This leads to the anthropomorphic concept of viscoelastic materials having a fading memory (like an aging professor), with $G_r(t)$ sometimes known as the *memory function*. (The concept, of course, is valid even in the absence of linear additivity—it is just much more difficult to quantify.)[1]

Example 15.5 A Maxwell element is initially free of stress and strain. At time $t = 0$, a strain of magnitude γ_o is suddenly applied and maintained constant until $t = \lambda/2$, at which time the strain is suddenly reversed to a value of $-\gamma_o$ and maintained at that value (Figure 15.13a). Obtain an expression for $\tau(t)$ and plot the result.

Solution. The relaxation modulus (memory function) $G_r(t)$ for a Maxwell element is given by Equation 15.7b. For this particular strain history, $\Delta\gamma(t_o) = +\gamma_o$ and $\Delta\gamma(t_1) = -2\gamma_o$ (remember, we need the increment, not the absolute value). Plugging these values into Equation 15.31 gives

$$\tau(t) = G_r(t - 0)\Delta\gamma(t_0) + G_r(t - t_1)\Delta\gamma(t_1)$$

$$= \gamma_0 G e^{-t/\lambda} - 2\gamma_0 G e^{-(t-t_1)/\lambda}$$

$$= \gamma_0 G[e^{-t/\lambda} - 2e^{-(t-t_1)/\lambda}]$$

This result is plotted in dimensionless form in Figure 15.13b. Keep in mind that the second term applies only at $t > t_1 = \lambda/2$.

Of course, not all strain histories consist of a nice series of finite step changes. No matter how an applied strain varies with time, however, it can always be approximated by a series of differential step changes, for which Equation 15.31 becomes

$$\tau(t) = \int_{\gamma(-\infty)}^{\gamma(t)} G_r(t - t')d\gamma(t') = \int_{-\infty}^{t} G_r(t - t')\frac{d\gamma(t')}{dt'}dt' = \int_{-\infty}^{t} G_r(t - t')\dot{\gamma}(t')dt'$$

(15.33)

---

[1] Note that this analysis is also applicable to material behavior when subjected to a tensile strain $\varepsilon(t_o)$, with tensile stress $\sigma(t)$ and relaxation modulus $E_r(t)$ or vice versa when an applied tensile stress results in material strain.

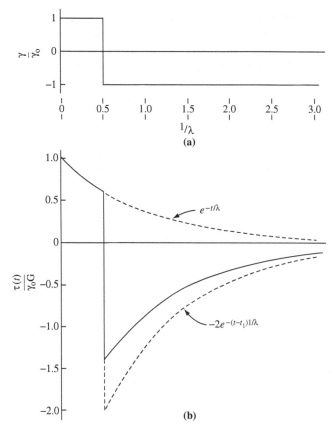

FIGURE 15.13   Stress response of a Maxwell element to a changing shear strain (Example 15.5).

where $t = present$ time and $t' = past$ time. A word is needed about the somewhat bizarre (but fairly standard) notation in Equation 15.33. The implication is that to evaluate the stress at the present time t, we must integrate over the entire past strain history of the sample; hence, the lower limit of $t' = -\infty$. In some (but not all) cases, it is convenient to assume that $\tau = \gamma = 0$ for $t' < 0$, in which case the lower limit on the integrals becomes zero.

Equation 15.33 allows calculation of $\tau(t)$ from stress-relaxation data $G_r(t)$ for any applied strain history as long as the response is linear. Furthermore, by inversion of Equation 15.31, it is possible (at least in principle) to obtain $G_r(t)$ from any test in which both $\tau(t)$ and $\gamma(t)$ are measured. When the independent variable is $\tau(t)$ and you wish to calculate $\gamma(t)$, analogs of Equations 15.31 and 15.33 may be written in terms of the creep compliance $J_c(t)$:

$$\gamma(t) = \sum_{i=0}^{n} \Delta\gamma_i = \sum_{i=0}^{n} J_c(t - t_i)\Delta\tau(t_i) \quad (\text{for } t > t_i) \tag{15.34}$$

$$\gamma(t) = \int_{\tau(-\infty)}^{\tau(t)} J_c(t - t')d\tau(t') = \int_{-\infty}^{t} J_c(t - t')\frac{d\tau(t')}{dt'}dt' = \int_{-\infty}^{t} J_c(t - t')\dot{\tau}(t')dt' \tag{15.35}$$

When all is said and done, probably the best definition of a linear material is simply one that follows Boltzmann's principle. Thus, spring–dashpot models, which are linear, automatically follow Boltzmann's principle. However, it is important not to infer a dependence of the Boltzmann principle on spring–dashpot models. The Boltzmann principle applies to a linear response regardless of whether it can be described with a spring–dashpot model. All that is needed are experimental $G_r(t)$ or $J_c(t)$ data. Models are used here simply as a matter of convenience to illustrate application of the principle.

Example 15.6 Solve Example 15.1 by applying the Boltzmann superposition principle, thereby demonstrating how the stress-time response in an engineering stress–strain test may be predicted from stress-relaxation data.

Solution. The *tensile* stress-relaxation modulus for a Maxwell element is

$$E_r(t) \equiv \frac{\sigma(t)}{\varepsilon_0} = Ee^{-t/\lambda}$$

In tensile notation, with the assumption that $\sigma = \varepsilon = 0$ for $t' < 0$, Equation 15.33 becomes

$$\sigma(t) = \int_0^t E_r(t - t') \frac{d\varepsilon(t')}{dt'} dt'$$

But for an engineering stress–strain test

$$\frac{d\varepsilon(t')}{dt'} = \dot{\varepsilon} \approx \text{constant} = \dot{\varepsilon}_0$$

Thus,

$$\sigma(t) = E\dot{\varepsilon}_0 \int_0^t e^{-(t-t')/\lambda} dt'$$

The integration is performed in the *present* time (i.e., t is a constant) over the material's past history, from $t' = 0$ to $t' = t$, with the result:

$$\sigma(t) = E\dot{\varepsilon}_0 \lambda [1 - e^{-t/\lambda}] = \eta_e \dot{\varepsilon}_0 [1 - e^{-(E/\eta_e)t}]$$

This was obtained by direct integration of the differential equation for the Maxwell element in Example 15.1.

Example 15.7 To demonstrate the fact that the stress in a viscoelastic material depends on its past strain history, calculate the stress $\tau(t_s)$ in a Maxwell element initially free of stress and strain that is brought to a strain γ_o at time t_s by three different paths:

(a)   For $t' < 0$, $\gamma = 0$; for $0 \leq t' \leq t_s$, $\gamma = \gamma_o$.
(b)   For $t' < t_s$, $\gamma = 0$; for $t' \geq t_s$, $\gamma = \gamma_o$.
(c) $\gamma(t') = (\gamma_o/t_s)t'$.

Solution.

(a) This is good old stress relaxation. From Equation 15.8*a*:

$$\tau(t_s) = G\gamma_0 e^{-t_s/\lambda}$$

(b) This corresponds to the initial extension in stress relaxation:

$$\tau(t_s) = G\gamma_0$$

(c) This is the shear analog of an engineering stress–strain test with the constant shear rate $\dot{\gamma}_o = \gamma_o/t_s$. By analogy to the solution of Example 15.6 above:

$$\tau(t_s) = G\dot{\gamma}_0\lambda(1 - e^{-t_s/\lambda}) = G(\gamma_0/t_s)\lambda(1 - e^{-t_s/\lambda})$$

15.7 DYNAMIC MECHANICAL TESTING

Creep and stress-relaxation measurements correspond to the use of step-response techniques to analyze the dynamics of electrical and process systems. Those familiar with these areas know that frequency-response analysis is perhaps a more versatile tool for investigating system dynamics. An analogous procedure, *dynamic mechanical testing*, is applied to the mechanical behavior of viscoelastic materials. It is based on the fundamentally different response of viscous and elastic elements to a sinusoidally varying stress or strain.

If a sinusoidal strain, $\gamma = \gamma' \sin \omega t$ (where ω is the angular frequency in radian/s) is applied to a linear spring, since $\tau = G\gamma$, the resulting stress $\tau = G\gamma' \sin \omega t$ is in phase with the strain. For a linear dashpot, however, because the stress is proportional to the rate of strain rather than the strain, $\tau = \eta\dot{\gamma} = \eta\omega\gamma' \cos \omega t$, the stress is 90° out of phase with the strain. These relations are sketched in Figure 15.14.

As might be expected, viscoelastic materials exhibit some sort of intermediate response, which might look like Figure 15.15b. This can be thought of as being a projection of two vectors, τ^* and γ^*, rotating in a complex plane (Figure 15.15a). The angle between these vectors is the *phase angle* δ ($\delta = 0$ for a purely elastic material and 90° for a purely viscous material). It is customary to resolve the vector representing the dependent variable into components in phase (designated by a prime) and 90° out of phase (designated by a double prime) with the independent variable. In this example, the applied strain is the independent variable, so the stress vector (τ^*) is resolved into its in-phase (τ') and out-of-phase (τ'') components, $|\gamma^*| = \gamma'$ and $\gamma'' = 0$. In complex notation,

$$\tau^* = \tau' + i\tau'' \tag{15.36}$$

where i is the out-of-phase unit vector.

An *in-phase* or *storage* modulus is defined by

$$G' \equiv \frac{\tau'}{\gamma'} \quad \text{storage modulus (in-phase component)} \tag{15.37}$$

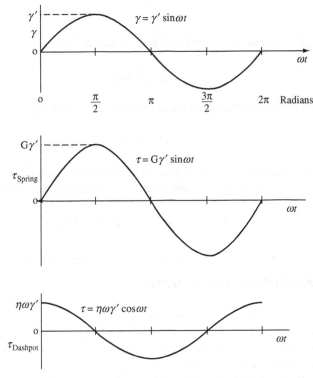

FIGURE 15.14 Stress in a linear spring and in a linear dashpot in response to a sinusoidal applied strain.

and an *out-of-phase* or *loss* modulus is defined by

$$G'' \equiv \frac{\tau''}{\gamma''} \quad \text{loss modulus (out-of-phase component)} \tag{15.38}$$

Both of these moduli have important physical significance to the design of materials. A material with a high storage modulus and low loss modulus will be springy—it can be used over and over again with little deformation. A material with a high loss modulus and low storage modulus will favor the dashpot, resulting in an easily deformed material (which is particularly useful in absorbing energy in a car collision).

The *complex modulus* G^* is the *vector sum* of the in-phase and out-of-phase moduli:

$$G^* \equiv G' + iG'' = \frac{\tau' + i\tau''}{\gamma'} = \frac{\tau^*}{\gamma^*} \quad \text{Complex modulus} \tag{15.39}$$

Additionally, a *complex viscosity*, η^*, may be defined as

$$\eta^* \equiv \eta' - i\eta'' = \frac{\tau^*}{\dot{\gamma}^*} \quad \text{Complex viscosity} \tag{15.40}$$

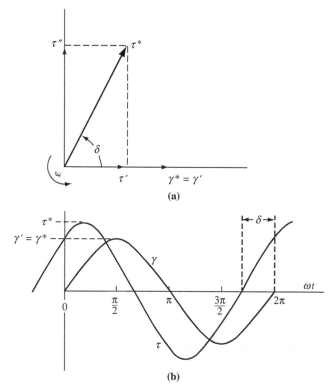

FIGURE 15.15 Quantities in dynamic testing: (a) rotating vector diagram; (b) stress and strain.

Also [10],

$$\dot{\gamma}^* = i\omega\gamma^* \tag{15.41}$$

By combining Equations 15.39–15.41 (and recalling that $i^2 = -1$), we get

$$\eta''\omega + i\omega\eta' = G' + iG'' \tag{15.42}$$

Comparison of the real (in-phase) and imaginary (out-of-phase) parts of Equation 18.40 gives

$$G' = \eta''\omega \tag{15.43}$$

and

$$G'' = \eta'\omega \tag{15.44}$$

Furthermore, combination of Equations 15.40–15.43 and 15.43 reveals that

$$G^* = G' + iG'' = i\omega\eta^* \tag{15.45}$$

From the geometry of Figure 15.15 and the relations above, the *loss tangent*, $\tan \delta$, is

$$\tan \delta = \frac{\tau''}{\tau'} = \frac{G''}{G'} = \frac{\eta'}{\eta''} \quad \text{Loss tangent} \tag{15.46}$$

The physical significance of the quantities just defined can best be appreciated by considering what happens to the energy applied to a sample undergoing cyclic deformation. From Equation 13.3, the work done on a unit volume of material undergoing a pure shear deformation is

$$W = \int \tau d\gamma \tag{15.47}$$

From Figure 15.15,

$$\gamma = \gamma' \sin \omega t \tag{15.48}$$

and

$$\tau = |\tau^*| \sin(\omega t + \delta) \tag{15.49}$$

Differentiating Equation 15.48 with respect to (ωt) gives

$$d\gamma = \gamma' \cos \omega t \, d(\omega t) \tag{15.50}$$

Inserting Equations 15.49 and 15.50 into Equation 15.47 gives

$$W = |\tau^*| \gamma' \int \sin(\omega t + \delta) \cos \omega t \, d(\omega t) \tag{15.51}$$

Let us first consider the work done on the first quarter-cycle of the applied strain, that is, integrate Equation 15.51 between 0 and $\pi/2$. Using appropriate trigonometric identities and a good set of integral tables gives

$$W(1\text{st} \tfrac{1}{4}\text{cycle}) = |\tau^*| \gamma' \left(\frac{\cos \delta}{2} + \frac{\pi}{4} \sin \delta \right) \tag{15.52}$$

Putting in terms of moduli or viscosities, using the trigonometry of Figure 15.15 and Equations 15.41 and 15.42, we find:

$$W(1\text{st} \tfrac{1}{4}\text{cycle}) = \frac{(\gamma')^2}{2} G' + \frac{\pi}{4}(\gamma')^2 G'' \tag{15.53a}$$

$$W(1\text{st} \tfrac{1}{4}\text{cycle}) = \frac{(\gamma')^2}{2} \omega\eta'' + \frac{\pi}{4}(\gamma')^2 \omega\eta' \tag{15.53b}$$

The first term on the right side of Equation 15.53a is simply the work done in straining a linear spring of modulus G' an amount γ' (the area under the spring's stress–strain curve).

It, therefore, represents the energy stored elastically in the material during its straining in the first-quarter cycle. Hence, G' is the storage modulus. If the applied mechanical energy (work) is not stored elastically, it must be "lost," converted to heat through molecular friction, that is, viscous dissipation, within the material. This is precisely what the second term on the right represents, so G'' is known as the loss modulus. Likewise, from Equation 15.53b, stored energy is proportional to η'' and the dissipated energy is proportional to η'.

Considering the second quarter of the cycle, integrating Equation 15.51 from $\pi/2$ to π gives results identical to Equation 15.53 except that the sign on the first (storage) term is negative. This simply means that the energy stored elastically in straining the material from 0 to γ' is recovered when it returns from γ' to 0. Thus, over a half cycle (0 to π or 180°) or a full cycle (0 to 2π), there is no net work done or energy lost by the elastic component. The sign of the second term, however, is positive for any quarter cycle, so the net energy loss (converted to heat within the material) for a full cycle (also obtainable by integrating Eq. 15.51 between 0 and 2π) is simply

$$W(\text{complete cycle}) = \pi(\gamma')^2 G'' = \pi(\gamma')^2 \omega \eta' \qquad (15.54)$$

The average power dissipated as heat within the material $\langle \dot{w} \rangle$ is obtained by dividing the energy dissipated per cycle by the period (time) of a cycle, $2\pi/\omega$:

$$\langle \dot{W} \rangle (\text{avg. power dissipation}) = \tfrac{1}{2}(\gamma')^2 \omega G'' = \tfrac{1}{2}(\gamma')^2 \omega^2 \eta' \qquad (15.55)$$

These results are of direct importance in the design of polymeric objects that are subjected to cyclic deformation. In a tire, for example, high temperatures contribute to rapid degradation and wear. A rubber compound with a low G'' (or η') therefore helps to minimize the energy dissipation and the resultant heat buildup. Moreover, dissipated energy wastes gasoline, therefore such a compound also contributes to better gas mileage. In the design of an engine mount, however, the goal is usually to prevent vibrations being transmitted from the engine. Here, a material with a large G'' (or η') would dissipate considerable vibrational energy as heat rather than transmit it to the passengers.

Dynamic mechanical analyses are also done to estimate a material's *fatigue*. The fatigue lifetime for a polymer is defined as the number of cyclic loading and unloading it can stand for a given stress before it fails. The higher the applied stress, the lower the fatigue lifetime.

Example 15.8 Obtain expressions for the quantities G', G'', $|G^*|$, $\tan \delta$, η', η'', and $|\eta^*|$ for a Maxwell element.

Solution. This problem is solved in Reference 10 (p. 56ff) by direct integration of the differential equation for the Maxwell element. Here, we will apply Boltzmann's superposition principle to obtain the results and, in doing so, again illustrate how information from one type of linear test (stress relaxation) may be used to predict the response in another (dynamic testing).

The shear stress-relaxation modulus (memory function) for a Maxwell element is

$$G_r(t) = Ge^{-t/\lambda}$$

We will assume that the element has *always* been subjected to a shear strain:

$$\gamma(t') = \gamma' \sin(\omega t')$$

which, when differentiated with respect to past time t', gives

$$\dot{\gamma}(t') = \gamma' \omega \cos(\omega t')$$

(Keep in mind that the prime here has two entirely different meanings: when applied to t, it designates past time; when applied to γ, it means the in-phase component of the strain.) Plugging these results into Equation 15.31 we get

$$\tau(t) = G\gamma' \omega \int_{-\infty}^{t} e^{-(t-t')/\lambda} \cos(\omega t') dt'$$

Evaluating the above integral is a nontrivial exercise, but again, with appropriate trigonometric identities and a good table of integrals, the result is

$$\tau(t) = \frac{G\gamma'(\omega\lambda)^2}{1 + (\omega\lambda)^2} \sin(\omega t) + \frac{G\gamma'(\omega\lambda)}{1 + (\omega\lambda)^2} \cos(\omega t)$$

(By assuming that the element has *always* been subjected to the periodic strain and integrating from $t' = -\infty$, we eliminate the transient part of the solution that would arise had we integrated from $t' = 0$.) It is clear that the first (sin) term in the above expression is in phase with the applied strain, while the second (cos) term is 90° out of phase with the applied strain. Therefore,

$$\tau' = \frac{G\gamma'(\omega\lambda)^2}{1 + (\omega\lambda)^2} \quad \text{and} \quad \tau'' = \frac{G\gamma'\omega\lambda}{1 + (\omega\lambda)^2}$$

From here on, its definitions and algebra are as follows:

$$G' = \frac{\tau}{'\gamma} = \frac{G(\omega\lambda)^2}{1 + (\omega\lambda)^2} \quad \text{and} \quad G'' = \frac{\tau''}{\gamma'} = \frac{G\omega\lambda}{1 + (\omega\lambda)^2}$$

$$|G^*| = [(G')^2 + (G'')^2]^{1/2} = \frac{G\omega\lambda}{[1 + (\omega\lambda)^2]^{1/2}}$$

$$\tan\delta = \frac{G''}{G'} = \frac{1}{\omega\lambda}$$

$$\eta' = \frac{G''}{\omega} = \frac{G\lambda}{1 + (\omega\lambda)^2} = \frac{\eta}{1 + (\omega\lambda)^2}$$

$$\eta'' = \frac{G'}{\omega} = \frac{G\lambda^2\omega}{1 + (\omega\lambda)^2} = \frac{\eta\omega\lambda}{1 + (\omega\lambda)^2}$$

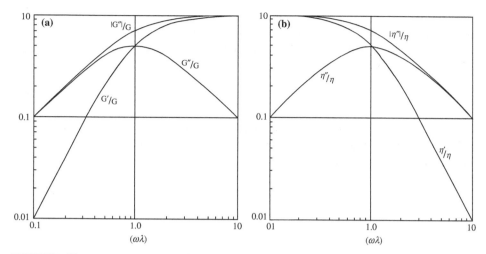

FIGURE 15.16 (a) Dynamic moduli of a Maxwell element (Example 15.8). (b) Dynamic viscosities of a Maxwell element (Example 15.8).

$$|\eta^*| = [(\eta')^2 + (\eta'')^2]^{1/2} = \frac{\eta}{[1 + (\omega\lambda)^2]^{1/2}}$$

The dynamic moduli are plotted in dimensionless form in Figure 15.16a and the dynamic viscosities are plotted in Figure 15.16b.

As when applied to other types of mechanical tests, the Maxwell element will not win any prizes for quantitatively fitting dynamic data for real materials. Nevertheless, Example 15.8 does serve to illustrate the frequency dependence of dynamic mechanical properties, and Figures 15.16a and 15.16b do in some ways that resemble the variation in isothermal dynamic data with frequency for real materials. In particular, the apparent "stiffness" $|G^*|$ increases to a limiting value with frequency. In the model, the dashpot simply cannot keep up with high frequencies, leaving only the spring to respond. Also, the maximum in G'' is usually observed at frequencies in the range where G' and $|G^*|$ are falling from their high-frequency limit. In the model, at low frequencies, the dashpot offers little resistance to motion and so dissipates little energy. At high frequencies, its high resistance prevents its motion, limiting response to the spring, and energy dissipation again falls off.

With real materials, the high-frequency limit of $|G^*|$ and G' corresponds quantitatively to the moduli obtained in the limit of short times in stress-relaxation and creep measurements, that is,

$$|G^*|(\omega \to \infty) = G'(\omega \to \infty) = G_r(t \to 0) = \frac{1}{J_c(t \to 0)} \qquad (15.56)$$

Furthermore, the measured low-frequency limit of $|\eta^*|$ and η' agrees with the zero-shear-rate steady-flow viscosity η_o:

$$|\eta^*|(\omega \to 0) = \eta'(\omega \to 0) = \eta_0 = \eta(\dot{\gamma} \to 0) \qquad (15.57)$$

The viscosity analogy has been pushed a bit further. The drop in $|\eta^*|$ with frequency resembles the variation in steady-flow viscosity η with shear rate. On a purely empirical basis, Cox and Merz [11] suggested that $|\eta^*|$ and η are the same when compared at equal values of frequency and shear rate:

$$|\eta^*|(\omega) \approx \eta(\dot{\gamma}) \quad \text{at } \omega = \dot{\gamma} \tag{15.58}$$

Although not exact, Equation 15.56 appears to be at least a reasonable approximation.

For a generalized Maxwell model consisting of n elements (Figure 15.10), the results obtained for a single Maxwell element in Example 15.8 are readily generalized to

$$G' = \sum_{i=1}^{n} \frac{G_i(\omega\lambda_i)^2}{1 + (\omega\lambda_i)^2} \tag{15.59}$$

and

$$G'' = \sum_{i=1}^{n} \frac{G_i\omega\lambda_i}{1 + (\omega\lambda_i)^2} \tag{15.60}$$

(the other dynamic properties may be obtained from the two above, as in Example 15.8). Presumably, the G_i and λ_i determined from, for example, stress-relaxation data can be used in Equations 15.59 and 15.60 to predict dynamic response, and vice versa. Equations 15.28 and 15.29 should be equally applicable to both. Also, the continuous distributions $G(\lambda)$ and $J(\lambda)$ as obtained from stress-relaxation and creep measurements are at least approximately interconvertible with $G'(\omega)$ and $G''(\omega)$ [2,4,8,9].

Techniques used to determine the complex behavior of real viscoelastic materials are covered in Chapter 16.

15.8 SUMMARY

This chapter described spring and dashpot models for the viscoelastic behavior of polymers. While a truly elastic material will stretch and contract back to its original form, most polymers will deform to an extent dependent on the magnitude and duration of an applied stress (shear or elongational). The ability of these materials to resist stretching or shearing is measured through the elastic or shear moduli (E or G). Dynamic mechanical testing breaks down these moduli into two components: a storage modulus that represents elastic behavior and a loss modulus that represents viscous flow or deformation. These moduli change with temperature, representing the material behavior below T_g (glassy), between T_g and T_m (rubbery) and above T_m (flow). Finally, time–temperature superposition was shown to be a useful tool by which the long-time behavior of a polymer can be predicted through measurements covering a range of temperatures.

PROBLEMS

1 Given enough springs and dashpots, it is possible in principle to fit any linear response to any desired degree of accuracy. For each of the individual springs, it is true by definition that $G_i = 1/J_i$ (the shear modulus of each individual spring is the reciprocal of its compliance).

Does it follow that $G_r(t) = 1/J_c(t)$? That is, is a material's stress-relaxation modulus always the reciprocal of its creep compliance? *Hint:* Examine this question for the simplest of materials, a Maxwell element.

2 Isothermal tensile creep data on polymers can sometimes be fit by an empirical equation of the form

$$\varepsilon = A(B + t^C)\sigma_o$$

where ε is the tensile strain, σ_o is the (constant) applied tensile stress, and t is the time. A, B, and C are positive constants.

(a) Is a material that follows this equation linear?

(b) What is the instantaneous elastic compliance?

(c) Is a material that follows this equation a liquid or a solid?

(d) Can this equation ever describe steady-state viscous flow? If so, under what conditions, and what is the equilibrium tensile viscosity under those conditions?

3 Example 15.1 analyzes the response of a Maxwell element in an engineering stress–strain test. Do the same thing for a Voigt–Kelvin element. Illustrate the effect of strain rate with a sketch. Do not expect great realism.

4 The four-parameter model discussed in the text consists of a series combination of Maxwell and Voigt–Kelvin elements. Here, consider a four-parameter model that consists of a *parallel* combination of Maxwell (G_1, η_1) and Voigt–Kelvin (G_2, η_2) elements.

(a) Does this model represent a fluid or a solid?

(b) Is the model suited for representing creep, stress relaxation, neither, or both?

(c) In terms of model parameters, write the steady-state ($t = \infty$) strain in response to an applied stress τ_o.

(d) Which of the following is the model capable of representing?
   1. Instantaneous elastic deformation
   2. Retarded elastic deformation
   3. Steady-state viscous flow
   4. Instantaneous elastic recovery
   5. Retarded elastic recovery
   6. Permanent set

5 Analyze the dynamic properties of a Voigt–Kelvin element, that is, obtain G' and G'' in terms of model parameters G and η and the frequency ω. *Hint:* Unless you are a masochist, do not use the Boltzmann principle here. Just examine the response to a sinusoidal strain.

6 Examine the response of a two-element generalized Maxwell model (Figure 15.10) with the following parameters:

$$G_1 = 10^9 \, \text{Pa} \qquad \eta_1 = 10^6 \, \text{P}$$
$$G_2 = 10^5 \, \text{Pa} \qquad \eta_2 = 10^6, \, 10^8, \, \infty \, \text{P}$$

(a)   Plot log $G_r(t)$ versus log t.

(b)   Plot log G' and log G'' versus log ω with $\eta_2 = 10^8$ P.

(c)   Assume that the above model parameters apply at $T_g + 20\,°C$, and further that the model's temperature shift factors a_T are given by the WLF equation with the "universal" constants. Plot the log of the 10-s relaxation modulus, $G_r(10)$ versus $(T - T_g)$.

Cover a wide enough range of the independent variable so that the G's range between 10^3 Pa and 10^9 Pa.

7   The text defines the Deborah number for transient and dynamic tests. How would you define De for an engineering stress–strain test?

8   A Maxwell element has $G = 10^7$ Pa and $\lambda = 1$ s at a temperature of $T = T_g + 20\,°C$. Assume that the element's shift factors are given by the WLF equation with "universal" constants. Calculate G' and G'' at a frequency of $\omega = 1$/s and a temperature of $T = T_g + 40\,°C$.

9   A three-parameter model, the Zener element or standard linear solid (SLS), has been used to represent viscoelastic behavior in certain solids. Two equivalent forms of the SLS are:

(a)   A spring (G_2) *in series* with a Voigt–Kelvin element (G_1, η_1) and

(b)   A spring (G_3) *in parallel* with a Maxwell element (G_4, η_2).

Obtain the equations necessary to relate the spring constants in one form of the SLS to the spring constants in the other, that is, given G_1 and G_2, how could you determine G_3 and G_4, or vice versa?

10   The shear creep compliance of a thermoplastic at $25\,°C$ is described by

$$J_{c25}(t) = 1.2 \times 10^{-3}\, t^{0.10} \quad (\text{m}^2/\text{N with } t \text{ in s})$$

(a)   This material has $T_g = 0\,°C$. Assume that its temperature shift factors are given by the WLF equation (Eq. 14.11) with the "universal" constants. Obtain an equation that gives its shear creep compliance at $35\,°C$, $J_{c35}(t)$.

(b)   The material is subjected to the following stress history at $25\,°C$:

$t < 0$ s	$\tau = 0$ Pa
$0 \leq t < 1000$ s	$\tau = 1000$ Pa
$1000 \leq t < 2000$ s	$\tau = 1500$ Pa
$2000 \leq t$ s	$\tau = 0$ Pa

Calculate the shear strain at 2500 s.

11   The four-parameter model (Figure 15.1f) is subjected to the following stress history, in which τ_o is a constant stress:

$t < 0$	$\tau = 0$
$0 \leq t < t_s$	$\tau = -\tau_o$
$t_s \leq t < 2t_s$	$\tau = +2\tau_o$
$2t_s \leq t$	$\tau = 0$

Write an expression for the permanent set $\gamma(\infty)$ that results from this stress history.

12 Consider the four-parameter model (Figure 15.1f). In terms of model and test parameters, answer the following.

(a) What is the initial modulus $G_r(0)$ in stress relaxation?

(b) What is the equilibrium modulus $G_r(\infty)$ in stress relaxation?

(c) Repeat (b) when the model is used to represent a lightly crosslinked polymer.

(d) In the shear analog of an engineering stress–strain test ($\dot{\gamma}_0$ constant), what is the limiting high strain rate ($\dot{\gamma}_0 \to \infty$) modulus.

(e) Repeat (d) for the limiting low strain rate ($\dot{\gamma}_0 \to 0$) modulus.

(f) In a dynamic test, what is the limiting low-frequency storage modulus, $G'(\omega \to 0)$?

(g) In a dynamic test, what is the limiting high-frequency storage modulus, $G'(\omega \to \infty)$?

(h) Repeat (f) for G''.

(i) Repeat (g) for G''.

13 Consider a four-parameter model made up of a Maxwell element (G_1, η_1) *in parallel* with a Voigt–Kelvin element (G_2, η_2). Obtain expressions for $G'(\omega)$ and $G''(\omega)$ for this model in terms of the model parameters.

REFERENCES

[1] Reiner, M., *Phys. Today*, Jan. 1964, p. 62.

[2] Ferry, J.D., *Viscoelastic Properties of Polymers*, 3rd ed., Wiley, New York, 1980.

[3] Eirich, F.R. (ed.), *Rheology*, Vols. **1–4**, Academic, New York, 1956–1964.

[4] Tobolsky, A.V., *Properties and Structure of Polymers*, Wiley, New York, 1960.

[5] Nielsen, L.E., *Mechanical Properties of Polymers and Composites*, Vol. 1, Dekker, New York, 1974.

[6] Spriggs, T.W., *Chem. Eng. Sci.*, **20**, 931 (1965).

[7] Rouse, P.E., Jr., *J. Chem. Phys.* **24**, 269 (1956).

[8] Schwarzl, F.R., *Pure and Appl. Chem.* **23**, 219 (1970).

[9] Schwarzl, F.R., *Rheol. Acta*, **8**, 6 (1969); **9**, 382 (1970); **10**, 166 (1971); **14**, 581 (1975).

[10] McKelvey, J.M., *Polymer Processing*, Wiley, New York, 1962.

[11] Cox, W.P. and Merz, E.H., *J. Polym. Sci.* **28**, 619 (1958).

CHAPTER 16

POLYMER MECHANICAL PROPERTIES

16.1 INTRODUCTION

This chapter covers some of the methods and instruments used to determine the mechanical properties of polymers. Examples of instrument designs and typical data generated in these measurements will be introduced. In particular, automated axial tensiometers (to find elastic modulus, yield stress, and ultimate stress), dynamic mechanical analyzers (to determine storage and loss moduli), and rheometers (to measure flow viscosity) will be introduced. This chapter considers the principles behind the devices used to establish and measure the properties of viscometric flows. One of the common techniques used to determine viscous flow properties, Poisueille (laminar) flow in cylindrical tubes, is also important in technical applications, as polymer melts and solutions are often transported and processed in this manner. The time–temperature superposition principle is also covered as a way to predict polymer behavior over long timescales by testing materials across a range of temperatures.

16.2 MECHANICAL PROPERTIES OF POLYMERS

A number of physical parameters can be measured as a way to characterize polymers. The specific tests that are done are normally dictated by the end use of a particular part. We have defined many of these parameters in the previous few chapters, such as elastic (or Young's) modulus, E, shear (or Hooke's) modulus, G, storage modulus (E' or G'), loss modulus (E" or G"), tan δ, which all apply to solids, as well as the viscosity used for polymer melts (and solutions) as described with polymer rheology. In addition to these terms, you may also run across hardness (how resistant the polymer is to penetration by a needle), toughness

Fundamental Principles of Polymeric Materials, Third Edition. Christopher S. Brazel and Stephen L. Rosen.
© 2012 John Wiley & Sons, Inc. Published 2012 by John Wiley & Sons, Inc.

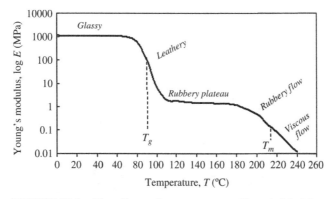

FIGURE 16.1   The effects of temperature on Young's Modulus.

(measures the total amount of energy that can be absorbed by a polymer before it fails or breaks), impact resistance (ability not to shatter upon sudden application of force), fatigue (ability to withstand cyclic forces and maintain mechanical integrity), and wear (how the surface of a polymer may be degraded or worn by friction).

While any individual polymer will have unique properties (due to sample geometry, polymer molecular weight, presence of crystals or crosslinks, etc.), some generalized characterizations can be made. We are well aware that temperature plays a large role in the state and behavior of polymers. Many mechanical properties follow a similar trend to that observed in Figure 16.1 for elastic modulus versus temperature. There are five general regions that describe the polymer depicted in this graph. At low temperatures (below T_g), the polymer is considered glassy. Closer to the T_g, the feel of the polymer softens slightly, and thus is referred to as leathery; above the T_g, the polymer reaches a rubbery plateau, where it is pliable and flexible. Nearing T_m, the polymer starts to melt and reaches a rubbery flow region, where the flow may be somewhat reversible if a force is removed. At the highest temperatures (assuming the polymer is not degraded), we have viscous flow, which is covered by the discussion on rheology in Chapter 14. While the absolute numbers on the y-axis will not be true for all polymers, it does show that temperature has a large effect on mechanical properties of the polymer.

16.3 AXIAL TENSIOMETERS

In many common applications of polymers, they are exposed to a uniaxial stress, by which they can stretch and rebound (if purely elastic) or stretch and deform with a smaller degree of rebound (for true viscoelastic materials). The equipment used for testing polymer mechanics are largely the same as those used for metals or ceramics, but the amount of force required to stretch a polymer will be (considerably) smaller. Automated materials testing apparatus (such as the one shown in Figure 16.2) have a range of designs, from the most common stretching, done by placing flat dogbone-shaped samples between clamps and measuring the force required to stretch, to instruments that can measure the force and deformation on samples subjected to compression or three-point bending and even instruments that can model human joints or test materials that are submerged in water.

For such axial tension experiments, the results are normally plotted as stress–strain curves. If the rate of strain is fixed and constant, the x-axis (independent variable) can be

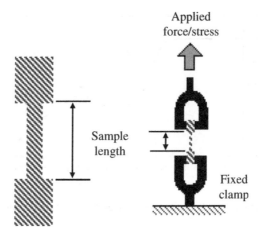

FIGURE 16.2   An automated materials testing apparatus typically applies a known stress on a dogbone-shaped sample and measures the resulting stretch (strain). A fixed or linear strain can also be programmed with the required force (or stress) measured as the sample is stretched.

either strain or time. The y-axis normally represents the stress on the given sample that is calculated by dividing the forces measured during the stretching experiment by the initial cross-sectional area of the sample, to give the typical pressure units of stress (Pa, or more commonly MPa or GPa).

A set of stress–strain curves are shown in Figure 16.3. At the lower temperature (the upper plot), the slopes of the linear portion of each plot are higher, as would be expected for

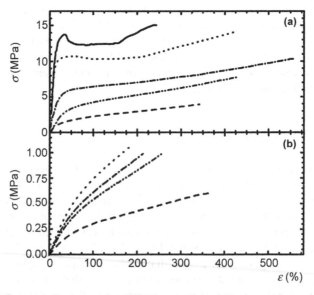

FIGURE 16.3   Stress–strain curves for different samples of block copolymers of caprolactone with n-butyl acrylate tested at (a) room temperature and (b) 70 °C. Polymer compositions were 0, 20, 39, 50 and 71 wt% n-butyl acrylate from top to bottom in (a). Reprinted with permission from Referenece 1. Copyright 2001 National Academy of Sciences, USA.

glassy polymers (compared to rubbery polymers at the higher temperature). The modulus of elasticity is easily found from the slope in the linear portion of the curve.

$$\sigma = E\varepsilon \tag{16.1}$$

The higher the slope, the higher the modulus, indicating that the material resists stretching with greater force. For most polymers, there exists a yield stress, σ_y, below which the deformation is elastic, as the material can snap back to its original conformation upon release of the force. The *yield stress* is most easily seen in Figure 16.3a, which marks the transition from linearity to uneven stretching. Beyond this yield stress, the material begins to deform permanently. As the force continues to stretch the sample, polymer chains physically move and rearrange the macromolecular structure. This causes necking and narrowing of the sample width and thickness. Because the polymer chains are slipping over each other above the yield stress, the sample is likely to "give" much easier, resulting in what can be significant, stretching without any additional force. This is the behavior that is described by the dashpot in viscoelastic models, but just because chains are able to slip over each other and rearrange does not mean that the material has failed ... just that it cannot snap back to its original configuration. By continuing to stretch the polymer, it may actually toughen a bit (often because the rearrangement of chains leads to an increase in crystallinity) before reaching the *ultimate stress*, σ_u, where the sample breaks. In Figure 16.3, the ultimate stress for each of the samples is the highest stress each sample can withstand, which is at the right end of each curve, just prior to the sample breaking (and the measured stress dropping to zero).

These experiments are great at picking out mechanical properties of materials subjected to single axial stresses. To study the effect of cyclic stresses, a dynamic mechanical analyzer (DMA) is recommended (next section).

16.4 VISCOSITY MEASUREMENT

As noted in Chapter 5, viscometry can be used to estimate polymer molecular weight, $\bar{M}_y$, but those solutions were required to be dilute. For polymer melts and more concentrated solutions, viscosity measurements require viscometers configured somewhat differently from the capillary viscometers introduced in Chapter 5. Details of such instruments and procedures have been reviewed [2,3].

16.4.1 The Couette Viscometer

One common device for measuring viscous properties is the cup-and-bob or Couette viscometer (Figure 16.4). The fluid is confined in the gap between two concentric cylinders, one of which rotates relative to the other at a known angular velocity while the torque on one is measured. This is a classic example of viscometric flow. In cylindrical coordinates, we assume only a tangential velocity component, so the 1 coordinate is the tangential or θ direction and the 2 coordinate is the radial direction.

Example 16.1

(a) Neglecting end effects, determine the shear stress as a function of radius in terms of the measured torque on the stationary inner cylinder (bob), $M(R_i)$, and the geometry

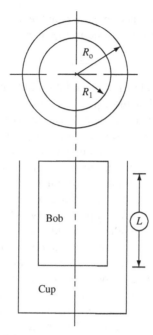

FIGURE 16.4   Schematic of Couette (cup-and-bob) viscometer.

of the apparatus as the outer cylinder (cup) is rotated with an angular velocity ω (radians/s).

(b) The is the classic example of a viscometric flow in which the shear rate is not equal to the velocity gradient dv_θ/dr. As a result, determining the shear rate as a function of radius is not easy. Bird *et al.* [4] did it by *assuming* a Newtonian relation between shear stress and shear rate. In general, however, the nature of the flow curve is not known *a priori*, and must be determined by viscometry. One means of doing this is to make the gap between the cylinders $(R_o - R_i)$ very small compared to the radius of either cylinder. Letting $R_o - R_i = \delta$ and $R_o \approx R_i = R$, obtain the expression for shear rate in terms of ω and the geometry.

Solution.

(a) In a rotating system at steady state, Σ torques $= 0$, or there would be angular acceleration. Consider a ring of fluid with inner radius R_i and outer radius r: $M(r) = M(R_i)$

$$\underbrace{\underset{\substack{\text{Surface} \\ \text{area}}}{2\pi r L} \quad \underset{\substack{\text{Moment} \\ \text{arm}}}{\tau(r)} \; r}_{\text{Force}} = \underset{\substack{\text{Measured} \\ \text{torque}}}{M(R_i)}$$

$$\tau(r) = \frac{M(R_i)}{2\pi r^2 L}$$

(16.2)

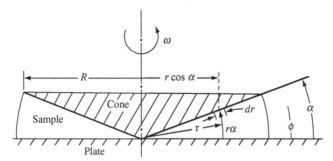

FIGURE 16.5   Schematic of cone-and-plate viscometer.

(b) This situation approximates the case of two flat plates separated by a distance δ (i.e., for a large enough cylinder, the fluid does not experience much curvature), one sliding past the other with a linear velocity equal to the tangential velocity $R\omega$:

$$\dot{\gamma} = \frac{R\omega}{\delta} \quad \text{(for } \delta \ll R_i \text{ only)} \tag{16.3}$$

Where the geometric approximations in the example above are not applicable, a more sophisticated analysis has been developed by Kreiger and Maron [5]. Another challenge to analyzing the Couette viscometer is accounting for the area below the bob at the bottom of the viscometer. This is best accounted for by making measurements with two fluid depths, the lower being well above the bottom of the bob, and using the differences between torques and depths in Equation 16.2, thereby subtracting out the effects of non-Couette flow at the bottom of the viscometer. Another approach is illustrated later in Example 16.3.

16.4.2 The Cone-and-Plate Viscometer

The cone-and-plate viscometer is another type of rotational viscometer. Here, the sample is sheared between a flat plate and a broad cone whose apex contacts the plate (Figure 16.5). For small cone-plate angles α, this approximates a viscometric flow with (in spherical coordinates) flow in the tangential θ or 1 direction and the gradient in the azimuthal ϕ or 2 direction. Here, the radial direction is the neutral or 3 coordinate.[1] It turns out that true viscometric flow of this type is inconsistent with the equations of motion if the inertial terms are included. (Formal flow solutions for rotational visc-ometers normally neglect the inertial terms in the equations of motion, as we tacitly do here.) There must, therefore, be radial and azimuthal velocity components. These are minimized in practice by keeping α quite small, often less than $1°$. The great advantage of this type of device is that (for small α) the shear rate, and hence the shear stress, is uniform throughout the material.

---

[1] At first glance, it might seem that there is a gradient component in the r direction because the tangential velocity increases with r. However, material points in a cone at constant ϕ do not move relative to one another, that is, they undergo *rigid-body rotation*, so there is no shearing in the r direction.

Example 16.2 Obtain the expressions for shear rate and shear stress in a cone-and-plate viscometer in terms of the rate of cone rotation ω, the measured torque M, and the geometry.

Solution. (See Figure 16.5). The tangential velocity v_θ of a point on the cone relative to the plate is $v_\theta = \omega r \cos \alpha$. Fluid is sheared between that point and the plate over a distance $\delta = \alpha r$:

$$\dot{\gamma} = \frac{v_\theta}{\delta} = \frac{\omega r \cos \alpha}{\alpha r} \xrightarrow{\text{small } \alpha} \frac{\omega}{\alpha} \text{(independent of } r) \tag{16.4}$$

Torque = (Shear stress) (Area) (Moment arm)

$$dM = (\tau)(2\pi r \cos \alpha \, dr)(r \cos \alpha)$$

Integrating over the cone face gives

$$M = 2\pi\tau \cos^2 \alpha \int_0^{R/\cos \alpha} r^2 \, dr = \frac{2\pi\tau R^3}{3 \cos \alpha} \xrightarrow{\text{small } \alpha} \frac{2\pi\tau R^3}{3}$$
$$\tau = \frac{3M}{2\pi R^3} \tag{16.5}$$

Note: Here, τ is constant, and can be taken outside the integral, because $\tau = \tau(\dot{\gamma}.)$, and $\dot{\gamma}$. is constant, as shown by Equation 16.4.

Cone-and-plate viscometers of the type shown here are usually limited to fairly low shear rates. At higher shear rates, solutions tend to be flung from the gap by centrifugal force, and melts tend to "ball up" (like rubbing a finger over dry rubber cement). These problems can be overcome by enclosing the fluid around a biconical rotor, giving, in effect, two cone-and-plate viscometers back-to-back [6]. The flow curves earlier in this chapter were obtained with such a device.

Example 16.3 One means of minimizing end effects in a Couette viscometer is to make the bottom of a bob a cone, the apex of which contacts the base of the cup, so that the area beneath the bob is a cone-and-plate viscometer. For a Couette geometry in which the gap δ is much smaller than the bob radius R_i, what must the cone angle α be to match the shear rates in the Couette and the cone-and-plate regions?

Solution. By equating the expressions for shear rate from the previous two examples, $\alpha = \delta/R$.

16.4.3 The Disk-and-Plate Viscometer

Another type of viscometer that finds occasional use is the disk-and-plate viscometer (Figure 16.6). A disk of radius R rotates with an angular velocity of ω relative to a parallel

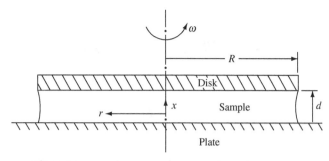

FIGURE 16.6 Schematic of disk-and-plate viscometer.

plate. The disk-and-plate are separated by a distance d ($d \ll R$), with the test fluid in between. The torque M on either the disk or the plate is measured. This is known as torsional (twisting) flow, and is another example of a viscometric flow.

Example 16.4

(a) Identify the 1 (flow), 2 (gradient), and 3 (neutral) directions with respect to the cylindrical coordinates in Figure 16.6.[2]

(b) Obtain an expression for $\dot{\gamma}$. Why is the cone-and-plate geometry usually preferred?

(c) Obtain an expression for the torque M needed to maintain a steady rate of rotation ω when a power-law fluid is in the gap.

Solution.

(a) 1 (flow) $= \theta$, 2 (gradient) $= z$, 3 (neutral) $= r$.

(b)

$$\dot{\gamma} = \frac{\text{Relative velocity, disk to plate}}{\text{Separation distance}} = \frac{r\omega}{d} \tag{16.6}$$

In this geometry, unlike the cone-and-plate, the shear rate increases linearly with radius, and so is not uniform throughout the fluid.

(c) Consider a differential ring of the disk (or plate) surface, at radius r with thickness dr:

$$dA = 2\pi r \, dr$$

$$dM = r\tau dA = 2\pi r^2 \tau dr$$

For a power law fluid, $\tau = K\dot{\gamma}^n = K(r\omega/d)^n$,

$$M = 2\pi K \left(\frac{\omega}{d}\right)^n \int_0^R r^{2+n} dr = \frac{2\pi K(\omega/d)^n}{3+n} R^{3+n}$$

---

[2] See the previous footnote. Here, points at constant z undergo rigid-body rotation like a phonograph record.

or

$$M = \frac{2\pi KR^3}{3+n} (\dot{\gamma}_R)^n \quad \text{(power-law fluid)} \qquad (16.7)$$

where

$$\dot{\gamma}_R = \frac{R\omega}{d} \qquad (16.8)$$

16.5 DYNAMIC MECHANICAL ANALYSIS: TECHNIQUES

Basically, three methods are available for determining dynamic properties: free oscillation, forced oscillation, and steady-state rotation. Experimental and analytical details for the first two are reviewed extensively by Ferry [7].

16.5.1 Free Oscillation

16.5.1.1 Torsion Pendulum   Free-oscillation measurements are made with a torsion pendulum (Figure 16.7). The sample is given an initial torsional displacement, and the frequency and amplitude decay of the oscillations are observed on release. G' is determined from the sample geometry, moment of inertia of the oscillating mechanism, and the observed period of oscillation. For example, with a cylindrical specimen of length L and radius R,

$$G' = \frac{8\pi LI}{R^4 P^2} \qquad (16.9)$$

where I is the moment of inertia of the oscillating mechanism and P is the observed period of oscillation ($P = 2\pi/\omega$). The damping or logarithmic decrement Δ is calculated from the amplitude decay of the oscillations (with a perfectly elastic material, there would be no damping of the oscillations):

$$\Delta = \ln\frac{A_1}{A_2} = \ln\frac{A_2}{A_3} = \ln\frac{A_i}{A_{i+1}} = \frac{1}{n}\ln\frac{A_i}{A_{i+n}} \qquad (16.10)$$

and for $\Delta < 1$:

$$\Delta \approx \pi\tan\delta \qquad (16.11)$$

Expressions for other specimen geometries and higher damping are reviewed by Nielson [9]. Solid or rubbery samples are twisted as illustrated in the form of rods, tubes, strips, etc. Liquids or soft solids may be contained in one of the geometries described for rotational viscometry earlier (Couette, cone-and-plate, etc.).

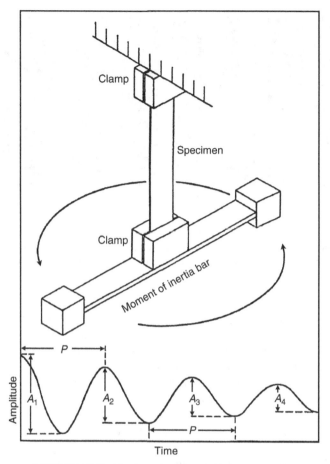

FIGURE 16.7 Torsion pendulum with its output [8].

16.5.1.2 *Torsional Braid Analysis*

In torsional braid analysis [10], a flexible, braided fiber, usually glass, is impregnated with the material to be studied. The impregnated fiber then becomes the torsion member in the pendulum. This type of device is useful for following the cure of a material that starts out as a liquid and cures to a solid (e.g., an epoxy). In analyzing the data from such a device, care must be exercised in separating interactions between the sample and the supporting fiber.

Although the frequency can be varied somewhat by changing the moment of inertia of the oscillating portion of the mechanism, torsion pendulums are usually intended to study only the temperature dependence of dynamic properties at a constant, relatively low frequency (≈ 1 cycle/s). On the other hand, they are inexpensive and rather simple to construct.

16.5.2 Forced Oscillation Devices

16.5.2.1 *Dynamic Mechanical Analyzers*

DMAs are forced-oscillation devices ("jiggle machines") that apply a sinusoidal stress or strain of known amplitude and frequency and measure the resulting strain or stress. The dynamic properties are calculated from the relation between the two. Solid materials are clamped into a DMA

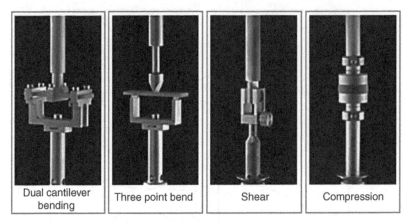

| Dual cantilever bending | Three point bend | Shear | Compression |

FIGURE 16.8   Dynamic mechanical analysis testing clamps (used with permission from TA Instruments, New Castle, DE). Different clamp setups can be used for tension, compression, three-point bending, and shear tests. The application of stress (or strain) is sinusoidal.

(Figure 16.8), where the cyclic load (stress or strain) can be applied, with the resulting strain or stress measured. For liquid samples, the geometries discussed in conjunction with rotational viscometry are often used with the drive system modified to produce sinusoidal rather than steady rotational deformation. Flexible samples such as fibers, films, and rubber are preloaded in tension and oscillated about a positive tensile strain so that they do not go slack at the "bottom" of the sine wave. Such tests give dynamic tensile properties, E', E'', etc., which are related to the corresponding shear properties by

$$|E^*| = 2|G^*|(1 + \nu) \tag{16.12}$$

where ν is Poisson's ratio ($\nu = 1/2$ for an incompressible material). Another type of forced-oscillation device applies a sinusoidal shear or compression wave to one end of a sample and monitors the attenuation of the wave as it progresses through the sample.

Forced-oscillation devices are generally intended to study dynamic properties as a function of frequency as well as temperature. Drive and detection systems for such devices may be strictly mechanical, but more sophisticated ones make use of piezoelectric crystals and, inorganic crystals (e.g., barium titanate) that change dimension in proportion to an applied voltage and, conversely, generate an output voltage proportional to an imposed deformation. Thus, the amplitude and, especially, the frequency of the applied strain can be conveniently controlled over a wide range in the form of electrical signals (the response of the crystals extends to very high frequencies, and electrons have very little inertia, unlike mechanical linkages). Input and output signals may be fed to a computer for online control and computations.

16.5.3   Steady Rotation Methods

The expense and complexity of oscillating devices and associated detection systems has led to the increased popularity of instruments that determine dynamic properties, but are driven in steady rotation and detect steady (nonoscillating) forces. This sounds like a contradiction in terms, but they really do work! We will attempt to describe the operation of one of the more popular of them. Mathematical details are provided in elegant fashion by Walters [11].

16.5.3.1 *Orthogonal Rheometry*

In the Maxwell orthogonal rheometer [12] (Figure 16.9), material is sheared between two parallel disks of radius R, separated from one another by a distance h, each rotating at the same steady angular frequency ω, but with their axes of rotation displaced by a distance d. Transducers are set up to measure three steady, orthogonal force components (hence, the name) on one of the disks, F_x, F_y, and F_z.

How, you ask, does this result in oscillating deformation? Well, the easiest way to see is to follow the motion of a point on the upper disk (point 2), relative to one on the lower disk (point 1) as the disks rotate. Here, we choose point 1 on the axis of rotation (center) of the lower disk and point 2 directly above it at the start of the analysis, $\omega t = 0$. This is the simplest choice, because point 1 remains stationary, but the same result will be obtained for any pair of points initially at the same values of x and y (you can prove this with a compass, ruler, and protractor). Figure 16.10a shows the relative displacement vector in the xy plane,

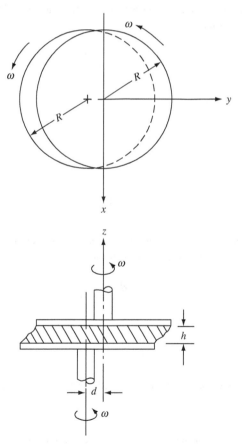

FIGURE 16.9 Orthogonal rheometer.

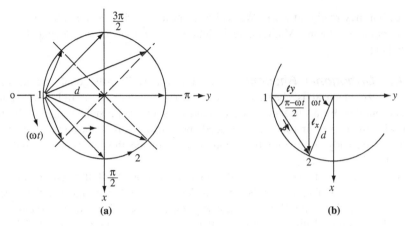

FIGURE 16.10   Analysis of the orthogonal rheometer: (a) relative displacement vector between point 1 on the center of the lower disk and point 2 immediately above it at $\omega t = 0$; (b) geometry of the displacement vector and its components.

l, and how it varies with the angle of rotation ωt. From Figure 16.10b, the magnitude of the relative displacement is then:

$$|\mathbf{l}| = 2d \ \sin\frac{\omega t}{2} \qquad (16.13)$$

and its x and y components are

$$l_x = d \ \sin(\omega t) \qquad (16.14)$$

and

$$l_y = |\mathbf{l}| \ \cos\left(\frac{\pi}{2} - \frac{\omega t}{2}\right) = d[1 - \cos(\omega t)] \qquad (16.15)$$

If we assume that the strain is simply the relative displacement divided by the disk separation h, the x and y components of strain are[3]

$$y_x = \frac{d}{h} \sin(\omega t) \qquad (16.16)$$

$$y_x = \frac{d}{h}[1 - \cos{(\omega t)}] \qquad (16.17)$$

Thus, both components of strain undergo simple harmonic oscillation as a result of the steady angular rotation.

---

[3] This requires that each fluid element move at fixed z in a circle that is centered on a line connecting the centers of the disks. There has been some controversy about this, but it is probably at least a good approximation for $d/h < 0.5$.

It is important to note that the x component of strain is symmetrical about 0, while the y component is not (it varies between 0 and $+2d/h$). For linear materials, because stress is directly proportional to strain, any shear stresses and, therefore, forces parallel to the xy plane produced by γ_x's in the region $0 < \omega t < \pi$ will be cancelled by those arising from the equal-and-opposite x-component strains in the region $(\pi < \omega t < 2\pi)$. As a result, the measured forces F_x and F_y can depend only on γ_y.

Looking at it another way, consider a purely elastic material in the rheometer to be represented by rubber bands stretched between points 1 and 2. Viewed from above, the rubber bands coincide with 1. It is obvious that the x components of the tug of the rubber bands will cancel, leaving only a net force (and, therefore, only a stress component τ_{zy}) in the y direction. Thus, the purely elastic stress is in phase with γ_y, so Equation 16.17 represents the variation of γ' with location in a disk as constant z, that is,

$$\dot{\gamma}(\omega t) = \gamma_y(\omega t) = \frac{d}{h}[1 - \cos(\omega t)] \tag{16.18}$$

and

$$\tau_{zy} = \tau' = \tau'(\gamma') \tag{16.19}$$

Any measured value of F_x, therefore, must arise from a viscous component, whose stress τ_{zx} will be 90° out of phase with γ_y, that is,

$$\tau_{zx} = \tau''(\gamma') \tag{16.20}$$

To determine the total force component on a disk, the stress-area product must be integrated over the surface of the disk:

$$F_y = \int_{\text{surface of disk}} \tau' \, dA \tag{16.21}$$

From Equation 15.37, $\tau' = G' \, \gamma'$ and, in polar coordinates, $dA = r \, d(\omega t) \, dr$, so that

$$F_y = G'\left(\frac{d}{h}\right) \int_0^R \int_0^{2x} [1 - \cos(\omega t)] \, d(\omega t) \, r \, dr = \pi R^2 \left(\frac{d}{h}\right) G' \tag{16.22}$$

or

$$G' = \frac{h/d}{\pi R^2} F_y \tag{16.23}$$

Similarly,

$$F_x = \int_{\text{surface of disk}} \tau'' \, dA \tag{16.24}$$

and since $\tau'' = G'' \, \gamma'$ (Eq. 15.38), it follows that

$$G'' = \frac{h/d}{\pi R^2} F_x \tag{16.25}$$

Equations 16.23 and 16.25 show how dynamic properties can be obtained from a device in steady rotation by measuring steady forces, which mechanically and analytically represents a great simplification over forced-oscillation techniques. It must be pointed out that there have been some questions as to whether a nonviscometric flow such as this can be used to determine quantities such as G', G'', η', η'', which are really defined in terms of viscometric deformations, but it is now pretty generally agreed that the technique is valid, at least in the limit of small strains, d/h [13,14].

Also, analyses are based on the assumption that both disks rotate at the same angular velocity, ω. In the usual instrument, one disk is driven and the other goes along for the ride; so with bearing friction and hydrodynamic effects, the assumption might not be strictly true. An analysis by Davis and Macosko [15] shows it to be a pretty good assumption under most conditions of interest.

16.5.3.2 Other Steady Rotation Devices   Other devices have been developed that operate along similar lines. The so-called balance rheometer confines a test fluid between two concentric hemispheres that rotate at the same rate but whose axes of rotation are at an angle to one another. Similarly, cone-and-plate geometry, in which the axis of the cone is not perpendicular to the plate, and Couette (cup-and-bob) geometry, with the bob not centered on the cup, can also be used to obtain dynamic data [11]. Dynamic tensile properties can be obtained for relatively rigid materials by subjecting a rotating cylindrical rod to a cantilever deflection [16].

Many of these dynamic tests can also be used to determine the thermal behavior of polymers. Figure 16.11 illustrates G' and damping $\approx \pi \tan \delta$ versus. T for poly(methyl methacrylate), an amorphous, linear polymer. The data were obtained with a torsion pendulum at about 1 cycle/s. At low temperatures, the typical glassy modulus 1–10 GPa is observed. In the vicinity of 110–130 °C, G' drops precipitously, ultimately reaching a plateau of 0.1–1 MPa, the typical rubbery modulus. Also, a sharp peak in the damping is observed in this region. Although the temperature at which this peak and drop are observed is frequency dependent, at low frequencies (such as obtained with the usual torsion pendulum), they are identified with the material's glass transition temperature, and the drop in G' is indicative of the decrease in "stiffness" at T_g, going from the straining of bond angles and lengths to coiling and uncoiling as the dominant response mechanism. The damping peak represents the onset of cooperative motion of 40–50 main-chain carbon atoms at T_g (Chapter 6).

In terms of the four-parameter model for viscoelasticity, below T_g, only spring 1 is operative, and the material is almost completely elastic (low damping). In the vicinity of T_g, the viscosity of dashpot 2 drops to the point where it can deform and dissipate energy, giving the damping peak. At higher temperatures, its viscosity drops to the point where it dissipates little energy, and the material is again highly elastic, mainly through spring 2. At still higher temperatures, the modulus drops off rapidly due to viscous flow, dashpot 1. The broad damping "hill" centered at about 40 °C (and the accompanying gradual drop in G') in Figure 16.11 has been shown to arise from the motion of the -CO-O-CH$_3$ side groups on the PMMA molecules.

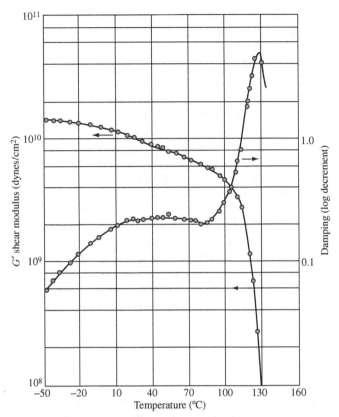

FIGURE 16.11 Dynamic mechanical properties of poly(methyl methacrylate) [8]. The data were obtained with a torsion pendulum at about 1 cycle/s.

Note that the dimensions of the angular frequency ω are per time. The angular frequency thus corresponds to a *reciprocal* timescale. For dynamic (oscillating) deformations, then, the Deborah number is

$$\text{De} = \lambda_c \omega \qquad (16.26)$$

In dynamic tests, a viscoelastic material becomes more like solid as the frequency is increased and the time-dependent response mechanisms (coiling and uncoiling, slip) are less able to follow the rapidly reversing stress. In the limit of very high frequencies, the straining of bond angles and lengths will be the only operative response mechanism and the polymer will exhibit the typical glassy modulus even though it may be well above its glass transition temperature.

16.6 TIME–TEMPERATURE SUPERPOSITION

One of the big challenges in developing materials that are expected to have a lifetime of years is developing laboratory tests to prove that the materials will last. One way to accelerate aging is to conduct tests at higher temperatures. Anyone who has ever wrestled with a cheap garden hose in cold weather appreciates the fact that polymers become stiffer and more rigid at lower temperatures, while at high temperatures they are softer and more flexible. In preceding examples, we have seen that the timescale (or frequency) of the application of stress has a similar influence on mechanical properties, short times (or high frequencies) corresponding to low temperatures and long times (low frequencies) corresponding to high temperatures. The quantitative application of this idea, *time–temperature superposition*, is one of the most important principles in polymer physics. It is based on the fact that the Deborah number determines quantitatively just how a viscoelastic material will behave mechanically. Changing either t_s (or ω) or λ_c can change De. The nature of the applied deformation determines t_s (or ω), while a polymer's characteristic time is a function of temperature. The higher the temperature, the more thermal energy the chain segments possess and the more rapidly they are able to respond, lowering λ_c. Thus, e.g., De can be doubled by halving t_s (or doubling ω in a dynamic test) or by lowering the temperature enough to double λ_c. The change in mechanical response will be the same either way, according to the time–temperature superposition principle.

Although time–temperature superposition is applicable to any viscoelastic response test (creep, dynamic, etc.), here, we will focus on its application to stress relaxation. Figure 16.12 shows tensile stress relaxation data at various temperatures for polyisobutylene, plotted in the form of a time-dependent tensile (Young's) modulus $E_r(t)$ versus. time on a log–log scale:

$$E_r(t) = \frac{\sigma(t)}{\varepsilon_0} = \frac{f(t)/A}{\Delta l / l} \qquad (16.27)$$

where $f(t)$ is the measured tensile force in the sample held at a constant strain $\varepsilon_o = \Delta l/l$ and A is its cross-sectional area. The technique is, of course, equally applicable to shear deformation. In stress relaxation, the lower measurement time limit is set by the assumption

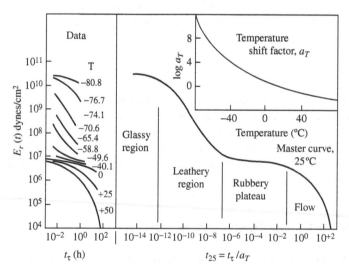

FIGURE 16.12 Time–temperature superposition for NBS polyisobutylene. Adapted from Tobolsky and Catsiff [17].

that the constant strain is applied instantaneously. In practice, inertia and other mechanical limitations make this impossible, so data are valid only at times an order of magnitude or so longer than it actually takes to apply the constant strain. The upper limit is set by the dedication (and patience) of the experimenter and the long-term stability of the sample and equipment. These data were obtained over a range of seconds to a couple of days. As might be expected, the modulus drops with time at a given temperature, and at a given time, it drops with increasing temperature.

Staring at the curves for a while indicates that they appear to be sections of one continuous curve, chopped up, with the sections displaced along the log-time axis. That this is indeed so is shown in Figure 16.12. Here, 25 °C has arbitrarily been chosen as a reference temperature T_o and the curves for other temperatures have been shifted along the log time axis to line up with it. The data below 25 °C are shifted to the left (shorter times) and those above 25 °C are shifted to the right (longer times), giving a *master curve* at 25 °C.

Sometimes, the relaxation moduli at each temperature T are corrected to the reference temperature T_o by multiplying be the ratio T_o/T before superimposing. This correction is based on the theory of ideal rubber elasticity (Chapter 13), which states that the modulus is proportional to the absolute temperature. This procedure is, however, open to question. Although ideal rubber elasticity might be a reasonable approximation when the major response mechanism is chain coiling and uncoiling, it certainly is not where response is dominated by straining of bond angles and lengths (glassy region) or molecular slippage (viscous flow region). In any event, such corrections are of minor practical significance.

Shifting a constant-temperature curve along the log-time axis corresponds to dividing every value of its abscissa by a constant factor (it is immaterial what kind of scale is used for the ordinate). This constant factor, which brings a curve at a particular temperature T into alignment with the one at the reference temperature T_o, is known as the temperature shift factor a_T:

$$a_T = \frac{t_T}{t_{To}} \quad \text{(for same response)} \tag{16.28}$$

where t_T is the time required to reach a particular response (E_r in this case) at temperature T and t_{To} is the time required to reach the same response at the reference temperature T_o. The time value for each relaxation datum on the abscissa of the master curve is from Equation 16.28 or $t_{To} = t_T/a_T$.

For temperatures above the reference temperature, it takes less time to reach a particular response (the material responds faster, i.e., has a shorter relaxation time), so a_T is less than one, and vice versa for temperatures below the reference temperature. The logarithm of the experimentally determined temperature shift factor is plotted as a function of temperature in the inset graph of Figure 16.12.

The master curve now represents stress relaxation at 25 °C *over 17 decades of time*. Since $t_{To} = t_T/a_T$, multiplication by the appropriate value of a_T (shifting along the log-time axis) establishes the master curve at any other temperature T, and can thus be used to predict material responses at that temperature over 17 decades of time.

Two additional aspects enhance the utility of the time–temperature superposition concept. First, the same temperature shift factors apply to a particular polymer regardless of the nature of the mechanical response, that is, the shift factors as determined in stress relaxation are applicable to the prediction of the time–temperature behavior in creep or dynamic testing. Second, *if the polymer's glass transition temperature is chosen as the reference temperature*, the shift factors are given by the Williams–Landel–Ferry (WLF) equation in the range $T_g < T < (T_g + 100\,\text{K})$:

$$\log a_T = \frac{-C_1(T - T^*)}{C_2 + (T - T^*)} \quad (\text{for } T_o = T_g) \tag{16.29}$$

With $T^* = T_g$, the "universal" constants $C_1 = 17.44$ and $C_2 = 51.6$ (with T's in K) give a rough fit for a wide variety of polymers. The WLF equation is most useful in this form because T_g is extensively tabulated [18]. Better fits can be obtained by using constants C_1, C_2, and T^* specific to the polymer, but these constants are not readily available. It has also been suggested that the fit can be improved by using $C_1 = 8.86$ and $C_2 = 101.6$, with T^* adjusted to fit specific data, if available. When this is done, T^* generally turns out to be $T_g + (50 \pm 5)\,\text{K}$ [19].

Example 16.5 The damping peak for poly(methyl methacrylate) in Figure 16.11 is located at 130 °C. Assuming the data were obtained at a frequency of 1 cycle/s, at what temperature would the peak be located if measurements were made at 1000 cycles/s? For PMMA, $T_g = 105$ °C.

Solution. Keep in mind that frequency is a *reciprocal* timescale, and therefore $a_T = \omega_{T_o}/\omega_T$. Applying the WLF equation with the "universal" constants gives

$$\log a_T = \log \frac{\omega_{T_o}}{\omega_T} = \frac{-17.44(T - T_g)}{51.6 + (T - T_g)}$$

Shifting the measurements at 1 cycle/s to T_g (i.e., finding the frequency at which the peak would be located at T_g) gives

$$\log \frac{\omega_{105\,°C}}{\omega_{130\,°C}} = \frac{-17.44(130 - 105)}{51.6 + (130 - 105)} = -5.69$$

$$\omega_{105\,°C}/\omega_{130\,°C} = 2.03 \times 10^{-6}$$

and

$$\omega_{105\,°C} = (1\,\text{cps})(2.03 \times 10^{-6}) = 2.03 \times 10^{-6}\,\text{cps}$$

Now, shifting from T_g to T:

$$\log \frac{2.03 \times 10^{-6}}{1000} = -8.69 = \frac{-17.44(T - 105)}{51.6 + T - 105}$$

$$T = 156\,°C$$

This quantitatively illustrates the statements made at the end of Section 16.4.

Example 16.6 The master curve for the polyisobutylene in Figure 16.12 indicates that stress relaxes to a modulus of 10^6 dyn/cm^2 in about 10 h at 25 °C. Using the WLF equation, estimate the time it will take to reach the same modulus at a temperature of −20 °C. For PIB, $T_g = -70$ °C.

Solution. To use the WLF equation, the reference temperature must be $T_g = -70$ °C:

$$\log \frac{t_T}{t_{T_g}} = \log \frac{t_{25\,°C}}{t_{-70\,°C}} = \frac{-17.44(25 - (-70))}{51.6 + (25 - (-70))} = -11.3$$

$$\frac{t_{25\,°C}}{t_{-70\,°C}} = 5.01 \times 10^{-12} \text{ and } t_{-70\,°C} = \frac{10}{5.01 \times 10^{-12}} = 2 \times 10^{12}\,\text{h}$$

$$\log \frac{t_{-20\,°C}}{t_{-70\,°C}} = \frac{-17.44(-20 + 70)}{51.6 - 20 + 70} = -8.59$$

$$t_{-20\,°C}/t_{-70\,°C} = 2.57 \times 10^{-9}$$

$$t_{-20\,°C} = (2 \times 10^{12})(2.57 \times 10^{-9}) = 5140\,\text{h}$$

This shows how lowering the temperature maintains mechanical "stiffness" for much longer periods of time.

Let us take a closer look at the stress–relaxation master curve. The one shown in Figure 16.12 is typical of linear, amorphous polymers, and illustrates the *five regions of viscoelastic behavior* [20]. At low temperatures, or short times (large De), only bond angles and lengths can respond, and so the typical glassy modulus of 10^{10}–10^{11} dyn/cm^2 (1–10 GPa) is observed. This is the so-called *glassy region*. At longer times or higher temperatures, the relaxation response is governed by the uncoiling of the chains, with the characteristic modulus 10^6–10^7 dyn/cm^2 (0.1–1 MPa) in the *rubbery plateau*. The intermediate region, where the modulus drops because of glassy to rubbery, is sometimes known as the *leathery region* because of the leather-like feel of materials with moduli in this range. At still longer times or higher temperatures (low De), the modulus drops from the rubbery plateau into the *rubbery-flow region*, where the material is still quite elastic but has a significant flow component and then falls off rapidly as a result of molecular slippage in the *viscous-flow region* (this last distinction is somewhat artificial; usually by this time (or temperature), the polymer is significantly deformed and for many polymers (and all thermosets) degradation begins before the polymer can flow).

Figure 16.13 illustrates the effects of molecular weight and crosslinking on the stress–relaxation master curve. Molecular weight should have no significant influence on straining of bond angles and lengths or on uncoiling, so the glassy and rubbery moduli are unchanged for these regions. Flow, however, is severely retarded by increasing molecular weight, which extends the rubbery plateau. For very low molecular weight polymers, the rubbery plateau is not even seen (they are liquids at room temperature). In the limit of infinite molecular weight (light crosslinking), flow is entirely eliminated and the curve levels off with the rubbery modulus. Higher degrees of crosslinking restrict uncoiling, ultimately leading to a material that responds only by straining of bond angles and lengths.

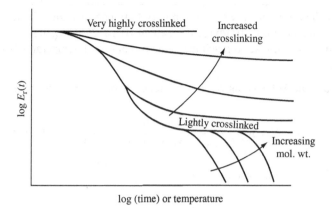

FIGURE 16.13 The effects of molecular weight and crosslinking on stress–relaxation master curves.

The effects of crystallinity on properties are similar to those of crosslinking. However, the applicability of time–temperature superposition in and across the region of T_m is open to question. In this region, the degree of crystallinity and crystalline morphology may change, and one would be, in effect, superimposing data for different materials. A second, vertical shift has been suggested to help superimpose data for crystalline polymers.

To illustrate the effect of temperature on mechanical properties, it is sometimes preferable to plot the property versus temperature for constant values of time. For example, data of the type shown in Figure 16.12 may be cross-plotted as $E_r[10]$ (the 10-s relaxation modulus) versus T. Such a plot is given in Figure 16.14 for several polystyrene

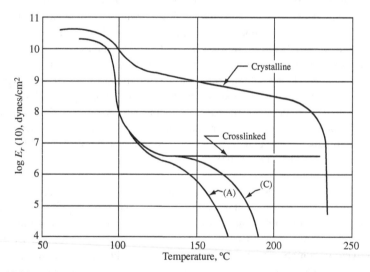

FIGURE 16.14 The 10-s tensile relaxation modulus $E_r[10]$ versus temperature for several polystyrenes [20]. Samples (A) and (C) are linear, amorphous (atactic) materials with narrow molecular weight distributions. $\bar{M}_n(A) = 140,000$; $\bar{M}_n(C) = 217,000$. The crosslinked sample is also amorphous. The crystalline material is isotactic.

samples [20]. The five regions of viscoelastic behavior are evident in the linear, amorphous (atactic) samples (A) and (C) along with the effect of molecular weight in the flow region. The drop in modulus in the vicinity of T_g (100 °C) is clearly seen. The crystalline (isotactic) sample maintains a fairly high modulus all the way up to T_m (~235 °C). Given values of a_T, one can convert data in the form E_r versus t at constant T (a master curve) to E_r versus T at constant t and vice versa.

16.7 SUMMARY

Material testing for polymers can be rather complex. Thus, a number of different methods is required for determining the mechanical properties of polymers when subjected to deformations, either glassy, rubbery, or solid (or even during a temperature ramp). A number of viscometers can be used to accurately determine the behavior of polymer solutions and melts, while dynamic mechanical analysis is used to determine the behavior of polymers to cyclic stresses, with the moduli separated into storage (or elastic) moduli and loss (or viscous) moduli. The time–temperature superposition principle is a valuable technique to take measurements for polymers at higher temperatures to estimate the behavior of a polymer sample at lower temperatures over long periods of time. The modern polymer characterization laboratory has rheometers, viscometers, dynamic mechanical analyzers, and techniques mentioned in earlier chapters such as differential scanning calorimeters and gel permeation chromatographs (or other equipment to measure molecular weights).

PROBLEMS

1 Data were collected from a mechanical stress/strain test on a strip of plastic, as below. The strip was initially 5 cm long, with a width of 1.40 mm and 0.20 mm thickness, when it was placed under axial tension.

Sample Length, cm	Force, N
5.00	0.0
5.03	6.8
5.08	14.3
5.12	22.0
5.23	41.0
5.25	38.0
5.29	39.5
5.33	41.5
5.36	42.5
5.39	44.9

(a) Plot the stress–strain curve and determine the ultimate stress, yield stress, and the elastic modulus.

(b) Draw a qualitative line on the same plot to show how a highly crystalline sample of the same polymer would behave.

2 By stretching, an initial biaxial stress of 0.76 MPa is placed on a rectangular polymer sample and is shown to decrease to 0.55 MPa in 133 days.

(a) What is the relaxation time for this rubbery polymer?

(b) Estimate how many days it will take for the stress to relax to 0.30 MPa.

3 An initial stress of 10.4 MPa is placed on a piece of rubber to strain it 50%. After the strain has been maintained for 40 days, the stress required is only 5.2 MPa. What stress will be required to hold the strain after 80 days?

4 The following data were obtained by Dr. N.D. Sylvester (Ph.D. thesis, Carnegie Institute of Technology, 1968) in the "world's largest capillary viscometer," an 8 ft long, $\frac{1}{2}$-in-inner diameter tube equipped with pressure taps to measure the pressure gradient down the tube. The fluid was a 1.34% solution of poly(ethylene oxide) in water.

Q, in^3/s	ΔP_{total} psi	ΔP_{ent} psi
2.10	1.898	0.036
3.11	2.510	0.039
3.49	2.732	0.081
4.48	2.947	0.082
5.04	3.330	0.123
6.16	3.599	0.132
7.87	4.238	0.211
10.74	4.965	0.341
14.31	6.046	0.620
15.49	6.249	0.675
15.85	6.451	0.726
20.13	7.597	1.183
23.35	8.670	1.607
29.09	10.130	2.335

Here, ΔP_{total} is the total pressure drop across the tube and ΔP_{ent} is the above-equilibrium pressure loss caused by entrance effects.

(a) Obtain the flow equation for this fluid, with constants in the CGS system. Note that at the higher flow rates, entrance losses are a significant portion of the total pressure drop.

(b) Check the Reynolds number at the highest flow rate to make sure the data are in laminar flow.

5 Show that the shear rate in Couette flow between two concentric cylinders is given by:

$$\dot{\gamma} = r\frac{d}{dr}\left(\frac{v_\theta}{r}\right)$$

where $v_\theta = \omega r$ is the tangential velocity.

6 Consider Couette flow between two concentric cylinders, the outer one at R_o fixed and the inner one at R_i rotating with a tangential velocity v_θ (R_i). Starting with Equation 16.2 and the equation in Problem 16.5, obtain an expression for the dimensionless tangential velocity profile $v_\theta(r)/v_\theta(R_i)$, for a power-law fluid. Write your expression in terms of the dimensionless quantities r/R_i, R_o/R_i, and the flow index n. Plot $v_\theta(r)/v_\theta(R_i)$ versus r/R_i for $R_o/R_i = 2$ and $n = 1$, $^1/_2$, and $^1/_4$.

7 A cylindrical shaft of radius R_i and weight W slides vertically downward through a fixed bearing of inner radius R_o and length L. Assume that the shaft remains centered in the bearing and that the annulus between them is filled with a power-law fluid. Neglect entrance and exit effects.

   (a) Is the flow in the annulus viscometric? If so, clearly identify the 1, 2, and 3 coordinate directions.

   (b) Obtain an expression for the shear-stress distribution in the annulus $\tau(r)$.

   (c) Obtain an expression for the velocity profile in the annulus.

   (d) Obtain an expression for V, the steady-state velocity with which the shaft falls through the bearing.

8 Consider the experiment in which the pressure drop ΔP is measured as a function of volumetric flow rate Q through a cylindrical tube of length L and inner radius R. Entrance and exit effects are negligible. The data may be summarized over a limited range of Q by:

$$\Delta P = AQ - BQ^2$$

where A and B are positive constants.

   (a) Obtain an expression for the zero-shear viscosity of the fluid in terms of geometry and quantities in the above equation.

   (b) Show how would you obtain the true shear stress–shear rate relation (flow curve) from the above data.

9 A power-law fluid is confined between two parallel, flat plates in simple shearing flow. The lower plate is fixed and the upper plate moves with a velocity V. The plates are separated by a distance δ. Calculate and sketch the velocity profiles for $n = 0.5, n = 1$, and $n = 1.5$.

10 Injection molding consists of forcing a heated thermoplastic into a cooled mold. It has been noticed that short shots (material freezing before the mold is filled) can sometimes be cured by forcing the material through a smaller gate (narrower diameter entrance to the mold). In fact, so-called "pin gates" (very small orifices at the mold entrance) seem very effective at this. Why?

11 Dynamic measurements taken on a polymer solution at a frequency 1 radian/s give values of $G' = 5.0 \times 10^3$ Pa and $G'' = 2.0 \times 10^3$ Pa. On the basis of these data, what quantitative statement (if any) can you make about the steady-state viscosity of this material at the same temperature?

12 A particular polystyrene sample shows a 10 s relaxation modulus $E_r(10)$ of 1.0×10^7 Pa at 100 °C. Estimate how long would it take for this same sample to reach the same modulus at a temperature of 115 °C. $T_g = 100$ °C.

13 Construct a master curve $\log E_r(t)$ versus $\log t$ for the linear polystyrene sample (C) in Figure 16.14. Assume that the shift factors are given by the WLF equation with the "universal" constants. $T_g = 100\,^\circ$C.

14 The Kepes balance rheometer is another device that determines dynamic properties in steady rotation. Material is confined between concentric hemispheres whose axes of rotation form an angle θ. Both hemispheres rotate with the same angular velocity ω. $R_o - R_i = \delta \ll R_i$ and θ is small.

   (a) Obtain an expression for the shear strain in this device.

   (b) A polymer melt is tested in such a device. Do you think the forces that arise (if any) will tend to increase, decrease, or not influence θ?

15 Dynamic measurements on a sample of poly(ethyl procrastinate) (PEP), give $G' = 1.0 \times 10^7$ Pa at $T_g + 20\,^\circ$C and a frequency of 1 cycle/s. At what frequency would this sample have the same G' at a temperature of $T_g + 40\,^\circ$C?

16 Crud Chemicals is enviously noticing the market for Tygon® flexible tubing. They have developed a linear, amorphous polymer with $T_g = -10\,^\circ$C, which they think may be a good competitor in this market. Their polymer has excellent clarity and chemical resistance, but they are not too sure about its mechanical suitability.

   To check it out mechanically, they extrude some tubing and subject it to rupture tests at 20 °C. These tests consist of taking a fixed length of tubing, pressurizing it to 50 psi, and determining the average burst time for a large number of samples. The average life of their specimens turns out to be 40 h at 20 °C.

   (a) To what common laboratory test of viscoelasticity does this test most closely correspond?

   (b) Estimate the average life of the same tubing in the same rupture test at 30 °C and 37 °C.

17 Figure 15.6 crudely illustrates the effect of strain rate on the engineering stress–strain test of a viscoelastic material at constant temperature.

   (a) Make a similar qualitative sketch that illustrates the effect of temperature at constant strain rate.

   (b) A stress–strain test is run on a sample of poly(vinyl acetate), $T_g = 24\,^\circ$C at $\dot{\varepsilon}_0 = 1$'s The initial slope (Young's modulus) is 5000 psi. At a temperature of 30 °C, what would $\dot{\varepsilon}_0$ have to be to give again an initial modulus of 5000 psi?

18 The 25 °C master curve in Figure 16.12 extends over the time range from about 10^{-14} to 10^{+2} h. What approximate time range would the master curve at $T_o = -40\,^\circ$C cover?

19 Given the statement that a polymer's mechanical behavior depends on the Deborah number, how is the material's characteristic time λ_c related to its temperature shift factor a_T?

REFERENCES

[1] Lendlein, A., A.M. Schmidt, and R. Langer, *PNAS* **98** (3), 842 (2001).

[2] Collyer, A.A. and D.W. Clegg (eds), *Rheological Measurement*, Elsevier, New York, 1988.

[3] Dealy, J.M., *Rheometers for Molten Plastics*, Van Nostrand Reinhold, New York, 1982.

[4] Bird, R.B., W.E. Stewart, and E.N. Lightfoot, *Transport Phenomena*, Wiley, New York, 1960, p. 94, Example 3.5-1.

[5] Krieger, I.M. and S.H. Maron, *J. Appl. Phys.* **25**, 72 (1954).

[6] Best, D.M. and S.L. Rosen, *Polym. Eng. Sci.* **8** (2), 116 (1968).

[7] Ferry, J.D., *Viscoelastic Properties of Polymers*, 3rd ed., Wiley, New York, 1980.

[8] Nielsen, L.E., *SPE J.* **16**, 525 (1960).

[9] Nielsen, L.E., *Mechanical Properties of Polymers and Composites*, Vol. 1, Dekker, New York, 1974.

[10] Gillham, J.K., *Am. Inst. Chem. Eng. J.* **20**, 1066 (1974).

[11] Walters, K., *Rheometry*, Halsted, New York, 1975.

[12] Maxwell, B. and R.P. Chartoff, *Trans. Soc. Rheol.*, **9**, 41 (1965).

[13] Bird, R.B. and E.K. Harris, *Am. Inst. Chem. Eng. J.* **14**, 758 (1968).

[14] Macosko, C.W. and W.M. Davis, *Rheol. Acta* **13**, 814 (1974).

[15] Davis, W.M. and C.W. Macosko, *Am. Inst. Chem. Eng. J.* **20**, 600 (1974).

[16] Maxwell, *B., J. Polym. Sci.* **20**, 551 (1956).

[17] Tobolsky, A.V. and E. Catsiff, *J. Polym. Sci.* **19**, 111 (1956).

[18] Brandrup, J. and E.H. Immergut (eds), *Polymer Handbook*, 3rd ed., Wiley-Interscience, New York, 1989.

[19] Bird, R.B., R.C. Armstrong, and O.A. Hassager, *Dynamics of Polymeric Liquids*, Vol. 1, Fluid Mechanics, Wiley, New York, 1977.

[20] Tobolsky, A.V., *Properties and Structure of Polymers*, Wiley, New York, 1960.

PART IV

POLYMER PROCESSING AND PERFORMANCE

This final part of the book covers how we get from producing and testing polymers to the processing methods used to make useful polymeric products, the additives that are often used, and includes a discussion of some of the challenges encountered in the various markets for polymeric materials. The thermal and mechanical behavior of polymers influences the selection of processing techniques. This section addresses the equipment used to transform polymers into a variety of useful forms including molded shapes, fibers, tubes, and thin (and very thin) sheets as well as a range of application areas for polymers, including thermoplastics, surface finishes, synthetic fibers, adhesives, and rubbers.

Fundamental Principles of Polymeric Materials, Third Edition. Christopher S. Brazel and Stephen L. Rosen.
© 2012 John Wiley & Sons, Inc. Published 2012 by John Wiley & Sons, Inc.

POLYMER PROCESSING AND PERFORMANCE

CHAPTER 17

PROCESSING

17.1 INTRODUCTION

In earlier chapters, we have discovered how to synthesize and characterize polymers. But how do we get the preponderance of polymeric products commonly manufactured? Here, we discuss polymer *processing*, which describes the technology used to convert raw polymer, or compounds containing raw polymer, to articles of a desired shape and size. Considering the wide variety of polymer types and the even wider variety of articles made from them, a complete description and analysis of the myriad of processing techniques that have sprung up through the years would be impossible here. There are references that provide detailed qualitative descriptions of polymer-processing operations [1–4] and many websites that provide photographs or videos of these operations. Similarly, quantitative treatment of processing operations has been the subject of a number of books [5–10]. Here, we outline some of the common processing techniques (Table 17.1), introduce terminology, and consider how the various techniques are based on fundamentals previously discussed. The reader wishing additional detail may consult these references and the more specialized ones cited in the sections to follow.

17.2 MOLDING

Molding consists of confining a material in the fluid state in a mold where it solidifies, taking the shape of the mold cavity. *Injection molding* [11,12] is the most common means of fabricating thermoplastic articles. Figure 17.1 illustrates a typical injection press. The molding compound, usually in the form of pellets of approximately 1/8-in. (3-mm) cubes or 1/8-in.-long cylinders (molding "powder") is fed from a hopper (which may be heated with

Fundamental Principles of Polymeric Materials, Third Edition. Christopher S. Brazel and Stephen L. Rosen.
© 2012 John Wiley & Sons, Inc. Published 2012 by John Wiley & Sons, Inc.

TABLE 17.1   Polymer Processing Techniques

Processing Technique	Brief Description	Example Products
Injection molding (Section 17.2)	Molten polymer injected into metal mold; solidifies by cooling	Lawn furniture, garbage cans
Foam molding (Section 17.2)	Molten polymer with volatile blowing agent that expands the polymer as it evaporates or degrades, producing a gas	Styrofoam® coffee cups, polyurethane seat cushions
Reaction injection molding (Section 17.2)	Starting prepolymers (usually condensation-type) are injected into a heated mold, where they react	Energy-absorbing front or rear car bumpers
Extrusion (Section 17.3)	Molten polymer is extruded through an opening with a cross-sectional shape matching the desired product	Hoses, tubing, vinyl house siding, wire coatings
Blow extrusion (Section 17.3)	Forms very thin sheets by expanding extruded cylindrical sheets with pressurized air before rolling	Plastic wrap
Blow molding (either injection or extrusion) (Section 17.4)	A parison of molten polymer is formed inside a mold, then filled with air to form the hollow insides	Soda bottles, detergent bottles
Rotational molding (fluidized bed or slush) (Section 17.5)	Powdered polymer is put in a heated mold that is rotated to melt the polymer into the desired shape	Irregular hollow objects, toys
Calendering (Section 17.6)	Polymer sheets are pulled and stretched between heated metal rolls (produces larger than 0.01 in. thick)	Shower curtains, vinyl upholstery, vinyl floor tile
Thermoforming (Section 17.7)	A polymer sheet is heated and forced into a cooled mold; vacuum can be used	Drinking cups, meat trays, boat hulls, corrugated sheeting
Stamping (Section 17.8)	Heated thin thermoplastic sheets stamped into a mold	Bucket seat bottoms, battery trays (in cars)
Solution casting (Section 17.9)	Polymer dissolved in solvent which evaporates leaving a thin film	Photographic film
Casting and crosslinking (Section 17.10)	Monomers or pre-polymers are cast into a mold and set by a crosslinking reaction	Electrical coatings, reproduction of carved objects
Reinforced thermoset molding (Section 17.11)	Fiberglass-reinforced plastics made by successive layers of fibers and polymer resin, which is crosslinked	Automotive body panels, bathtubs, golf clubs
Fiber spinning (Section 17.12)	Polymer extruded in thin fibers through a spinnerette and solidified by cooling, drying, or solvent exchange	Fibers for most kinds of clothing; stretchy spandex, acrylic yarns
Compounding (Section 17.13)	Intensive mixing, similar to kneading dough, used to mix additives in polymers	Automotive tires; composites

TABLE 17.1 *(Continued)*

Processing Technique	Brief Description	Example Products
Lithography (photo-) (Section 17.14)	A mask is used to shield light from a pre-polymer solution; forms highly-complex products with small features	Electronic materials such as integrated circuits
3-D printing (Section 17.15)	Adding polymer layer-by-layer using computer-aided design to form complex geometries	Architectural prototypes, medical (tissue scaffolds, teeth)

circulating hot air to dry the material) to an electrically heated barrel. The pellets are conveyed forward through the barrel by a rotating screw. The material is melted as it goes by a combination of heat from the barrel and the shearing (viscous energy dissipation) of the screw.

Older machines had a reciprocating plunger in place of the screw. All the heat for melting had to be supplied by conduction from the barrel, and mixing was very poor. This resulted in low plasticizing (melting) rates, and a thermally nonuniform melt. The screw generates heat *within* the material and provides some mixing, thereby increasing plasticizing rates and giving a more uniform melt temperature.

Molten material passes through a check valve at the front of the screw, and as it is deposited ahead of the screw, it pushes the screw backward while material from the previous shot is cooling in the mold. The cooled parts are ejected from the mold, the mold closes, and the screw is pushed forward hydraulically, injecting a new shot of molten plastic into the mold.

Some machines have a two-stage injection unit. A rotating-screw preplasticizer feeds molten polymer into an injection cylinder, from which it is injected into the mold by a hydraulically driven plunger. This configuration is claimed to provide better control of shot size.

The molten polymer flows through a *nozzle* into the water-cooled mold, where it travels in turn through a *sprue, runners,* and a narrow *gate* into the *cavity* (Figure 17.2). When the part has cooled sufficiently, the mold opens and knockout pins eject the parts, with material cooled in the sprues and runners attached. This material is removed by hand or by robots (or left intact in the case of many toy model kits). With thermoplastics, this material is chopped and recycled back to the hopper as *regrind*. Depending on the

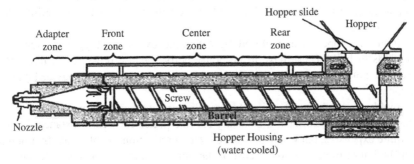

FIGURE 17.1 Injection molding machine. From *Modern Plastics Encyclopedia*, McGraw-Hill, New York, 1969–1970 ed.

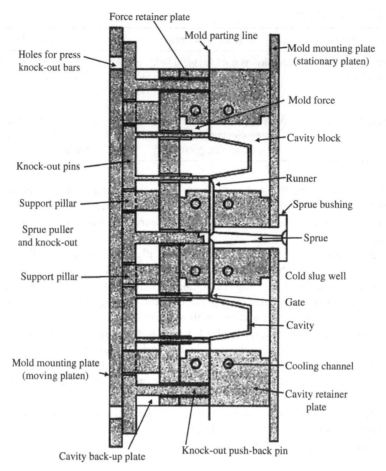

Force retainer plate

Mold parting line

Holes for press knock-out bars

Mold mounting plate (stationary platen)

Mold force

Cavity block

Knock-out pins

Runner

Support pillar

Sprue bushing

Sprue puller and knock-out

Sprue

Support pillar

Cold slug well

Gate

Cavity

Mold mounting plate (moving platen)

Cooling channel

Cavity retainer plate

Cavity back-up plate

Knock-out push-back pin

FIGURE 17.2 Two-cavity injection mold. From *Modern Plastics Encyclopedia*, McGraw-Hill, New York, 1969–1970 ed.

extent of material degradation in the cycle and the property requirements of the finished part, injection-molding operations may tolerate up to 25% regrind in the hopper feed. *Hot-runner molds* contain heaters that maintain the material in the sprue and runners in the molten state, and often a shutoff valve ("valve gate") at the gate. This eliminates the need for regrind.

The molds are opened and closed by hydraulic cylinders, toggle mechanisms, or combinations of both. These have to be pretty hefty because pressures in the mold can reach several thousand psi and the projected surface area of some parts may be hundreds of square inches.

The molds themselves often involve much intricate hand labor and can be quite mechanically complex. Surfaces are usually chrome-plated for wear resistance. Molds can thus be quite expensive, but when amortized over a production run of many thousand parts, the contribution of mold cost to the cost of the finished item may be insignificant.

Injection-molding presses are rated in terms of tons of mold-clamping capacity and in ounces (of general-purpose polystyrene) of shot size. They range from 2-ton, 0.25-oz laboratory units to 6600-ton, 2400-oz monsters that are used to mold garbage cans, TV cabinets, dishwasher tubs, lawn furniture, etc. They are usually completely automated. Cycle times (and therefore production rates) sometimes depend on plasticizing capacity (the rate at which the material can be melted), but more often than not are limited by the cooling time in the mold, which, in turn, is established by the thickness of the part and the (usually very low) thermal diffusivity of the material. Typical cycle times are between 1/4 and 2 min. As usual, there are compromises involved. It is often tempting to try to reduce cooling time by lowering the mold temperature, and thereby increasing the cooling rate, and/or lowering the temperature at which the material is injected into the mold, reducing the amount it must be cooled before solidifying enough to allow removal from the mold without distortion. The higher material viscosity that results, however, can give rise to short shots (incomplete mold filling), poor surface finish, and part distortion from frozen-in strains (see below).

Injection-molding machines now are commonly equipped with feedback control systems that monitor cavity pressure, screw position, and/or screw velocity and control these variables through the hydraulic system to provide high-quality, uniform parts, using a minimum amount of material and minimizing the number of rejects.

Example 17.1 Use the four-parameter model (Figure 15.1f), to explain the presence of frozen-in strains in injection-molded parts.

Solution. When a polymer melt is squirted into a mold, it is subjected to a combination of volumetric (from the high pressures), shear, and elongational (from flow-induced orientation) strains. On a molecular level, these all arise because of deviations from the most-random, highest entropy configuration, and therefore can be represented by extension of the Voigt–Kelvin element, spring 2 and dashpot 2 of the four-parameter model. When the mold fills and the gate freezes, these strains begin to relax (the Voigt–Kelvin element begins to contract). However, before retraction is complete, the rapidly dropping temperature in the mold raises η_2 to the point where the rate of retraction becomes very low, and the part is ejected from the mold with the strain frozen in. Subsequent recovery outside the mold causes distortion of the part. The problems are compounded by the fact that temperatures and flow fields are not uniform within the mold, so the frozen-in strains vary from point to point in the part, giving rise to nonuniform distortions (warping). (It is for this reason that the thermally uniform melt from a screw machine is beneficial.) These problems can, of course, be minimized by using higher melt temperatures and/or mold temperatures, but both lengthen the cycle time. It might also be pointed out here that part failure can result if these frozen-in strains are suddenly released, as, for example, by the concentration of stress by a scratch or sharp corner on the part, or by plasticization by a liquid (stress cracking).

Foamed plastic articles are molded by incorporating a blowing agent (Chapter 7). These blowing agents may be inert but volatile liquids, compounds that decompose chemically at elevated temperatures to liberate a gas such as nitrogen or carbon dioxide, or simply nitrogen dissolved in the polymer. In any case, the blowing gas is kept in solution by the high pressures ahead of the screw. The mold is partially filled with a shot from the cylinder. The reduced pressure within the mold allows the blowing agent to vaporize, expanding the shot to fill the cavity and giving a foamed part.

One drawback of traditional injection molding was its inability to produce parts with complex internal passages. To be sure, simple holes could be incorporated by using core pins that are retracted mechanically either parallel or perpendicular to the mold-opening direction. To permit retraction, however, these pins (and the resulting holes) had to be straight and if not of constant cross section, tapered appropriately with regard to the direction of retraction.

The *lost-core* process allows the molding of parts with curved internal passages of varying cross section. The core is cast in a low melting bismuth–tin alloy. This cast core is placed in the injection mold and the plastic is molded around it. After the part is ejected from the mold, the metal core is melted out. The high thermal diffusivity of the metal core prevents its melting during the molding step. This process can be used for automotive intake manifolds, where it provides the necessary curved passages with a nice, smooth interior finish.

Injection molding was at one time confined exclusively to thermoplastics. It is now also used for thermosets, which are injected into a heated mold, where they solidify through a curing (crosslinking) reaction. This is a tricky operation. The compound must be heated just enough in the barrel to achieve fluidity and injected into the mold before it begins to cure appreciably, otherwise the entire machinery becomes a sticky mess. This requires precise control of temperatures and cycle timing to prevent premature cure in the barrel and the resulting shutdown-and-cleanout operation. Thermoset sprues and runners cannot be recycled, of course, and must be minimized to cut waste.

Thermosetting compounds are traditionally *compression* molded. The molds are mounted in hydraulic presses on steam-, electric-, or oil-heated platens. The molding compound is fed to the heated mold, which closes, maintaining the material under pressure until cured. The part is then ejected from the mold.

Molding compound in the form of granules or powder may be fed to the mold automatically in weighed shots, or as pre-formed (by cold pressing) tablets. The charge is often preheated to reduce heating time in the mold and thereby cycle time.

Mold temperature is a critical variable in compression molding. The higher it is, the faster the material cures, but if it cures too fast, it will not have enough time to fill thin sections and far corners of the mold, so a compromise must be reached. Material suppliers attempt to optimize the cure characteristics to provide minimum cycle times.

Transfer molding is a variation of compression molding. Here, the material is melted in a separate transfer pot, from which it is squirted into the mold. This can give faster cycle times and since no solid material is pushed around in the mold cavity itself, damage to delicate inserts (e.g., metal electrical contacts molded into a part) is minimized, mold wear is reduced, and greater ease in filling intricate molds and more uniform cures are obtained. However material cured in the transfer pot and sprue is wasted.

Reaction injection molding (RIM) [13] (Figure 17.3) is a process developed to mold large polyurethane (Example 2.4Q) or polyurea (Example 2.4) materials directly from the starting chemicals with cycle times comparable to injection molding. Between shots, the highly reactive liquid polyol (for polyurethanes) or polyamine (for polyureas) and isocyanate components are circulated from separate tanks through heat exchangers that regulate their temperatures. During the shot, the streams are pumped at about 2500 psi through nozzles in a mixing head, where they are impinged at high velocity and thoroughly mixed by the resulting turbulence. The mixed stream begins to react as it flows into the mold, where the polymerization reaction is completed. The mold is maintained at about 150 °F, initially heating the reacting system but later removing the exothermic heat of

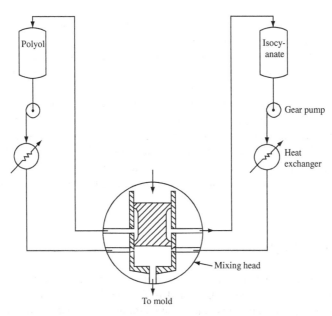

FIGURE 17.3   RIM. The Kraus–Maffei mixing head is shown in the shot position. At the end of the shot, the plunger moves down, packing the mold, cleaning the mixing chamber, and recirculating the reactant streams back to their storage tanks.

polymerization as the temperature rises while the reaction approaches completion. Since the unreacted system is relatively low in viscosity (about like pancake syrup), unlike the molten polymer in an injection-molding process, it flows easily to fill large molds with narrow clearances at relatively low pressures (100 psi, or so, in contrast to the thousands of psi in injection molding). These low pressures allow the use of relatively inexpensive molds and low-force mold clamping systems (compared to injection molding).

The polyurethanes and polyureas may be formulated to be flexible or rigid, solid or foamed. Fillers and/or reinforcing agents (e.g., glass fibers or flake) may be added to one or both components (in which case the process is sometimes known as RRIM-reinforced RIM). A major application of RIM is to produce energy-absorbing front and rear ends and body panels for automobiles. Cycle times of 2 min or less are feasible for such large parts.

RIM has achieved considerable success, particularly in the automotive industry, where the drive to cut vehicle weight is intense. RIM panels have replaced stamped steel in a variety of automotive applications. However, RIM faces stiff competition from other plastics and processes, injection-molded thermoplastics, for example, for this large market.

A somewhat similar process known as *resin-transfer molding* (RTM) is used to produce highly reinforced parts from low-viscosity, reactive starting materials. Reinforcing preforms, usually consisting of layers of fiberglass cloth or mat, are placed in the mold. When the mold is closed, they pretty much extend throughout the cavity and therefore the finished part. The resin and catalyst components are forced by positive-displacement piston pumps through a static or motionless mixer (see Figure 17.12) into the closed mold, where they cure. To ensure rapid and complete impregnation of the reinforcing material, care must be taken to vent the mold properly to allow the escape of air as the resin is injected. In some cases, the air is sucked out of the mold with a vacuum pump prior to resin injection.

RTM has been mainly used with unsaturated polyesters (Example 2.3), but it is also applicable to epoxies, polyurethanes, etc. Parts weighing hundreds of pounds have been produced (e.g., truck air deflectors). As with RIM, because the viscosity of the flowing material is low, relatively modest pressures and inexpensive molds are required.

Some newer methods of forming complex three-dimensional shapes combine computer-aided three-dimensional design with printing technology that build an object by adding polymer pellets and growing a shape. This has been done to make replicas of fragile artifacts, develop medical implants with precise geometries, and construct complex three-dimensional scaffolds for seeding cells for tissue engineering. Alternatively, shapes can also be carved from a polymer block (see Section 17.14 on lithography), similar to sculpting, but the size of any interior cavities is dependent on the size of the tool used in carving.

17.3   EXTRUSION [14–17]

Thermoplastic items with a uniform cross section are formed by extrusion. This includes many familiar items such as pipe, hose and tubing, gaskets, wire and cable insulation, sheeting, window-frame moldings, and vinyl siding for houses. Molding powder is conveyed down an electrically or oil-heated barrel by a rotating screw. It melts as it proceeds down the barrel and is forced through a *die* that gives it its final shape (Figure 17.4). Vented extruders incorporate a section in which a vacuum is applied to the melt to remove volatiles such as traces of unreacted monomer, moisture, solvent from the polymerization process, or degradation products.

The design of extruder screws is an interesting and complex technical problem and has received considerable study. Screws are optimized for the particular polymer being extruded. Basically, a screw consists of three sections: *melting, compression,* and *metering.* The function of the melting section is to convey the solid pellets forward from the hopper and convert them into molten polymer. Its analysis involves a combination of fluid and solid mechanics and heat transfer. The compression section, in which the depth of the screw flight decreases, is designed to compact and mix the molten polymer to provide a more-or-less homogeneous melt to the metering section, where the molten polymer is pumped out through the die.

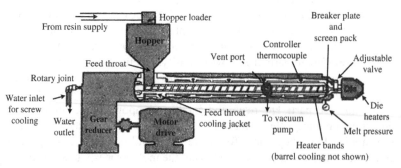

FIGURE 17.4   Vented extruder. From *Modern Plastics Encyclopedia*, McGraw-Hill, New York, 1969–1970 ed.

This last section is well understood. Analysis of the metering section is an interesting application of the rheological principles discussed previously (Chapter 14), as is much of die design. The determination of the die cross section needed to produce a desired product cross section (other than circular) is still pretty much a trial-and-error process, however. Viscoelastic polymer melts swell upon emerging from the die (i.e., die swell, which is caused by the recovered stored elastic energy after squeezing through the die) and the degree of die swell cannot be reliably predicted.

In addition to die swell, as the extrusion rate is increased, the extrudate begins to exhibit roughness, and then an irregular, severely distorted profile. This phenomenon is known as *melt fracture*. It is generally attributed to melt elasticity, but there is currently no way of quantitatively predicting its onset or severity. It can be minimized by increasing die length, smoothly tapering the entrance to the die, and raising the die temperature.

Extruders are normally specified by screw diameter and length-to-diameter ratio. Diameters range from 1 in (2.5 cm) in laboratory or small production machines to 1 ft (30 cm) for machines used in the final pelletizing step of production operations. Typical *L/D* ratios seem to grow each year or so, with values in the 20/1 to 36/1 range.

Single-screw extruders pump the polymer based on the drag flow of material between the rotating screw and the stationary barrel. As a result, they are not positive-displacement pumps and tend to give a rather broad residence time distribution. Moreover, they are not particularly good mixing devices. Counter-rotating *twin-screw* extruders are true positive-displacement pumps, capable of generating the high pressures needed in certain profile extrusion applications. Co-rotating twin-screw extruders, though not positive displacement pumps, can give excellent mixing and a narrow residence time distribution with proper screw design (subjecting all the material to essentially the same shear and temperature history). They are, therefore, used extensively in polymer compounding (mixing) operations and, to a certain extent, as continuous polymerization reactors.

In steady-state extruder operations, most or all of the energy needed to plasticize the polymer is supplied by the drive motor through viscous energy dissipation. The heaters on extruders are needed mainly for startup and because enough heat cannot always get from where it is generated (the compression and metering zones) to where it is needed (the melting zone). In fact, cooling through the barrel walls and/or screw center is sometimes necessary. A steady-state energy balance on a typical operating extruder is shown below [17]:

Extruder:	3.5-in (8.9-cm) barrel 32/1 *L/D*
Material:	high-impact polystyrene
Operating Conditions:	screw speed 121 rpm throughput 566 lb/h (257 kg/h)
Energy Inputs	
Shaft work (viscous energy dissipation)	+90 %
Heaters	+10%
Energy Outputs	
Polymer enthalpy increase	−65%
Losses to ambient	−15%
Cooling water	−10%

The complex interplay between the rate of viscous energy dissipation, temperature, and material viscosity in an extruder sometimes results in a variation in output (*surging*) even at constant screw speed. Where it is necessary to minimize surging, positive-displacement *gear pumps* are often added after the extruder.

Most packaging film, such as Saran® wrap, is produced by *blow extrusion* (Figure 17.5). A thin-walled, hollow cylinder is extruded vertically upward. Air is introduced to the interior of the cylinder, expanding it to a tube of film (less than 0.01 in thickness). The tube is grasped between rolls at the top, preventing the escape of air and flattening it for subsequent slitting and windup on rolls. The expanded tube is rapidly chilled by a blast of air from a chill ring as it proceeds upward. With crystalline polymers such as polyethylene and polypropylene, this rapid chilling produces smaller crystallites and enhances film clarity.

"Cast" film is produced by extruding the polymer from a slot die onto polished metal chill rolls. The rolls maintain a more uniform thickness (gage) and the more rapid cooling and smoother surface finish provide superior film clarity. On the other hand, cast film is oriented (stretched) only in the machine direction, whereas blown film is biaxially oriented, and therefore has superior mechanical properties in equivalent thickness (or equivalent properties in a thinner film).

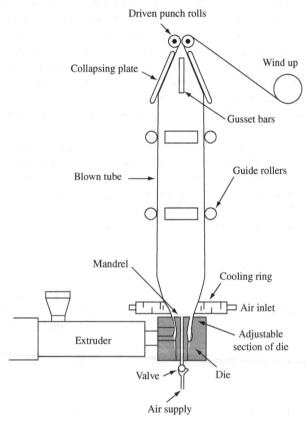

FIGURE 17.5 Blow extrusion film line. From *Modern Plastics Encyclopedia*, McGraw-Hill, New York, 1969–1970 ed.

17.4   BLOW MOLDING [18]

A quick walk through a supermarket provides convincing proof of the economic importance of plastic bottles. They are nearly all made by blow molding. In one form of this process, *extrusion blow molding* (Figure 17.6), a hollow cylindrical tube or *parison* is extruded downward. The parison is then clamped between halves of a water-cooled mold. The mold pinches off the bottom of the parison and forms the threads on the neck of the bottle. Compressed air expands the parison against the inner mold surfaces, and when the part has cooled sufficiently, the mold opens and the part is ejected. In one form of machine, the mold "shuttles" aside for blowing and ejection as a new parison is being formed beneath the die. High-production rotary machines may have two parison heads and index as many as 20 molds past them on a rotating table.

The rheological properties of the parison are important. If it sags too much before being grasped by the mold, the walls of the bottle will be too thin in places. Sag is minimized by using high molecular weight compounds with high viscosities (high "melt strength").

Many blow-molding extruders are equipped with *parison programmers*, which vary the orifice diameter as the parison is being extruded, minimizing variations in wall thickness in the blown product. The parisons may be "ovalized" to reduce variations in wall thickness in objects of noncircular cross section.

Extrusion blow molding has passed well beyond the bottle-production stage. It is now being used to produce large items such as drums (to replace the familiar 55-gallon steel drum) and truck, automobile, and recreational-vehicle gasoline tanks. In such metal-replacement applications, blow-molded containers offer light weight and great design flexibility.

In *injection blow molding,* the parison is formed by injection molding rather than extrusion. A variation known as *stretch* (or *orientation*) *blow molding* is responsible for the now ubiquitous plastic soda-pop bottle. In this process, parisons (often called *preforms* in this case) are injection molded with the bottom end closed and the threads and neck molded on the open top. They are allowed to cool to room temperature. Prior to blowing, they are reheated in a radiant-heat oven in which close control is exercised over the temperature profile of the parison. When introduced to the blow mold, the parison is normally a fraction of the length of the mold (Figure 17.7), but before blowing, a rod

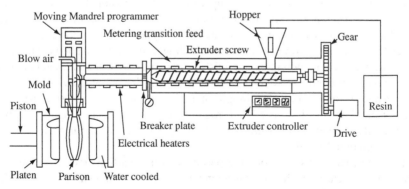

FIGURE 17.6   Extrusion blow molding. From *Modern Plastics Encyclopedia,* McGraw-Hill, New York, 1969–1970 ed.

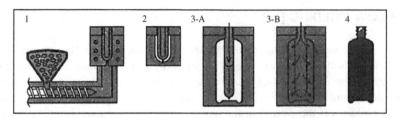

FIGURE 17.7 Stretch blow molding: 1, parison injection molded; 2, parison is reheated; 3-A, a rod stretches the parison, imparting axial orientation; 3-B, air expands parison against mold walls, imparting tangential orientation; 4, finished bottle ejected from mold. From *Modern Plastics* **55**(10), 22 (1978).

rapidly stretches the parison to nearly the full length of the mold, orienting the polymer molecules in the axial direction. This is followed by blowing with compressed air, which imparts tangential (hoop) orientation. The resulting *biaxial* orientation improves the toughness, creep resistance, clarity, and barrier properties (resistance to permeation) of the bottle material. In this case, the frozen-in strains that are normally detrimental to thick, injection-molded parts (Example 17.1) are deliberately introduced to thin-walled blown bottles with highly beneficial benefits. The biaxial orientation provided by stretch blow molding is absolutely essential to the success of the poly(ethylene terephthalate) soda-pop bottles.

17.5 ROTATIONAL, FLUIDIZED-BED, AND SLUSH MOLDING

Molding techniques have been developed to take advantage of the availability of finely powdered plastics, mainly polyethylene and nylons. In *rotational molding*, a charge of powder is introduced to a heated mold, which is then rotated about two mutually perpendicular axes. This distributes the powder over the inner mold surfaces, where it fuses. The mold is then cooled by compressed air or water sprays, opened, and the part is ejected.

Rotational molding is capable of producing extremely irregular hollow objects. The molds are inexpensive, often simply sheet metal, because no elevated pressures are involved, and can be heated in simple hot-air ovens. Thus, capital outlay is relatively low. The process is in many ways competitive with extrusion blow molding for the production of large, hollow items such as drums and gasoline tanks. Blow molding requires a much larger initial investment, but is capable of higher production rates.

These two processes (blow and rotational molding) also provide a good example of how the processing operation, polymer, and finished properties are often intimately connected. Blow molding can handle very high molecular weight linear polyethylenes, and in fact usually *requires* high molecular weight material to prevent excessive parison sag. Rotational molding, on the other hand, requires a low molecular weight resin because a low viscosity is needed to permit fusion of the powder under the influence of the low forces in a rotational mold.

When subjected to stresses for long periods of time, particularly in the presence of certain liquids, linear polyethylene has a tendency to fail through stress cracking. It turns out that high molecular weight resins are much more resistant to stress cracking. Thus, although the parts might appear similar, those produced by extrusion blow molding will ordinarily have superior resistance to stress cracking. The difference can be narrowed or eliminated by using more material (thereby lowering stress for a given load) or by using a crosslinkable polyethylene in rotational molding. Either of these solutions increases material cost, however.

Polymer powders are also used in a process known as *fluidized-bed coating.* When a gas is passed up through a bed of particles, the bed expands and behaves much like a boiling liquid. When a heated object is dipped into a bed of fluidized polymer particles, those that contact it fuse and coat its surface. Such 100% *solids* coating processes are increasing in importance because they eliminate the pollution often caused by solvent evaporation when ordinary paints are used.

Similar procedures have been in use for years with plastisols (Chapter 7). Liquid platisol is poured into a heated female mold. The plastisol in contact with the mold surface fuses and the remainder is poured out for reuse. This slush-molding process is used to produce objects such as doll's heads. Dipping a heated object into a liquid plastisol coats it with plasticized polymer. Vinyl-coated wire dishracks are familiar products of this process.

17.6 CALENDERING

Polymer sheets (greater than 0.01 in thickness) may be produced either by extrusion through thin, flat dies or by *calendering* (Figure 17.8). Basically, a calender consists of a series of rotating, heated rolls, between which the polymer compound (most often plasticized PVC) is squeezed into sheet form. The thickness of the sheet is determined by the clearance between the rolls. Commercial calenders may be very large (rolls up to 3 ft

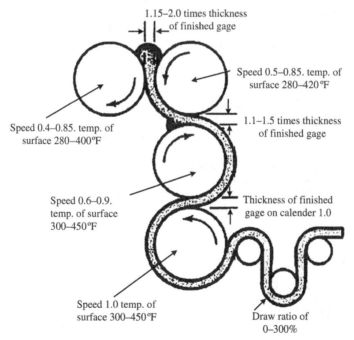

FIGURE 17.8 Inverted "L" calender, illustrating process variables for the production of PVC sheet. From *Modern Plastics Encyclopedia,* McGraw-Hill, New York, 1969–1970 ed.

(1 m) in diameter by 8 ft (3 m) long) and may represent a big capital outlay, but are capable of tremendous production rates (up to 100 yards/min). The polymer may be laminated to a layer of fabric between two rolls to give a supported sheeting. The final rolls may also be embossed to impart a pattern to the sheet. Shower curtains, vinyl upholstery materials (Naugahyde), and vinyl floor tile are produced by calendering.

17.7 SHEET FORMING (THERMOFORMING) [19, 20]

Thermoplastic sheet is converted to a wide variety of finished articles by processes generically known as *sheet forming* or *thermoforming*. Although the details vary considerably, these processes all involve heating the sheet above its softening point and forcing it to conform to a cooled mold (Figure 17.9).

In *vacuum forming*, for example, the heat-plasticized sheet (either directly from an extruder or preheated in an oven) is drawn against the mold surface by the application of a vacuum from beneath the surface. Similarly, a positive pressure may be used to force the sheet against the mold surface, and where very deep draws are required, mechanical assists, *plug forming*, are used. In *drape forming*, the plasticized sheet is draped over a male mold, perhaps with a vacuum assist.

From a material standpoint, polymers used for sheet forming should have high "melt strengths," that is, high melt viscosities so that they do not draw down or thin out excessively or perhaps even tear in the forming operation. Thus, high molecular weight resins are preferred. One of the great advantages of sheet forming is that relatively inexpensive molds are required since no high pressures are involved. Epoxy molds are often used because they can be easily cast to shape from a handmade pattern.

Among the many familiar items made by sheet forming are drinking cups, meat trays, wraps for cigarette packs, aircraft canopies, corrugated plastic sheets for roofing, lighting globes, and even items as large as the hull of a boat.

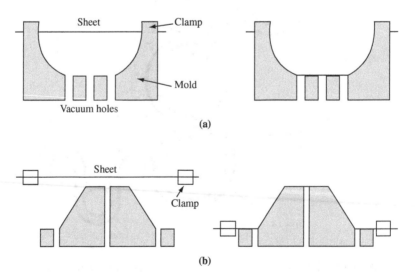

FIGURE 17.9 Sheet-forming process: (a) vacuum forming; (b) drape forming.

A similar technique is used to make sheet molding compound, but requires the addition of several additives, so the discussion is included in Chapter 18.

17.8   STAMPING

The plastics industry has long envied the ability to stamp sheet-metal parts with cycle times of a few seconds. Because of the high production rates possible with a stamping process (particularly for forming auto bodies from stamped steel) and the large existing investment in stamping presses, the stamping process has been adapted to form thin-walled (0.1-in) parts from glass-reinforced thermoplastic sheet (Azdel®). The sheet is heated in an infrared oven and then stamped in modified metal-forming equipment, with a dwell time of about 8 s as the material cools in the mold. Because of the thin sections, cooling is rapid and cycle times as low as 15–20 s can be achieved with automated feeding and part-removal systems [21,22]. So far, the process has been limited to rather simple shapes and shallow draws (crankcase oil pans, battery trays, bucket seat bottoms and backs, etc.), but because of its economic potential, stamping has been combined with other polymer processes (such as thermoforming) in the development of relatively inexpensive high-throughput processes.

17.9   SOLUTION CASTING

Solvent-evaporation casting (commonly known as solution casting) produces plastic sheets or films by dissolving the polymer in an appropriate solvent, spreading the viscous solution onto a polished surface, and evaporating the solvent. Additives such as plasticizers and pigments can be easily added to the solution, with a good dispersion achieved in the dried material. In the manufacture of photographic film base, the solution is spread with a "knife" onto a slowly rotating wheel about 2 ft wide and 20 ft in diameter. As the wheel revolves, heated air evaporates the solvent, which is recovered for reuse, and the dried film is stripped from the wheel before the casting point is reached again.

17.10   CASTING

The raw materials for many thermosetting polymers are available as low molecular weight liquids that are converted to solid or rubbery materials by a crosslinking reaction. Similarly, liquid vinyl monomers can be converted to solid, linear polymers through an addition reaction. These materials are easily cast to shape at atmospheric pressure in inexpensive molds. A good example of this process is the reproduction of carved wood furniture panels. Only the original is actually carved, which is on expensive step. A mold is made by pouring a room-temperature-curing silicone rubber over the original. When cured, the rubber mold is stripped from the original and used to cast multiple copies from a liquid, unsaturated polyester (Example 2.3). A free-radical initiator causes crosslinking to the final solid object, which faithfully reproduces the detail of the carved original at a fraction of the cost.

Such furniture components provide a good example of how the processing method of choice often depends on the production volume. They can also be injection molded from, for example, high-impact polystyrene. Because the panels are usually large, they require a large and very expensive molding machine. Similarly, the mold, with its intricate carving and wood-grain detail in hardened steel, is very expensive to produce. Thus, the capital outlay for injection-molded panels is huge, while the reverse is true for cast panels. Nevertheless, parts can be injection molded with cycle times under a minute and very little hand labor involved, and when the capital investment is amortized over a large number of panels, it becomes a small contribution to the total panel cost. On the other hand, the long cure times and hand labor required for the casting process make it prohibitively expensive for large production runs, but it is the preferred process for small numbers of parts.

Delicate electrical components are often encapsulated or "potted" by casting a thermosetting liquid resin (usually an epoxy) around them. The casting of acrylic sheet from monomer or syrup is described in Chapter 12.

Example 17.2 A polymer of divinyl benzene $H_2C{=}CH{-}\phi{-}HC{=}CH_2$ is to be used to make rotameter bobs (1-in-long $\times$ $^1/_2$-in-diameter cylinders). What processing technique should be used?

Solution. The monomer is tetrafunctional (Chapter 2) and hence will be highly crosslinked in the polymerization reaction; therefore, it cannot be softened by heat or melted. The only suitable technique is casting directly from the monomer, either to final shape or casting rods that can then be cut and machined to the final shape.

17.11 REINFORCED THERMOSET MOLDING

Many plastics, when used by themselves, do not possess enough mechanical strength for structural applications. When reinforced with high-strength fibers, however, the composites have high strength-to-weight ratios and can be fabricated into a wide variety of complex shapes. Glass and other fibers are used extensively to reinforce thermosetting plastics, mostly polyesters and epoxies. Fiberglass-reinforced plastics (FRPs) are now used for just about all boats under 40 ft in length, truck cabs, automobile body panels, structural panels, aircraft components, and bathtubs.

Many such objects are fabricated by a process known as *hand layup*. The mold surface is often first sprayed with a pigmented but nonreinforced *gel coat* of the liquid resin to provide a smooth surface finish. The gel coat is followed up by successive layers of fiberglass, either in the form of woven cloth or random matting, impregnated with the liquid resin, which is then cured (crosslinked) to give the finished product.

This process is tremendously versatile. The molds are relatively inexpensive, because no high pressure is required. Objects may be selectively reinforced by adding extra layers of material where desired. The major drawback of the process is the expense of hand labor involved, which makes it uneconomical for large production volumes. For this reason, a sprayup process is often used. A special gun chops continuous fibers into approximately 1 in lengths. The chopped fibers are combined with a stream of liquid resin and sprayed directly onto the mold surfaces. Although the random chopped fibers do not reinforce quite as well as woven cloth or random mat, the labor savings are substantial. Sprayed-on polyester-fiberglass backings are also applied to vacuum-formed poly(methyl methacrylate) sheets, thus combining a smooth, hard, strain- and UV-resistant acrylic surface skin with the lightweight strength of a FRP core. Such composites are used for sinks, bathtubs, and recreational vehicle bodies.

In many objects, stresses are not uniformly distributed, so the reinforcing fibers may be arranged to support the stress most efficiently. In "fiberglass" or "graphite" fishing rods, vaulting poles, golf club shafts, etc., the fibers are arranged along the long axis to resist the bending stresses applied. The process of *filament winding* extends this principle to more complex structures. Continuous filaments of the reinforcing fiber are impregnated with liquid resin and then wound on a rotating *mandrel*. The winding pattern is designed to resist most efficiently the anticipated stress distribution. The range of the Polaris submarine ballistic missile was increased by several hundred miles by replacing the metallic rocket casing with a filament-wound reinforced plastic system. This technique is also used to produce tanks and pipe for the chemical process industries, and even gun barrels, by filament winding about a thin metal core that provides the necessary heat and abrasion resistance.

More recently, nanofibers (particularly carbon nanotubes) have found their way into fibrous polymer composites, offering even greater strength per mass when compared to composites made using the larger fibers above.

Pultrusion [23] is used to produce continuous lengths of objects with a constant cross section, for example, structural beams. Continuous fibers (such as roving that is made into yarn) and/or mat are impregnated by passing them through a tank of liquid resin and pulling slowly through a heated die of the desired cross section. The resin cures to a solid as it passes through the die.

17.12 FIBER SPINNING

The polymers used for synthetic fibers are similar, and in many cases, identical to those used as plastics, but for fibers, the processing operation must produce an essentially infinite length-to-diameter ratio. In all cases, this is accomplished by forcing the plasticized polymer through a *spinnerette*, a plate in which a multiplicity of holes has been formed to produce the individual fibers, which are then twisted together to form a thread for subsequent weaving operations. The cross section of the spinnerette holes obviously has a lot to do with the fiber cross sections, which, in turn, greatly influences the properties of the fiber. The three basic types of spinning operations differ mainly in the method of plasticizing and deplasticizing the polymer.

Melt spinning is basically an extrusion process. The polymer is plasticized by melting and pumped through the spinnerette. The fibers are usually solidified by a cross-current blast of air as they proceed to the drawing rolls. The drawing step stretches the fibers, orienting the molecules in the direction of stretch and inducing the high degrees of crystallinity necessary for good fiber properties. Nylons are commonly melt spun.

In *dry spinning*, a solution of the polymer is forced through the spinnerette. As the fibers proceed downward to the drawing rolls, a countercurrent stream of warm air evaporates the solvent (Figure 17.10). In this process, the cross section of the fiber is determined not only by the shape of the spinnerette holes but also by the complex nature of the diffusion-controlled solvent evaporation process, because there is considerable

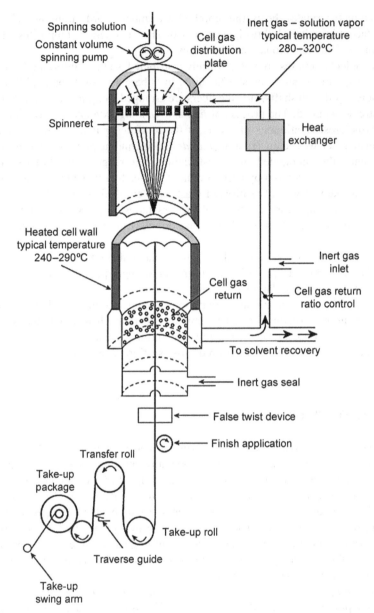

FIGURE 17.10 Dry spinning of polymer fibers [24]. Reprinted with permission of John Wiley and Sons, Inc.

shrinkage as the solvent evaporates. The acrylic fibers (Orlon®, Creslan®, Acrilan®, etc.), mainly polyacrylonitrile, are produced by dry spinning.

Wet spinning is similar to dry spinning in that a polymer solution is forced through the spinnerette. Here, however, the solution strands pass directly into a liquid bath. The liquid might be a nonsolvent for the polymer, precipitating it from solution as the solvent diffuses outward into the nonsolvent. The bath might also contain a substance that precipitates a polymer fiber by chemically reacting with the dissolved material. Here again, the process as well as the spinnerette influences the fiber cross section. Kevlar is an example of a wet-spun fiber (Section 4.8). The polymer is dissolved in concentrated sulfuric acid and the solution is spun into a water bath. The water leaches out the acid, causing the polymer to precipitate in fiber form.

Example 17.3 Suggest processing techniques for the manufacture of the following:

(a)   100,000 ft of plasticized PVC garden hose

(b)   50,000 polystyrene pocket combs

(c)   100,000 polyethylene detergent bottles

(d)   5000 phenolic (phenol-formaldehyde) TV knobs

(e)   six souvenir paperweights of poly(methyl methacrylate) containing a coin

(f)   a strip of chlorinated rubber (a linear, amorphous polymer) and white pigment (60:40), roughly 0.001 in thick, 4 in wide, running 20 miles down the center of a highway

(g)   1,000,000 hard (not foamed) polystyrene meat trays, 0.005 in thick

Solution.

(a)   Extrusion

(b)   Injection molding

(c)   Blow molding

(d)   Compression, transfer, or thermoset injection molding

(e)   Although these objects could be injection molded, the small number of articles required would make injection molding uneconomical. They can easily be cast from the monomer.

(f)   Dissolve rubber in solvent, add pigment, and spray on highway. Evaporation of solvent leaves the desired strip.

(g)   Extrude sheet, then vacuum form trays.

17.13 COMPOUNDING

Polymers are almost always used in combination with other ingredients. These ingredients are discussed in subsequent chapters, but they must be combined with the polymer in a *compounding* operation.

Occasionally, if they do not interfere with the polymerization reaction, such ingredients may be incorporated at the monomer or low molecular weight polymer (e.g., unsaturated polyester, Example 2.3) stage and carried through the polymerization and/or crosslinking reaction. In such cases, viscosities are low enough to permit the use of standard mixing equipment. Similarly, powdered PVC and thermosets are compounded with other ingredients in the usual tumbling-type of blending equipment.

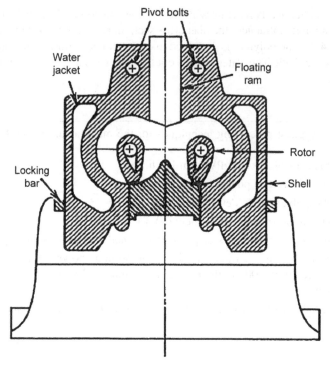

FIGURE 17.11   Banbury mixer. From L.K. Arnold, *Introduction to Plastics,* Iowa State UP, Ames (1968).

Because of their extremely high melt viscosities, specialized equipment, however, is usually needed to compound ingredients with high molecular weight thermoplastics. In general, high shear rates and large power inputs per unit volume of material are required to achieve a uniform and intimate dispersion of ingredients in the melt. Single- and twin-melt extruders (Section 17.3) are used extensively for continuous compounding. The latter, with screws often modified to incorporate special mixing sections, are better mixers and provide a narrower, more uniform distribution of residence times, while the formed offer lower cost and greater mechanical simplicity. *Intensive mixers*, such as the Banbury mixer (Figure 17.11), subject the material to high shear rates and large power inputs in a closed, heated chamber containing rotating, intermeshing blades.

A *two-roll mill* (Figure 17.12) generates high shear rates in a narrow nip between two heated rolls that counter-rotate with slightly different velocities. In commercial mills, the rolls are about 1 ft in diameter and 3 ft long. Once the polymer has banded on one of the

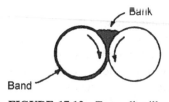

FIGURE 17.12   Two-roll mill.

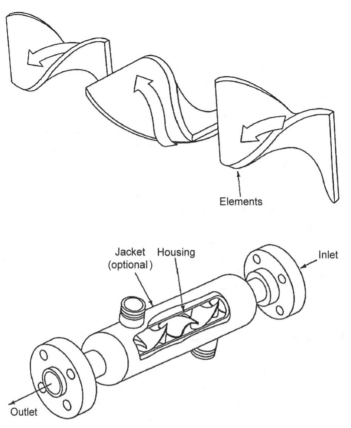

Elements

Jacket Housing   Inlet
(optional)

Outlet

FIGURE 17.13   The Kenics Static Mixer. From McCabe W.L., J.C. Smith, and P. Harriott, *Unit Operations of Chemical Engineering,* 4th ed., McGraw-Hill, New York, 1985.

rolls, ingredients are added to the bank between the rolls. The band is cut off the roll with a knife, rolled up, and fed back to the nip at right angles to its former direction. This is done several times to improve mixing. Despite extensive safety precautions, operators of two-roll mills often have n fingers ($n < 10$). To the author's knowledge, no systematic studies of the effect of the additional ($10-n$) ingredients on the properties of the compounded polymers have been reported.

Motionless or *static mixers* are a more recent development for continuous compounding [25–27]. One example of this type of device, the Kenics mixer (Figure 17.13), consists of a series of alternating right- and left-handed helices that continuously divide the melt and cause it to rotate around its own hydraulic axis. Each element in a Kenics mixer breaks the stream into two parts. Therefore, an n-element mixer generates 2^n layers, so n does not have to be very large to achieve intimate mixing. The obvious advantage of this type of device is that it contains no moving parts (although the material must be pumped through it). It also requires relatively low power input per unit of material processed. Such mixers provide good radial mixing and relatively narrow residence-time distributions, despite the fact that the flow is invariably laminar in the processing of polymer melts. They are also incorporated in extruders, between the screw and die, and in injection-molding machines ahead of the nozzle to improve the thermal homogeneity of the melt.

17.14   LITHOGRAPHY

Lithography, literally stone writing, has been used to etch patterns into solid materials (primarily harder materials than polymers), and has been used extensively in microelectronics to etch fine and complex patterns into silicon wafers to make integrated circuits. A photoresist, usually made of a polymer, is made to block an intense light, resulting in the reverse pattern being etched into the wafer. For polymer processing, similar photoresists (or masks) are placed between a light source and a monomer or pre-olymer solution containing a photoinitiator. After polymerization, unreacted solution behind the mask can be removed, resulting in polymer films with features that approach submicron sizes that match the holes in the mask. The mask is recycled, similar to the molds used in injection molding. Photolithography is best suited for small parts (films) that require complex two-dimensional features, but can be used for high-volume production.

Electron beam lithography is a similar process, but instead of using light, a beam of electrons is shot at a surface to create the surface architecture with the advantage of being able to produce even smaller features (into the nanometer region, as electron beams can be focused to dimensions smaller than the wavelength of light).

17.15   THREE-DIMENSIONAL (RAPID) PROTOTYPING

A more recent addition to the methods of polymer processing is inkjet printing or three-dimensional prototyping. Here, polymer droplets can be placed using a machine similar to an inkjet printer to build complex three-dimensional architectures, usually with the assistance of computer-aided design (Figure 17.14). This is a rather expensive technique in that (even with rapid prototyping) the time required to make each part can be fairly lengthy. However, there is one really great advantage: the interior architecture of a polymeric part can be designed with a great deal of geometric specificity, much more complex than can be achieved by blow molding or extrusion. This allows void spaces to be built into materials, such as conduits for flow, but in three dimensions. One area of application is in light-emitting diode (LED) displays used in televisions and computer screens. They are also useful in three-dimensional reconstructions.

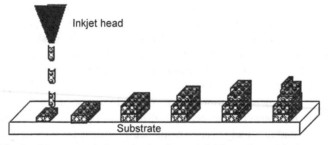

FIGURE 17.14   Rapid prototyping method using an inkjet printer head to construct three-dimensional objects. After leaving the head, the droplets coalesce, and by rastering over a region multiple times, a three-dimensional object can be made.

17.16 SUMMARY

Because polymer properties can be easily modified by addition of solvents, plasticizers, or heat, they are quite versatile when selecting a production method. This gives rise to a wide range of processing techniques (summarized in Table 17.1), but also presents challenges. Not every technique will work for each type of polymer, e.g., thermosets and polymers that degrade easily cannot be melt processed. On the other hand, starting with a liquid (monomer, prepolymer, polymer solution, or polymer melt) allows polymers to be somewhat easily formed into rather complex (sometimes large) shapes with a great degree of reproducibility. Although the major techniques used in polymer processing have been presented in this chapter, numerous variations and combinations continue to evolve to allow processing of new polymers and develop economical methods to create even more complex products. Chapters 18 to 22 include some additional processing techniques, but is focused on additives (as in sheet molding compound) or on more specifically applied areas (such as surface finishing and microencapsulation).

PROBLEMS

1 Using the Internet, identify the commercial suppliers of any three pieces of equipment used to process polymers. List the supplier, approximate capacity of the equipment (lbs/day or kg/day), materials the equipment is made of, and an example of a product made by each piece of equipment.

2 What processing methods are used to make the following polymeric products:

(a) computer keyboards   (b) rubber tires
(c) contact lenses   (d) mylar balloons
(e) acrylic fibers for clothing (f) car windshields
(g) a bottle for liquid detergent

3 Explain the difference between thermoforming and compression molding.

4 What is the function of a spinnerette in wet and dry spinning? Define denier.

5 You are using an extruder to make rubber tubing, but one of the heating elements near the nozzle does not work. What kind(s) of problems might this cause in the process? What problems might arise in the quality control of the product?

6 Explain why injection molding techniques can only process thermoplastics, but can also be used to make thermosets.

7 In making 50,000 plastic shower liners (approximately 6 ft by 8 ft sheets with thickness of 1/16 in) out of poly(vinyl chloride), what process would you recommend using? What temperature should be used for the process (approximately)? What would happen if the temperature was too high or too low? What additive(s) would you use, and why?

8 You discover that the plastic PET bottles made using blow molding in your plant have a large rate of rejection due to problems with holes in the threaded top part of the bottles (and that the threaded part is often very thin).

(a) Give a likely explanation of what is causing this problem.

(b) Suggest two changes in processing that will help resolve the problem.

REFERENCES

[1] Kroschwitz, J. (ed.), *Encyclopedia of Polymer Science and Engineering*, 2nd. ed., Wiley, New York, 1985.

[2] *Modern Plastics Encyclopedia*, McGraw-Hill, New York (yearly).

[3] Rosato, D.V. and D.V. Rosato, *Plastics Processing Data Handbook*, Van Nostrand Reinhold, New York, 1989.

[4] Morton-Jones, D.H., *Polymer Processing*, Chapman and Hall, London, 1989.

[5] Han, C.D., *Rheology in Polymer Processing*, Academic, New York, 1976.

[6] McKelvey, J.M., *Polymer Processing*, Wiley, New York, 1962.

[7] Middleman, S., *Fundamentals of Polymer Processing*, McGraw-Hill, New York, 1977.

[8] Pearson, J.R.A., *Mechanics of Polymer Processing*, Elsevier, New York, 1985.

[9] Pearson, J.R.A. and S.M. Richardson (eds), *Computational Analysis of Polymer Processing*, Applied Science, New York, 1983.

[10] Tadmor, Z. and C.G. Gogos, *Principles of Polymer Processing*, Wiley-Interscience, New York, 1979.

[11] Rosato, D.V. and D.V. Rosato, *Injection Molding Handbook*, Van Nostrand Reinhold, New York, 1985.

[12] Rubin, I.I., *Injection Molding: Theory and Practice*, Wiley-Interscience, New York, 1973.

[13] Macosko, C.W., *Fundamentals of Reaction Injection Molding*, Hanser, Munich, 1988.

[14] Stevens, M.J., *Extruder Principles and Operation*, Elsevier, New York, 1985.

[15] Tadmor, Z. and I. Klein, *Engineering Principles of Plasticating Extrusion*, Van Nostrand Reinhold, New York, 1970.

[16] Rauwendaal, C., *Polymer Extrusion*, Hanser, Munich, 1986.

[17] Private communication to S.L. Rosen, NRM Corporation.

[18] Rosato, D.V. and D.V. Rosato (eds), *Blow Molding Handbook*, Hanser, Munich, 1989.

[19] Gruenwald, G., *Thermoforming: A Plastics Processing Guide*, Technomic, Lancaster, PA, 1987.

[20] Throne, J.L., *Thermoforming*, Hanser, Munich, 1987.

[21] Sikes, S., *Mod. Plast.* **51**(1), 70 (1974).

[22] Ward, L.G., *Plast. Eng.* **35**(4), 47 (1979).

[23] Meyer, R.W., *Handbook of Pultrusion Technology*, Chapman and Hall, London, 1985.

[24] Boliek, J.E. and A.W. Jensen, Fibers, Elastomeric, in *Kirk-Othmer Encyclopedia of Chemical Technology*, Wiley, 2009, p. 15.

[25] Skoblar, S.M., *Plast. Technol.* October 1974, p. 37.

[26] Schott, N.R., B. Weinstein, and D. LaBombard, *Chem. Eng. Prog.* **71**(1), 54 (1975).

[27] Chen, S.J., *Chem. Eng. Prog.* **71**(8), 80 (1975).

CHAPTER 18

POLYMER APPLICATIONS: PLASTICS AND PLASTIC ADDITIVES

18.1 INTRODUCTION

As mentioned in the introduction to this book, there are five major applications for polymers: plastics, rubbers, synthetic fibers, surface finishes, and adhesives. Previous sections have dealt with the properties of polymers themselves. Although these properties are undoubtedly most important in determining the ultimate applications, polymers are rarely used in a chemically pure form; so in a discussion of the technology of polymers, it is necessary to mention the properties of polymers used for these industrially important applications. There are also numerous modifiers and additives commonly used, which are introduced in Section 18.4, but are useful in many products that fit any of the five applications discussed in the next five chapters. This chapter will focus on simple plastics.

18.2 PLASTICS

Plastics are normally thought of as being polymer compounds possessing a degree of structural rigidity— in terms of the usual stress–strain test, a modulus on the order of 10^8 Pa (N/m^2) (15, 000 psi) or greater. The *molecular requirements* for a polymer to be used in a plastic compound are as follows: (1) if linear or branched, the polymer must be below its glass transition temperature (if amorphous) and/or below its crystalline melting point (if crystallizable) at use temperature; otherwise (2) it must be crosslinked sufficiently to restrict molecular response essentially to straining of bond angles and lengths (e.g., ebonite or "hard" rubber).

Fundamental Principles of Polymeric Materials, Third Edition. Christopher S. Brazel and Stephen L. Rosen.
© 2012 John Wiley & Sons, Inc. Published 2012 by John Wiley & Sons, Inc.

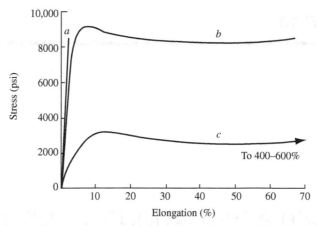

FIGURE 18.1   Typical stress–strain curves for plastics: (a) rigid and brittle; (b) rigid and tough; (c) flexible and tough.

18.3   MECHANICAL PROPERTIES OF PLASTICS

The engineering properties of commercial plastics vary considerably within the broad definition given above. Figure 18.1 sketches some representative stress–strain curves for three types of plastics. As discussed previously, for any given material, the quantitative nature of the curves depends markedly on the rate of strain and the temperature. In general, faster straining and lower temperatures give higher moduli (slopes) and smaller ultimate elongations.

Plastics with stress–strain curves of the type in Figure 18.1a are *rigid* and *brittle*. The former term refers to the high initial modulus. The latter refers to the area under the stress–strain curve, which represents the energy per unit volume required to cause failure. These materials usually fail by catastrophic crack propagation at strains in the order of 2%. Since hardness correlates well with tensile modulus, which is another valuable property of this type of plastic. Examples of this class are polystyrene, poly(methyl methacrylate), and most thermosets.

Curve *b* represents *rigid* and *tough* materials, sometimes known as *engineering plastics*. In addition to high modulus, tensile strength and hardness, these materials undergo *ductile deformation* or drawing beyond the yield point, evidence of considerable molecular orientation before failure. The latter confers toughness or *impact resistance*, that is, the ability to withstand shock loading without brittle failure. Examples of this class of materials are polycarbonates, cellulose esters, and nylons. Although the reasons are not entirely clear, such plastics generally exhibit a secondary dynamic damping peak (Chapter 15) well below their use temperature.

Curve *c* represents flexible and tough plastics, as typified by low- and medium-density polyethylenes. Here, the ductile deformation leading to very high ultimate elongations results from the conversion of low-crystallinity material to high-crystallinity material. The tensile samples "neck down" with the density, percent crystallinity, and modulus of the material in the neck being appreciably greater than those of the parent material. Despite their good toughness, these materials are limited in their structural applications by their low moduli and tensile strengths.

TABLE 18.1 Common Additive Types for Plastics

Additive Type	Example(s)
Reinforcing agents	Graphite fibers
Fillers	Wood flour, calcium carbonate
Coupling agents	Bifunctional silanes
Stabilizers	Quinones and other free-radical scavengers
Pigments	Titanium dioxide, carbon black
Dyes	Various colored dyes
Plasticizers	Phthalates, medium MW organics
Lubricants	Stearic acid, low MW organics
Processing aids	Low MW polymers (to reduce viscosity)
Curing agents	Sulfur, crosslinking agents (often with a promoter)
Blowing agents	Pentane and volatile organics, CCl_3F (makes foam)
Flame retardants	Phosphorus, organics with Br or Cl groups

18.4 CONTENTS OF PLASTIC COMPOUNDS

In addition to the polymer itself, which is seldom used alone, plastics usually contain at least small amounts of one or more of the following additives (summarized in Table 18.1) [1,2].

18.4.1 Reinforcing Agents [3]

The function of these is to enhance the structural properties of the compound, in particular, properties such as modulus (stiffness), strength, and the retention of these properties at higher temperatures. The use of long fibers to reinforce epoxy and polyester thermosets was mentioned in the previous chapter. While glass is by far the most common reinforcing fiber, largely due to its strength, ready availability, and relatively low cost, other, more exotic fibers are used where very high strength-to-weight ratios are required and cost is of secondary concern. Carbon or "graphite" fibers, made by pyrolysis of pitch or polyacrylonitrile fibers, offer superior performance at a cost premium over glass. They are used in many aerospace applications and in more down-to-earth objects such as golf-club shafts, tennis-racket frames, racing-car bodies, and automotive springs and driveshafts. Aramid (*aro*matic poly*amid*) fibers (Kevlar) are also used to reinforce plastics for premium-property applications, sometimes in combination with other fibers. And, more recently, nanofibers (including carbon nanotubes) and nanoclays have been employed as additives to strengthen plastics with even smaller amounts of additive (see Figure 18.2).

Short (1/8 to $^1/_2$ in.) (3–13-mm) glass and other fibers are used extensively to reinforce thermoplastics [5]. Addition of up to 40 wt% short glass fibers to nylon, polypropylene, poly(butylene terephthalate), and others, provides a relatively inexpensive way of greatly improving the structural strength of the plastic, as illustrated in Table 18.2. The glass-reinforced materials also have less mold shrinkage and, therefore, give parts with better dimensional control and stability. The down side is that they are more difficult to process and often have a rough surface finish.

Note that major gains in heat-distortion temperature are obtained when using reinforcing agents with the crystalline polymers nylon 6/6, poly(butylene terephthalate), and polypropylene, but not with the amorphous polystyrene and polycarbonate. This is a

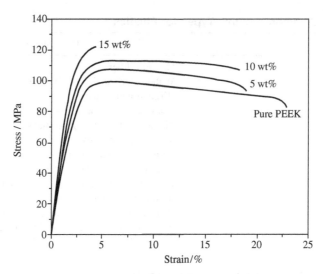

FIGURE 18.2   Stress–strain curves showing the effect of adding carbon nanofibers to poly(ether ether ketone), PEEK, a semicrystalline thermoplastic. Reprinted from [4] with permission from Elsevier.

general result. The crystallites "lock on to the fibers, maintaining mechanical integrity well above T_g.

Another example of reinforcement is the use of up to 20% dispersed rubber particles, on the order of 1 μm in diameter, to increase the *toughness* or *impact strength* of a normally brittle plastic. Figure 18.3 illustrates the effect of adding increasing levels of a micron-sized rubber to polystyrene in a high rate-of-strain tensile test (to simulate shock loading). The addition of the rubber particles decreases the modulus (slope) and ultimate tensile strength but introduces considerable ductile deformation (elongation beyond the "knee"), thereby increasing the area under the stress–strain curve. It also imparts a secondary dynamic damping peak below use temperature. In effect, it converts a type *a* material (Figure 18.1) to a type *b* material. For many applications, the improvement in impact strength considerably outweighs the slight decreases in modulus and tensile strength produced by the addition of rubber. *Impact modifiers* are rubbery additives that impart such toughness when blended with plastics. Proper compatibility (good adhesion but not solubility) between the phases is essential in these applications. This is often achieved with graft and block copolymers (*compatibilizers*), which have blocks of repeat units at each end that are compatible with each polymer phase.

Commercially important examples of *rubber-toughened* plastics include the high-impact polystyrenes (HIPS), in which polystyrene is toughened with a polybutadiene rubber, and the acrylonitrile-butadiene-styrene (ABS) plastics, in which a polybutadiene or poly(butadiene-*co*-acrylonitrile) rubber toughens a poly(styrene-*co*-acrylonitrile) glassy phase. More complete discussions of these two-phase polymer systems are available elsewhere [7,8].

18.4.2   Fillers [9]

Fillers are particulate materials whose major function often is simply to extend the polymer and thereby reduce the cost of the plastic compound. Fillers can take up a substantial

TABLE 18.2 Some Properties[a] of Short Glass Fiber-Reinforced Thermoplastic [6]

Plastic[b]	PP		Nylon 6/6[c]		PC		PBT		PS	
wt% glass	0	40	0	30	0	30	0	30	0	30
Ultimate tensile strenght, psi ASTM D638	5300	11,700	12,400	25,000	9500	19,500	8200	16,500	6400	12,000
Ultimate elongation, %	350	2.8	300	4	110	3.5	175	3	1.9	1.1
Tensile modulus, 10³ psi ASTM D638	195	1300	390	1300	345	1300	360	1380	400	1250
Izod impact, 1/8-in ft-lb/in ASTM D256A	0.8	1.7	1.3	3.0	16	2.3	0.85	1.3	0.4	2.0
Rockwell hardness ASTM D758	R91	R106	R120	R110	M70	M92	M73	M90	M67	M90
Linear coefficient of expansion, 10⁻⁶/°C ASTM D696	90	30	80	35	68	22	67	25	77	20
264 psi DTUL,[d] °F ASTM D648	130	315	179	371	270	297	154	430	186	217
Density, g/cm³ ASTM D792	0.91	1.22	1.14	1.30	1.20	1.41	1.34	1.51	1.05	1.29

[a]Where a range is given, the average is listed here.
[b]PP, polypropylene; PC, polycarbonate; PBT, polybutylene terephthalate; PS, polystyrence.
[c]Nylon properties are very sensitive to moisture content.
[d]Deflection temperature under load.
Source: Reference 6.

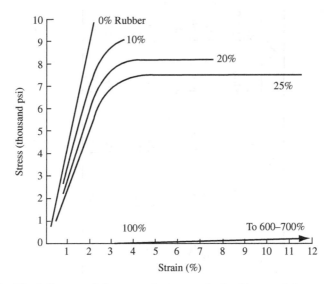

FIGURE 18.3   The influence of discrete, micrometer-sized rubber particles on the high-speed (133 in/in min) stress–strain curves of polystyrene [7].

portion of the volume of polymeric products (often approaching 50%), particularly for items where low cost is more important than product performance. One of the earliest examples of a filler is the wood flour (fine sawdust) long used in phenolic polymers and other thermosets. Calcium carbonate is used in a variety of plastics and polypropylene is often filled with talc. Even water has been used to extend polyester casting compounds.

Fillers may also be used to improve certain properties of the compound. Almost all of them reduce mold shrinkage and thermal expansion coefficients. They reduce warping in molded parts. Mica and other fillers can increase heat resistance. Hollow glass or phenolic microspheres reduce the density of the composite. *Conductive fillers*, such as certain carbon blacks and aluminum flake, impart electrical conductivity that bleeds off static charge (important in car tires) and provides plastics with electromagnetic interference (EMI) shielding capability. Just about any filler whose hardness and modulus are greater than those of the polymer will increase the hardness and *initial* modulus of a mixture with the polymer, but generally, elongation, ultimate strength, toughness, and processability suffer.

18.4.3 Coupling Agents

For a reinforcing fiber, such as glass, to be of maximum benefit (or a filler not to be too detrimental to mechanical properties), stress must be efficiently transferred from the polymer to the reinforcing agent or the filler. Unfortunately, most inorganics have hydrophilic surfaces, while most polymers are hydrophobic, so interfacial adhesion is poor. This problem is exacerbated by the tendency of the surface of many inorganics (particularly glass) to adsorb water, which further degrades adhesion by reducing the amount of hydrogen bonds between the fibers and the polymer. Coupling agents are intended to help overcome these difficulties.

The most common coupling agents are silanes, with the general formula $YRSi(OR')_3$. The (OR') group reacts with the inorganic substrate and the Y group reacts (or at least forms strong secondary bonds) with the polymer, thereby enhancing interfacial adhesion. The reaction sequence by which vinyl-triethoxysilane is believed to couple to a substrate containing hydroxyl surface functionality (e.g., glass) is shown below.

Step 1: Hydrolysis (most likely with water adsorbed on the substrate):

$$H_2C=CH-Si(OC_2H_5)_3 + 3H_2O \rightarrow H_2C=CH-Si(OH)_3 + 3C_2H_5OH$$

Step 2: Reaction with substrate:

$$H_2C=CH-Si(OH)_3 + HO{-} \rightarrow H_2C=CH-Si(OH)_2-O{-} +H_2O$$

The "dangling double bonds" so introduced can then participate in the cure of an unsaturated polyester resin (Example mprb2.3), covalently bonding the polymer to the surface.

A wide variety of silane coupling agents is available, with the functional groups optimized for use with various polymers and substrates. Titanate coupling agents perform in a similar fashion. Fillers and reinforcing fibers are supplied pretreated with coupling agent or the agent may be added during compounding (in which case, presumably most of it migrates, or blooms, to the interface).

In the case of fillers, coupling agents have an added benefit. By converting an otherwise hydrophilic surface to a hydrophobic surface, compatibility with the polymer is improved. This results in easier and more uniform dispersion of the filler in the polymer, and at a given filler, loading can reduce the viscosity substantially, thereby improving processability.

Graft and block copolymers often perform a coupling function in the case of rubber-toughened plastics mentioned in the discussion of reinforcing agents. In the manufacture of high-impact polystyrene, for example, the polybutadiene is dissolved in the polymerizing styrene monomer. As a result, some of the styrene is grafted to the rubber. The grafted polystyrene chains are physically compatible with the surrounding homopolystyrene and are chemically bound to the rubber, thereby enhancing adhesion between the dispersed polybutadiene particles and the polystyrene matrix. The graft copolymer is thought to be essential for effective impact enhancement in this particular system, because simple mechanical blends of polystyrene and polybutadiene tend to be less homogeneous and show little impact enhancement. However, minor amounts of styrene–butadiene *block* copolymers do act as compatibilizers and, therefore, improve the impact strength of such blends, much as a soap or a surfactant that emulsifies oil in water.

18.4.4   Stabilizers

Most polymers are susceptible to one or more forms of degradation, usually as a result of environmental exposure to oxygen or ultraviolet (UV) radiation, or to high temperatures during processing operations. Stabilizers inhibit degradation of the polymer. In the case of poly(vinyl chloride), for example, the major product of thermal degradation is HCl, which, besides being a strong acid, catalyzes further degradation. Compounds such as metal oxides, which react with the HCl to form stable products, are used as stabilizers. Oxidative degradation of polymers is thought to take place by a free-radical mechanism involving crosslinking and/or chain scission initiated by free radicals from peroxides formed in the initial oxidation step. Similarly, these reactions can be initiated by free radicals produced by UV radiation (a serious threat to polymers used in cars, particularly dashboard materials). Stabilizers against such reactions are generally quinone-type organics that are effective free-radical scavengers.

18.4.5 Pigments

Plastics are often colored by the addition of pigments, which are finely divided solids. If the polymer is itself transparent, a pigment imparts opacity. Titanium dioxide is a common pigment where a brilliant, opaque white is desired. Sometimes, pigments perform other functions. For example, calcium carbonate acts as both a filler and a pigment in many plastics, and carbon black is often a stabilizer as well as a pigment. It prevents degradation by absorbing and preventing the penetration of UV light beyond the surface of the material.

18.4.6 Dyes

Dyes are colored organic chemicals that dissolve in the polymer to produce a transparent compound (assuming the polymer is transparent to begin with). Many polycarbonate drinking bottles that are colored, yet transparent, use dyes. Some thermoplastics, though transparent, develop a slight yellow tinge from minor degradation during processing that causes selective absorption of light toward the blue end of the spectrum. The yellow can largely be "canceled out" by the addition of a blue dye that reduces the transmission of yellowish wavelengths. This technique unavoidably results in a slight lowering of the total light transmission.

18.4.7 Plasticizers

The addition of a relatively low molecular weight organic plasticizer (often exceeding 10–20% of the polymer mass) to a normally glassy polymer will progressively reduce its modulus by bringing the compound's glass transition temperature down closer to use temperature. The level at which the modulus is reduced to the point where the compound is considered an elastomer rather than a plastic is somewhat arbitrary, but plastics sometimes contain small amounts of a plasticizer to reduce brittleness. Plasticizers must be thermodynamically compatible with the host polymer to avoid phase separation.

Common plasticizers, such as those based on phthalates or trimellitates, have been used frequently, particularly with poly(vinyl chloride), allowing an otherwise hard material (e. g., drain pipes) to be used in flexible products (e.g., shower curtains or water hoses). Several recent studies showing migration of plasticizers out of materials during their lifetime has led to the development of more stable or less toxic plasticizers, such as citrates that are used in medical plastics (e.g., IV bags and tubing). A recent review paper covers the major uses of common plasticizers and discusses the numerous applications and research trends [10].

18.4.8 Lubricants

External lubricants are low molecular weight organics that are relatively insoluble in the polymer and "plate out" or migrate to the surface of the compound and form a slippery coating during processing operations. They produce smoother extrudates and molded articles, and minimize sticking in the mold by acting as mold release agents. Where subsequent painting or printing of the surface is required, they can reduce adhesion, and so must be used with care. Stearic acid and its metal salts are common external lubricants. *Internal* lubricants are soluble in the polymer, and ease processing by lowering the compound's viscosity.

18.4.9 Processing Aids

Various compounds (often a second polymer) are used to modify the rheological properties of a polymer during processing to provide higher outputs, better surface finish, and easier handling in general. For example, low molecular weight polyethylene is added to PVC to improve extrusion behavior.

18.4.10 Curing Agents

These are chemicals whose function is to produce a crosslinked, thermosetting plastic from an initially linear or branched polymer. For example, a vinyl monomer, such as styrene, a free-radical initiator, and sometimes a *promoter* (e.g., cobalt naphthenate, to speed up the reaction) can be dissolved in a low molecular weight unsaturated polyester resin, which forms crosslinks by an ordinary addition mechanism involving the double bonds in the polyester (Example mprb2.3). Curing is also important in forming tires, as sulfur is compounded with natural, butadiene, or isoprene rubber to form a crosslinked rubber.

The polymer for a *two-stage* thermosetting compound is deliberately produced with a stoichiometric shortage of one of the reagents to give a highly branched, but not yet crosslinked structure, which is later cured in the mold with a curing agent. An "A-stage" phenolic polymer is produced by reacting phenol and formaldehyde in about a 1.25:1 mol ratio (a molal excess of formaldehyde is required for crosslinking). This still-thermoplastic product (a *novolac*) is compounded with fillers, pigments, reinforcing agents, and a curing agent, hexamethylenetetramine, to give a "B-stage" resin. The curing agent decomposes in the presence of moisture (a product of the condensation reaction) upon heating in the mold, giving the additional formaldehyde required for crosslinking and ammonia which acts as a basic catalyst for the reaction:

$$+\ 6H_2O \longrightarrow 6CH_2=O\ +\ 4NH_3$$

Hexamethylenetetramine Water Formaldehyde Ammonia

Single-stage resins (*resoles*) are made using the final desired reactant ratio, about 1:1.5 phenol:formaldehyde, but the reaction is stopped short of crosslinking. Crosslinking is completed in the mold without the necessity of a curing agent.

Free-radical initiators (usually organic peroxides) are used as curing agents for saturated thermoplastic polymers. The curing agent must be compounded with the polymer at temperatures low enough to prevent appreciable decomposition to free radicals. The compound is then heated in the mold, producing free radicals that abstract protons from the polymer, leaving unshared electrons on the chains, which combine to form crosslinks:

$$R{:}R \xrightarrow{\text{Heat}} 2R\bullet$$

Initiator   Free radicals
(curing agent)

R• +   ∿C–C–C∿   ⟶   ∿C–C–C∿   +   R:H

Polymer chain

Crosslinked chains

Sometimes, sulfur or multifunctional vinyl monomers are added to improve cross-linking efficiency by helping to "bridge the gap" (i.e., produce longer crosslinks) between the chains. In this manner, polyethylene is crosslinked with dicumyl peroxide, converting it from a thermoplastic to a thermoset with much greater heat resistance to stress cracking and abrasion. It can also tolerate much higher levels of carbon black filler without becoming excessively brittle.

18.4.11 Blowing Agents

Foamed plastics must contain a material that generates a gas to produce foaming. They are *chemical blowing agents* (CBAs), which generate gas through a chemical reaction, and physical blowing agents, inert but volatile chemicals that are dissolved in the polymer or its precursors and simply vaporized upon heating or a reduction in pressure. As an example, polyurethanes (Example 2.4Q) may be "water blown" by the reaction of excess diisocyanate with water, generating CO_2:

$$O{=}C{=}N{-}R{-}N{=}C{=}O \; + \; 2H_2O \rightarrow R(NH_2)_2 + 2CO_2$$

Diisocyanate   Diamine

This is not often done in practice because of the expense of the diisocyanate and the difficulty in controlling foam structure. Alternatively, they can be foamed with a physical blowing agent, traditionally CCl_3F or other chlorofluorocarbons (CFCs). The latter gives excellent control of foam structure and, because of its low thermal conductivity, makes good insulating foams. Unfortunately, it poses a threat to the atmospheric ozone layer and has been phased out for many operations. Carbon dioxide and liquefied gases are among several alternatives to replace the CFCs.

CBAs must be compounded with a polymer below their decomposition temperatures and then decomposed to generate the gas at temperatures low enough to prevent degradation of the polymer. A variety of CBAs is available, covering a range of decomposition temperatures, for use with most common plastics [11,12].

In the manufacture of polystyrene foam molding beads, pentane is added to the monomer in a suspension polymerization or absorbed by the beads afterward. When the beads are placed in a mold and heated (usually with steam), the pentane volatilizes, expanding the beads against each other and the mold walls. These beads are used in many familiar applications, such as drinking cups, picnic coolers, and packaging peanuts.

18.4.12   Flame Retardants [13–15]

Most synthetic polymers, being composed largely of carbon and hydrogen, are flammable (although no more so than natural materials such as wood and cotton—both natural polymers, which synthetic polymers often replace—a fact sometimes overlooked). Plastics are increasingly being compounded with flame retardants to reduce their flammability. The most common flame-retarding additives for plastics contain large proportions of chlorine or bromine. These elements are believed to quench the free-radical flame propagation reactions. They may be compounds that are simply mixed with the plastic, for example, decabromodiphenyl ether, or they may be reactive monomers that become part of the polymer. Examples of the latter include tetrabromobisphenol-A substituted for some of the normal bisphenol-A in epoxies (Example 2.4O) or polycarbonates (Example 2.4N) and tetrabromophthalic anhydride used in polyesters (Example 2.3). Antimony trioxide is often used in synergistic combination with halogenated compounds. Organic phosphates are thought to function as flame retardants by forming a char that acts as a barrier to flames. A compound such as *tris*(2,3-dibromopropyl) phosphate combines halogen and phosphate. Poly(vinyl chloride), because of its chlorine content, is inherently flame resistant, but if it is to maintain this valuable property in plasticized form, it must be compounded with halogenated or phosphate plasticizers.

   The safety of halogen and antimony flame-retardant systems has been questioned, both the compounds themselves during processing operations and their combustion products. They have largely been replaced with phosphorus-based systems [16]. Hydrated alumina, another alternative, is a particulate filler that contains 35% water of hydration, the evaporation of which absorbs energy and inhibits flame spread.

18.4.13   Miscellaneous

Various other materials are added to plastics to provide certain end-use properties. For example, where a polymer will be subjected to a warm, humid environment (e.g., vinyl shower curtains and silicone-rubber bathtub caulks), it often contains a *biocide* to inhibit the growth of mold, mildew, and fungus. These biocides are generally organic copper, mercury, nanosilver, or tin compounds. Fatty-acid amines are added to compounds for use in bottles and phonograph records as antistatic agents. These chemicals, because of their limited compatibility with the polymer, migrate (or *bloom*) to the surface of the material and, because of their polarity, attract moisture from the atmosphere. The moisture bleeds off static electricity charges, which otherwise would attract considerable dust over time. These same fatty-acid amines also function as *slip* or *antiblock* agents, which prevent layers of plastic film and sheet from sticking to each other. *Nucleating agents*, which promote more but smaller crystallites, were discussed in Chapter 4.

18.5   SHEET MOLDING COMPOUND FOR PLASTICS

The subject of sheet molding compound (SMC) is introduced here because: (1) they are commercially important, (2) they make use of many of the additives discussed in the previous section, and (3) they do not seem to fit logically anywhere else. Sheet molding compounds provide a high strength-to-weight ratio, with low cost and easy, rapid processing. SMCs are easily-moldable polymer composites that are formed into sheets with a curing agent included (which is activated during a delayed molding step). They are used for automotive body parts such as front ends and fender extensions and for other structural parts in such items as business copy machines and air-conditioner hosings. They, along with RIM polyurethanes and polyureas and injection-molded thermoplastics, are in the competition for increased use in auto body panels (doors, hoods, fenders, etc.).

SMCs consist mainly of unsaturated polyester resin (Example mprb2.3), filler (usually $CaCO_3$), and chopped-glass reinforcing fiber. Early SMCs contained roughly a third of each, but newer materials are closer to 25% resin, 30% glass, and 45% filler [17]. The liquid polyester and filler are compounded in a high-shear mixer along with minor amounts of a curing agent (usually an organic peroxide), a thickener (MgO or CaO), an external lubricant as a mold-release agent (zinc stearate), and sometimes a low-shrink additive.

The material leaves the mixer looking (but not smelling) like pancake batter, and is spread continuously onto layers of polyethylene film (Figure 18.4). Chopped glass fibers (about 1 in long) are sprinkled on, the layers are combined, and the sheet is kneaded by rollers to wet out the glass. The sheet is then rolled up and allowed to age, which during time. The thickens to the consistency of cardboard. This thickening is believed to involve interaction of the MgO or CaO with the carboxylic acid groups on the polyester, with water also playing a role, but it is not a true polymerization.

Before use, the sheets are cut to size and the polyethylene stripped off. They are then placed in a heated compression-type mold (several layers may be stacked for greater thicknesses) where true cure (crosslinking) of the polyester takes place.

The choice of a curing agent for a SMC involves a compromise between an active initiator, which will give quick cures, but conversely, imparts a short shelf life to the stored SMC, and a less active initiator with a longer shelf life but slower cure. A drawback of early SMCs was the wavy surface and sometimes visible glass-fiber texture produced by shrinkage of the polyester during cure. This necessitated hand-finishing operations, which are not at all popular in the automotive industry. The low-shrink additives that counteract this usually involve a second, linear polymer dissolved in the polyester, which precipitates

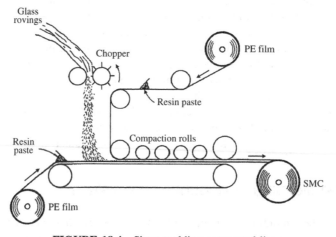

FIGURE 18.4   Sheet molding compound line.

out as a micrometer-sized dispersed phase as the polyester cures. Why this should reduce shrinkage is not at all clear, but if the low-shrink additive is a rubber, it may act as a toughening agent as well (see Section 18.4.1).

Bulk molding compounds (BMCs) are similar in composition to SMCs, with somewhat shorter fibers ($^1/_4$ to $^1/_2$ in), and perhaps less glass (15%) and more filler. They are produced in chunk form and compression molded like other thermosetting compounds.

18.6 PLASTICS RECYCLING

The recycling and reuse of plastics is important for both environmental and monetary reasons. Just as metals are separated prior to recycling, polymers must be separated as well. Typical recycling symbols indicate which type of plastic an item is made from (Figure 18.5). The additives covered in the chapter make recycling difficult, as they are generally well-mixed with the polymer host, and thermodynamically difficult to remove. Some plastics are copolymers or blends of polymers that can also be challenging to recycle. Impurities (i.e., the additives) can significantly alter the properties of a material, and most of the reprocessing methods used to make new products from recycled plastics reduce the molecular weight of the materials. Thus, recycled plastics do not find many uses in products requiring high mechanical strength or that have a narrow range of acceptable properties. The U.S. Food and Drug Administration requires the use of virgin plastic materials for linings that touch a food product, thus recycled plastics are limited to outer packaging for food applications. One reason for this is the potential for harmful chemicals or bacteria to find their way into the void space of the original material. Recycled plastics more often than not are used in lower-end products like garbage bags and park benches.

The recycling process normally involves several steps of chopping, grinding, and melting to form pellets, similar to those used for the original hopper feed for thermoplastic processing. Thermosets, on the other hand, cannot be reprocessed by melting without degrading the material, so after chopping, they are generally used only as filler material.

FIGURE 18.5 Plastic recycling symbols for (1) poly(ethylene terephthalate), (2) high density polyethylene, (3) vinyl polymers, (4) low density polyethylene, (5) polypropylene, (6) polystyrene, and (7) all other polymers.

PROBLEM

1 Many consumer products sold these days are advertised as being nanofiber composites (e.g., golf clubs and tennis rackets). Give an example of the nanotechnology inside these composites and explain why this improves the performance of the object. (Use the Internet for assistance.)

REFERENCES

[1] Seymour, R.B. (ed.), *Additives for Plastics, Vol. 1, State of the Art, Vol. 2, New Developments*, Academic, New York, 1978.

[2] Greek, B.F., *Chem. Eng. News*, June 13, 1988, p. 35.

[3] Milewski, J.V. and H.S. Katz (eds), *Handbook of Reinforcements for Plastics*, Elsevier, New York, 1987.

[4] Sandler, J., P. Werner, M.S.P. Shaffer, V. Demchuk, V. Altstadt, and A.H. Windle, *Composites-Part A: Appl. Sci. Manufac.* **33**, 1033 (2002).

[5] Clegg, D.W. and A.A. Collyer (eds), *Mechanical Properties of Reinforced Thermoplastics*, Elsevier, New York, 1986.

[6] Kline, G.M., *Modern Plastics Encyclopedia* McGraw-Hill, New York, 1990.

[7] Rosen, S.L., *Trans. N.Y. Acad. Sci.* **35**(6), 480 (1973).

[8] Riew, C.K. (ed.), *Rubber-Toughened Plastics*, American Chemical Society, Washington, DC, 1989.

[9] Katz, H.S. and J.V. Milewski (eds), *Handbook of Fillers for Plastics*, Elsevier, New York, 1987.

[10] Rahman, M. and C.S. Brazel, *Prog. Polym. Sci.* **29**, 1223 (2004).

[11] Heck, R.L., III, *Plast. Compd.*, November/December 1978, p. 52.

[12] Hallans, R.S., *Plast. Eng.*, December 1977, p. 17.

[13] Schongar, L.H., *Plast. Compd.*, May/June 1978, p. 44.

[14] Blake, W.P., *Plast. Compd.*, July/August 1978, p. 26.

[15] Nametz, R.C., *Plast. Compd.*, January/February 1979, p. 31.

[16] Wood, S.A., *Mod. Plast.* **67**(5), 40 (1990).

[17] Wigotsky, V., *Plast. Eng.* **47**(6), 8 (1991).

CHAPTER 19

POLYMER APPLICATIONS: RUBBERS AND THERMOPLASTIC ELASTOMERS

19.1 INTRODUCTION

A rubber is generally defined as a material that can be stretched to at least twice its original length, and that will retract rapidly and forcibly to substantially its original dimensions on release of the force. In contrast, an *elastomer* is a rubber-like material from the standpoint of modulus but one that has limited extensibility and incomplete retraction. The most common example is highly plasticized poly(vinyl chloride).

Rubber has the following characteristics. (1) From the molecular standpoint, a rubber must be a high molecular weight polymer, since rubber elasticity is mainly due to the coiling and uncoiling of long chain segments (Chapter 13). (2) For the molecules to be able to coil and uncoil freely, the glass-transition temperature of the polymer must be above its use temperature. (3) Furthermore, the polymer must be amorphous in its unstretched state because crystallinity would hinder coiling and uncoiling. (4) There used to be an additional requirement that the polymer be crosslinked (although physical entanglements act as effective crosslinks). If it were not crosslinked, the chains would slip past one another (undergo viscous flow) under stress and recovery would be incomplete; however, if they were chemically crosslinked, they would be thermosets rather than thermoplastics.

19.2 THERMOPLASTIC ELASTOMERS [1,2]

The introduction of the so-called thermoplastic elastomers (actually rubbers) has gotten around the requirement for crosslinking in the strictest sense, that of covalent bonding between chains. They are commonly found for applications requiring flexibility in automotives, household appliances, and soft-grip tools. The most common thermoplastic

Fundamental Principles of Polymeric Materials, Third Edition. Christopher S. Brazel and Stephen L. Rosen.
© 2012 John Wiley & Sons, Inc. Published 2012 by John Wiley & Sons, Inc.

elastomers are styrene-*b*-butadiene-*b*-styrene (SBS) block copolymers produced in solution by anionic polymerization (Example 10.1). Since most polymer pairs are mutually insoluble, the polystyrene chain ends aggregate together in microscopic domains. These polystyrene domains, being below their glass transition temperature (100 °C) at normal use temperatures, are rigid and act to tie together the long, flexible polybutadiene segments (which are above their T_g at normal use temperature) as do ordinary crosslinks. Unlike the usual covalent crosslinks, however, the polystyrene domains soften above the T_g of polystyrene, and the polymer behaves as a true thermoplastic.

All thermoplastic elastomers have in common a continuous rubbery phase of "soft segments" tied together by glassy or crystalline "hard segments" at use temperature, which soften at elevated temperatures. Thus, thermoplastic elastomers do not have to be vulcanized and can be processed by the usual economical thermoplastic techniques.

19.3 CONTENTS OF RUBBER COMPOUNDS

Natural rubber and, with a few exceptions, the many synthetic rubber polymers commercially available are unsaturated; that is, they contain double bonds that provide sites for vulcanization (crosslinking) reactions. The polymers are mostly linear or branched, but some also contain (either intentionally or unintentionally) minor amounts of gel (crosslinked particles) prior to vulcanization. This can have a profound effect on processing properties [3,4]. The following substances are often included in rubber compounds.

19.3.1 Reinforcing Fillers

Most rubber compounds that are intended to develop a reasonable tensile strength, abrasion, and tear resistance will contain up to 50 phr (parts per hundred parts rubber by weight) or more of a reinforcing filler, nearly always a carbon black (thus the color of car tires). The use of carbon black in rubbers is quite different than in plastics, where it is strictly a pigment and limited to much lower loadings.

Carbon blacks are not simply carbon. Basic blacks have hydroxyl groups at the particle surfaces and acid blacks have carboxylic acid functionality. It has been shown that the rubber polymer forms strong secondary and primary covalent bonds with the carbon black, which accounts for its reinforcing ability. A wide variety of carbon blacks is available [5]. In addition to chemical functionality, they differ in factors such as particle size, degree of aggregation, and surface area, and different types of rubber polymers require particular kinds of black for optimum reinforcement. Silicone rubbers are sometimes reinforced with finely divided silica (SiO_2).

Natural rubber (*cis*-1,4-polyisoprene) and its synthetic counterpart and butyl rubber are among the few rubber polymers that can develop reasonable mechanical strength without reinforcement. This is because (especially for natural rubber that has a regular tacticity) they crystallize with molecular orientation at high elongations and the crystallites function as a reinforcing agent, causing the sharp upturn in the stress–strain curves at high elongations shown in Figure 19.1 [6].

19.3.2 Fillers

As with plastics, the function of fillers in rubber is mainly to reduce the cost of the compound. Most rubber fillers are finely divided inorganics such as $CaCO_3$. The addition of

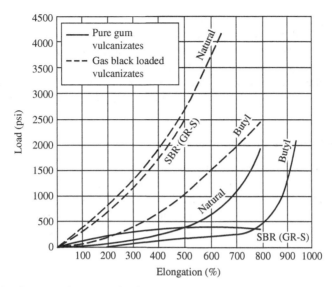

FIGURE 19.1 Stress–strain curves for filled and unfilled rubber vulcanizates [6]. Used with permission of McGraw-Hill Book Company.

such high modulus fillers raises the modulus (stiffens the compound), though too much will cause a loss of rubbery properties, all other things being equal. Carbon blacks sometimes perform as fillers and as reinforcing agents in rubber compounds.

19.3.3 Extending Oils

Hydrocarbon oils are often used in rubber compounds. Their function is twofold. First, they plasticize the polymer, making it softer and easier to process. This is particularly important with very high molecular weight polymers. Second, since they are usually cheaper than the rubber polymer, they act like fillers in reducing the cost of the compound. Extending oils are available in various degrees of aromaticity and the properties of the compound depend on the type of oil in relation to the polymer as well as oil level.

Carbon black and extending oils have opposite effects on the modulus of a rubber compound. One trick for producing low-cost compounds from synthetic polymers is to polymerize to higher-than-normal molecular weight. The modulus is cut down by the addition of an extending oil and then brought back up to the desired level by the addition of large amounts of black. With these ever-increasing levels of black and oil in rubber compounds, it appears that there soon will not be any polymer left! Seriously, there is a limit to this technique, because other properties are soon excessively degraded.

19.3.4 Vulcanizing or Curing Systems

The function of these is to crosslink the polymer. The most common curing systems are based on sulfur. Although sulfur alone can cure unsaturated rubbers upon heating, the process is slow and inefficient in its use of sulfur. The mechanisms of sulfur curing are not well understood, but are thought to include (among other things) the formation of sulfide or disulfide links between chains and the abstraction of protons from adjacent chains to form H_2S, with the chains crosslinking at the remaining unshared electrons.

To speed up the vulcanization process, *accelerators* are generally used. These are usually complex sulfur-containing organic compounds, often of proprietary composition. *Promoters* or activators are used to improve the cure still further. Zinc oxide is a common example, particularly in conjunction with stearic acid.

Nonsulfur cures are used occasionally. Free-radical initiators can provide crosslinks, as discussed in Section 18.4, and zinc oxide can crosslink polymers that contain chlorine.

19.3.5 Antioxidants or Stabilizers

These are particularly important with natural rubber and the many synthetic rubbers that contain a major proportion of butadiene or isoprene. These polymers are highly unsaturated and the double bonds are extremely susceptible to attack by oxygen and ozone, resulting in embrittlement, cracking, and general degradation. Since the degradation reactions take place by free-radical mechanisms, antioxidants work by scavenging free radicals.

19.3.6 Pigments

Where great mechanical strength is not required, and thus carbon black not used, rubbers can be colored with pigments, just as plastics. A typical radial passenger-tire tread formulation is shown in Table 19.1. Notice that the additives (everything except SBR polymer and *cis*-polybutadiene) make up 122.25 parts per hundred parts polymer or more than *half* of the tire!

TABLE 19.1   Radial Passenger-Tire Tread Formulation

Ingredient	phr[a]	Function
SBR 1712[b]		
Polymer	45	Rubber polymer
Aromatic oil	37.5	Extending oil
cis-Polybutadiene 1252	55	Rubber polymer
Carbon black N-234	70	Reinforcing agent
Oil-soluble sulfonic acid	1	Processing aid
Sulfur	1.75	Vulcanizing agent
Stearic acid	2	Promotor
Zinc oxide	3	Promotor
N-Cyclohexyl-2-benzothiazole sulfenamid	1	Accelerator
Polymerized 1,2-dihydrotrimethylquinoline	2	Antioxidant
N-(1,3-dimethylbutyl)-N'-Phenyl-p-phenylene diamine	1	Antiozonant
Blended petroleum wax	3	Crack inhibitor and antiozonant
Total	222.25	

[a]Parts per hundred parts rubber, by weight.
[b]Oil-extended master batch. Cold emulsion polymer $\approx$75% butadiene, 25% styrene extended with highly aromatic oil.
Source. H. L. Stephens [7].

19.4 RUBBER COMPOUNDING [7,8]

The ingredients added to a rubber (such as those in Table 19.1) must be *compounded* with the rubber polymer to produce the final rubber compound for molding, extrusion, etc. This is usually done in *two-roll* mills or *Banbury* mills (Chapter 17). Both devices are driven by relatively large electric motors and put a lot of energy per unit time and volume into the polymer. This energy input serves two purposes. First, it breaks down the polymer and reduces its "nerve" to the point where it can be easily compounded and processed. Nerve is a term used to describe the difficult-to-handle highly elastic response caused by a molecular weight too high. This mastication step mechanically degrades the polymer, lowering its molecular weight and, therefore, increasing its viscous response. Premastication is particularly important with natural rubber, since its molecular weight cannot be controlled during polymerization. Second, the high energy input breaks up and disperses the compounding ingredients evenly throughout the polymer.

REFERENCES

[1] Legge, N.R., G. Holden, and H.E. Schroeder (eds), *Thermoplastic Elastomers*, Hanser, Munich, Germany, 1987.

[2] Walker, B.M., and C.P. Rader (eds), *Handbook of Thermoplastic Elastomers*, 2nd ed., Van Nostrand-Reinhold, New York, 1988.

[3] Rosen, S.L. and F. Rodriguez, *J. Appl. Polym. Sci.* **9**, 1601 (1965).

[4] Rosen, S.L. and F. Rodriguez, *J. Appl. Polym. Sci.* **9**, 1615 (1965).

[5] Byers, J.T., Fillers, Part I: Carbon black, Chapter 3 in *Rubber Technology*, 3rd ed., M. Morton (ed.), Van Nostrand-Reinhold, New York, 1987.

[6] Schmidt, A.X. and C.A. Marlies, *Principles of High Polymer Theory and Practice*, McGraw-Hill, New York, 1948, p. 537.

[7] Stephens, H.L., The compounding and vulcanization of rubber, Chapter 2 in *Rubber Technology*, 3rd ed., M. Morton (ed.), Van Nostrand-Reinhold, New York, 1987.

[8] Barlow, F.W., *Rubber Compounding, Principles, Materials and Techniques*, Dekker, New York, 1988.

CHAPTER 20

POLYMER APPLICATIONS: SYNTHETIC FIBERS

20.1 SYNTHETIC FIBERS

Although many of the polymers used for synthetic fibers are identical to those in plastics, the two industries grew up separately with completely different terminologies, testing procedures, etc. Many of the requirements for fabrics are stated in nonquantitative terms such as "hand" and "drape" that are difficult to relate to normal physical property measurements, but which can be critical from the standpoint of consumer acceptance, and therefore the commercial success, of a fiber.

A fiber is often defined as an object with a length-to-diameter ratio of at least 100. Synthetic fibers are spun (Chapter 17) in the form of continuous *filaments*, but may be chopped to much shorter *staple*, which is then twisted into thread before weaving. Natural fibers, with the exception of silk, are initially in staple form. The thickness of a fiber is most commonly expressed in terms of *denier*, which is the weight in grams of a 9000-m length of the fiber. Stresses and tensile strengths are reported in terms of *tenacity*, with units of grams/denier.

20.2 FIBER PROCESSING

The polymer molecules in synthetic fibers are only slightly oriented by flow as they emerge from the spinnerette. To develop the tensile strengths and moduli necessary for textile fibers, the fibers must be drawn (stretched) to orient the molecules along the fiber axis and develop high degrees of crystallinity. All successful fiber-forming polymers are crystallizable and so, from a molecular standpoint, the polymer must have polar groups, between which strong hydrogen bonding holds the chains in a crystal lattice (e.g., polyacrylonitrile,

Fundamental Principles of Polymeric Materials, Third Edition. Christopher S. Brazel and Stephen L. Rosen.
© 2012 John Wiley & Sons, Inc. Published 2012 by John Wiley & Sons, Inc.

nylons), or be sufficiently regular to pack closely in a lattice held together by dispersion forces (e.g., isotactic polypropylene).

In Chapter 17, it was pointed out that the cross section of fibers is determined by the cross section of the spinnerette holes and the nature of the spinning process. This plays an important role in establishing the properties of the fiber. For certain applications, the spun fibers are textured after spinning. Carpet fibers, for example, are often given a heat twist and/or are crimped by passing them through a pair of gear-like rollers.

20.3   FIBER DYEING

The dyeing of fibers is a complex art in itself. A successful dye must either form strong secondary bonds to polar groups on the fiber or react to form covalent bonds with functional groups on the polymer. Furthermore, since the fibers are dyed after spinning, the dye must penetrate the fiber diffusing into it from the dye bath. The dye molecules cannot penetrate the crystalline areas of the polymer, so it is mainly the amorphous regions that are dyed. This often conflicts with the requirement of high crystallinity. The chains of polyacrylonitrile, for example, while possessing the necessary polar sites for dye attachment in abundance, are so strongly bound to each other that it is difficult for the dye to penetrate. For this reason, acrylic fibers usually contain minor amounts of plasticizing comonomers to enhance dye penetration. Nonpolar, nonreactive fibers such as polypropylene, on the other hand, have no sites to which the dye can bond even if it could penetrate. This was a problem long with polypropylene fibers and was overcome by incorporating a finely divided solid pigment in the polymer before melt spinning. Many of these dyes are also subject to leaching out, as you may have observed when washing a new bright red sweater with white clothes.

20.4   OTHER FIBER ADDITIVES AND TREATMENTS

Static electricity can be a big problem with carpets. Many carpet fibers therefore incorporate an antistatic agent (such as quaternary ammonium salts or alkyl esters of poly(ethylene glycol) to bleed off static charge. These additives are designed to bloom to the surface of the host polymer reducing the build-up of static charge.

The same polar bonding sites used for dyeing fibers also make them stainable. In the past, the finished item, a carpet, for example, would be treated with an anti-staining agent such as a fluorocarbon telomer (such as 3M's Scotchguard) or the anti-staining agents could be applied to the fibers before weaving. However, many of these have gone out of use due to toxicity and bioaccumulation problems of the fluorocarbons. Another approach is to use bicomponent fibers that mix the feel and texture of the original fiber (e.g., using nylon 6/6 as a core fiber) with a highly stain-resistant sheath (e.g., polypropylene or Teflon). These additives are also commonly found in many stain-resistant clothing lines.

20.5   EFFECTS OF HEAT AND MOISTURE ON POLYMER FIBERS

The polarity of the polymer also directly influences its degree of water absorption. Other things being equal, the more polar the polymer, the higher its equilibrium moisture content under any given conditions and humidity. As with dyes, however, moisture content is reduced by strong interchain bonding. The moisture content exerts a strong influence on the feel and comfort of fibers. Hydrophobic fibers tend to have a "clammy" feel in clothing and can build up static electricity charges. Recent advances in athletic apparel has led to clothing that actively wicks away moisture, using fibers that are "breathable."

Perhaps the most important effect of moisture on polar polymers is as a plasticizer. Since fiber-forming polymers are linear, heat acts basically as a plasticizer. This explains why suits wrinkle on hot, humid days, and why the wrinkles can be removed by steam pressing. "Wash-and-wear" and "permanent-press" fabrics are produced by operations that crosslink the fibers by reacting with functional groups on the chains, such as the hydroxyls on cellulose. The more hydrophobic polymers are inherently more wrinkle resistant because they are not plasticized by water. Wash-and-wear shirts, therefore, usually are made of blends of poly(ethylene terephthalate), a polyester, and cotton, about 65%/35%.

CHAPTER 21

POLYMER APPLICATIONS: SURFACE FINISHES AND COATINGS

21.1 SURFACE FINISHES

Nearly all surface finishes and coatings, with the exception of ceramic types for high-temperature applications, are based on a polymer film of some sort. They account for the use of a lot of polymer, but determining just how much and which polymers is not easy because most formulations are proprietary, and production figures do not always separate polymer and nonpolymer components. Five traditional types of surfaces finishes, lacquers, oil paints, varnishes, enamels, and latex paints, will be discussed, along with the role the polymers play in the finish.

21.1.1 Lacquers

One of the more widely used lacquer products is fingernail polish. A lacquer consists of a polymer solution to which a pigment has been added. The film is formed simply by evaporation of the solvent, leaving the pigment trapped in the polymer film. Since no chemical change occurs in the polymer, it retains its original solubility characteristics. Hence, a major drawback of lacquers is their poor solvent resistance.

Much of the technology of lacquers involves the development of polymers that form tougher, more adhesive, and more stable films, and the choice of solvent systems that provide the optimum application viscosity and minimum cost, and meet environmental constraints. The volatility of the solvent system is also important. If volatility is too high, the solvent will evaporate before the film has had a chance to "level," leaving brush marks or an "orange peel" or rough surface when sprayed. If too low, the coating will "sag" excessively after application.

Fundamental Principles of Polymeric Materials, Third Edition. Christopher S. Brazel and Stephen L. Rosen.
© 2012 John Wiley & Sons, Inc. Published 2012 by John Wiley & Sons, Inc.

A wide variety of polymers is used for lacquers. Newer systems favor acrylic polymers for their superior chemical stability. Acrylic lacquers may also incorporate a curing agent [1] such as hexamethoxymelamine, which crosslinks the film in a subsequent baking operation. Crosslinking toughens the film and improves solvent resistance.

The term "spirit varnish" is an old and imprecise name for a lacquer in which the solvent is alcohol. One example is shellac, a natural polymer secreted by the female lac bug, found in India and Thailand. Shellac can be dissolved in alcohols and then brushed on as a glaze and for finishing wood. Since such polymers are soluble in highly polar solvents, spirit varnishes have rather poor water resistance.

21.1.2 Oil Paints

These popular and widely used finishes typically consist of a suspension of pigment in a *drying oil*, an ester of glycerin, and unsaturated fatty acid such as linseed oil. The film is formed by a free-radical reaction involving atmospheric oxygen that polymerizes and crosslinks the drying oil through its double bonds. Sometimes an inert solvent (mineral spirits, turpentine) is added to control viscosity, and catalysts (e.g., cobalt naphthenate) are used to promote the crosslinking reaction. Oil paints, once cured, are no longer soluble, although they may be swelled and softened considerably by appropriate solvents (as used in paint removers).

21.1.3 Oil Varnish or Varnish

Varnishes are commonly used to coat wood, and provide protection from weather and humidity. These coatings consist of a polymer, either natural or synthetic, dissolved in a drying oil, with perhaps an inert solvent to control viscosity and a catalyst to promote the crosslinking reaction with oxygen. When cured, they produce a clear, tough, solvent-resistant film.

21.1.4 Enamel

An enamel is a pigmented oil varnish. It is much like an oil paint except that the added polymer provides a tougher, glossier film.

21.1.5 Latex Paints

These versatile finishes have largely replaced oil paints and enamels for home use because of their quick drying, low odor, and water cleanup properties. Basically, they are polymer latexes, produced by emulsion polymerization (Chapter 12), often of acrylic polymers, to which pigments and rheological-control agents have been added.

The paint film is formed by coalescence of the polymer particles upon evaporation of the water. The polymer itself is not water soluble. Although they are sometimes termed "water-soluble" paints, this is a serious misnomer, as is well known to anyone who has tried to clean a paintbrush after the water has evaporated. As long as the individual latex particles have not coalesced, they are water-dispersable.

In order that the particles coalesce to form a film when the water evaporates, the polymer must be deformable under the action of surface-tension forces. Thus, polymers for latex paints must be near or above their glass transition temperatures at use temperature. This formerly resulted in a rather soft paint film, which inherently lacked "scrubability."

Reasonable resistance to abrasion can be achieved by the incorporation of fairly large particle size inorganic filler-pigments such as $CaCO_3$. The rough paint film that results scatters reflected light and therefore has a "flat" finish, a characteristic of all early latex paints. By incorporating a low-volatility leveling solvent, latex "enamels" (they are not true enamels in the traditional sense) are produced. The polymer is formulated to have a T_g above use temperature. It is plasticized by the solvent so that its T_g is below application temperature, allowing the particles to coalesce to form a film. Over a period of a day or so after application, the solvent evaporates, leaving the hard, scrub-resistant, high-T_g polymer film, with a reasonable gloss if desired.

Early latex paints were based on styrene–butadiene copolymers or poly(vinyl acetate). These have been largely supplanted by paints based on acrylic latices (acrylates, $ROOCH=CH_2$, or methacrylates, $ROOCCH_3=CH_2$), because of the acrylics' superior chemical stability and, therefore, resistance to color change and degradation.

Much work has gone into adjusting the formulations of latex paints to allow high pigment contents and thick films for one-coat coverage, together with reasonable ease of application. The main rheological property desired is *thixotropy* (Chapter 14), which gives a high viscosity to prevent the settling of pigments and sagging and dripping of the applied film, together with a lower viscosity under the shearing of brushing or rolling for easy application. These properties are achieved by the addition of finely divided inorganics and water-soluble polymers.

21.2 SOLVENTLESS COATINGS

Traditionally, the application of most surface finishes has been accompanied by the evaporation of an organic solvent. These evaporated solvents are now recognized as a significant source of air pollution. Moreover, since the hydrocarbon crunch, it has become economically undesirable simply to lose them to the atmosphere. Although the water-based latex paints have gone a long way toward eliminating these problems, the coatings produced are sometimes lacking in thickness, uniformity, and durability. As a result, there is considerable interest in solventless or 100% *solids* coatings.

Two techniques for producing surface coatings without solvent evaporation, plastisol coating and fluidized-bed coating, were discussed earlier. Like the latter, *electrostatic spraying* makes use of a polymer powder. The powder is sprayed past an electrode, charging the particles, and onto a heated substrate with the opposite charge. The powder contacting the substrate melts to form a continuous film. Thermosets may be further oven-cured, if necessary. The resulting film is usually quite uniform. Another advantage of this process over traditional spraying of either latex- or solvent-based paints is that the material that is sprayed but not actually deposited on the surface (*overspray*) can be recovered and recycled back to the same process. Since overspray may account for much of the material sprayed, this can be economically significant.

The other approach to 100% solids coatings is to carry out the polymerization reaction right on the surface. Reactive liquid monomer or oligomer (low molecular weight polymer) is deposited on the surface and polymerized there. The reactants may be two components or a condensation pair (e.g., epoxies, Example 2.4O, or polyurethanes, Example 2.4Q), which are mixed just prior to the application and heat-cured on the surface, or they may be unsaturated materials that undergo addition polymerization (acrylates, $ROOC-CH=CH_2$, because of their high reactivity, are favored here).

In *radiation curing* processes, addition reactions are activated by various forms of radiation: infrared, microwave, radio frequency, gamma rays, ultraviolet (UV), and electron beam [2,3]. Basically, the energies of the first three are such that they simply thermally activate the system, that is, heat it, and are used in conjunction with the usual free-radical initiators.

Gamma rays generate free radicals in polymer solutions (or in any solution for that matter) and are generated by the radioactive decay of certain elements, such as ^{60}Co. Due to the radioactivity, these systems are not frequently used, but the newly formed free radicals can effectively crosslink polymers in solution. However, the gamma rays also cause some degradation, so the resulting material may have reduced mechanical properties compared to other curing methods.

UV-cured systems have achieved significant commercial importance, not only as coatings but also as printing inks, as a result of their low energy requirements (as compared to heat-cured systems) and freedom from pollution. They are also used in dental fillings, where multifunctional prepolymers are cured by UV light. They are highly crosslinked to prevent shrinkage within the tooth. The polymeric portion of a typical UV-cured coating has three main ingredients: (1) a reactive oligomer, (2) a multifunctional acrylate monomer, and (3) a photoinitiator. The reactive oligomer can be just about any low molecular weight polymer containing at least a couple of alkene double bonds, preferably acrylate double bonds because of their high reactivity. A simple example would be the system formed by reacting an excess of 1,6-hexanediol, $HO\text{-}(CH_2)_6\text{-}OH$, with adipic acid, $HOOC\text{-}(CH_2)_4\text{-}COOH$. The resulting hydroxyl-capped polyester is then condensed with acrylic acid, $HOOC\text{-}CH=CH_2$, to give the diacrylate polyester:

$$H_2C=\overset{\overset{\displaystyle H}{|}}{C}-\overset{\overset{\displaystyle O}{||}}{C}-O(CH_2)_6 \left[O-\overset{\overset{\displaystyle O}{||}}{C}(CH_2)_4\overset{\overset{\displaystyle O}{||}}{C}-O(CH_2)_6 \right]_x O-\overset{\overset{\displaystyle O}{||}}{C}-\overset{\overset{\displaystyle H}{|}}{C}=CH_2$$

This material, being tetrafunctional (two double bonds), will cure in an addition polymerization, but the crosslink density tends to be too low and, even though x is kept low, the viscosity is too high for many applications, so it is diluted with a multifunctional acrylate monomer such as trimethylol propane triacrylate:

$$H_3C\text{-}CH_2\text{-}C(CH_2\text{-}OOC\text{-}CH=CH_2)_3$$

This reduces the viscosity of the mix for easier formulation and increases the crosslink density in the cured film.

The film is cured by free radicals generated by the UV-induced decomposition of a photoinitiator, for example, benzoin methyl ether:

Another common UV-curing agent is dimethoxy phenyl acetophenone.

Electron-beam curing systems are similar in composition. Although they have been around a long time, they have the disadvantage of a high capital investment required to produce the electrons and the need to shield the equipment (stray X-rays are generated). Nevertheless, for highly pigmented systems, where UV radiation tends to be absorbed preferentially near the surface, they offer more uniform cures. The economics are reasonable for high-volume applications; therefore, as equipment continues to improve, electron beam systems should increase in importance.

The chemistry of polymers for high-solids coatings has been extensively reviewed [4].

21.3 ELECTRODEPOSITION [5,6]

The process of electrodeposition has been developed to provide uniform, highly adhesive, corrosion-resistant primer coats, particularly for the automotive industry. (The ultimate solution to the corrosion problem is simply to replace metal with plastic, as is done for many applications.) In the most common form of the process, polymers that contain carboxylic acid functionalities are produced. These polymers are then "solubilized" in an aqueous medium by partial neutralization with a base to give macroanions (P represents a polymer backbone): the macroanions are believed to exist as micelles.

$$P(\overset{O}{\overset{\|}{C}}-OH)_n + mBOH \rightarrow (HO-\overset{O}{\overset{\|}{C}})_{\overline{n-m}}P(\overset{O}{\overset{\|}{C}}-O^-)_m + mB^+ + mH_2O$$

Macrocarboxylic   Base   Macroanion
        acid

The part to be coated is made an anode ($+$) and immersed in a tank containing the macroanion solution plus pigment, plasticizer, curing agent, and other additives. The applied electric field draws the macroanions to the surface where they lose their charge and are deposited as a film. From the standpoint of forming a continuous, uniform, and, therefore, highly corrosion-resistant film, this process has significant advantages. The electric field draws the particles into all the nooks and crannies of the part (good "throwing power"). Sharp exterior corners, which in other processes tend to be thinly coated because of the surface tension that draws the material away from the corner, produce a stronger electric field and, thus, attract more polymer. Because the deposited coating is a dielectric, the field is stronger at holes in the film, preferentially attracting polymer to seal the hole. Furthermore, pollution and fire hazards are minimized and the process wastes very little material (unlike spraying), offering economic advantages.

A cationic process can also be used. Theoretically, it offers the best potential for rust resistance. This process is based on amine-functional polymers treated with an acid:

$$P(NR_2)_n + mHX \rightarrow (R_2N)_{\overline{n-m}}P(NR_2H^+)_m + mX^-$$

Acid   Macrocation

The macrocations are then deposited on a cathodic substrate.

Following electrodeposition, objects are usually rinsed to remove any non-adhering material and then baked to dry and fuse the film, and perhaps crosslink it.

21.4 MICROENCAPSULATION

Microencapsulation is a way of coating materials while ending up with a powder or very fine particles. Polymers are particularly useful, as they can form continuous thin films while in the liquid state and then harden upon cooling or drying of a solvent. Microcapsules are perhaps best known in the field of pharmaceuticals, where a drug can be trapped inside a polymer shell that is used to prevent release of the drug until the capsule reaches a desired location for best pharmacological effect. They are also commonly used in agriculture (for controlled release of fertilizers and pesticides), the food industry (e.g., to keep ingredients such as omega-3 fatty acids shelf-stable without spoiling prior to consumption), cosmetics, and industrial processes.

Nearly any polymer can be used to form a coating, although the food and pharmaceutical fields have a narrow list of materials approved for human ingestion. Many such microcapsules are made using natural polymers or derivatives of natural polymers, such as gelatin, cellulose derivatives, and alginates (from seaweed). The process of forming microcapsules requires three main steps:

- dispersion,
- film coating, and
- hardening.

In the dispersion phase, the core ingredient to be encapsulated must be separated into tiny droplets or solid particles through methods such as high-shear mixing, atomization with a

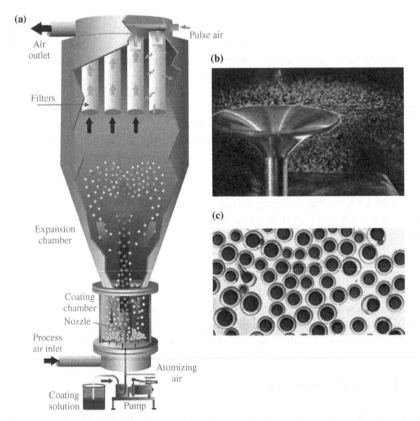

FIGURE 21.1   Microencapsulation methods include (a) fluidized bed coating and (b) rotating disk dispersion. Uniform coating around a core can be seen in (c). Image in (a), courtesy of Coating Place, Inc. Images in (b) and (c), courtesy of Southwest Research Institute®.

gas, or using rotational energy to form droplets. One method of coating solid core materials is fluid-bed coating (Figure 21.1a), where the solid particles are dispersed by a stream of air and then a polymer solution is sprayed into the unit. The polymer sticks to the surfaces of the particles and the air (which is often heated) evaporates the solvent to harden the coating; the longer the polymer is sprayed into the fluidized bed coater, the thicker the coating (as shown in Figure 21.1c). The dispersion phase can also be done at the same time as the film coating, as shown in a rotating disc encapsulation system (Figure 21.1b), where a usually molten polymer surrounds the core and is flung off a fast-spinning plate after which the polymer hardens by cooling in air before being collected as a powder. Besides fluidized bed coating and rotating disc technologies, other methods used for microencapsulation include spray drying, coacervation, solvent evaporation during high shear mixing, coextrusion, and pan coating [7].

PROBLEM

1 One method of making microcapsules is fluidized bed coating. If 10 kg of iron oxide (Fe_3O_4) magnets with 25 μm diameter are fluidized with air and a 10 wt% gelatin solution is sprayed in constantly at 50 g/min, how long will it take to coat all of the magnets with a 1 μm thick coating of gelatin (when dry)? (*Note*: You will need to look up the densities of the materials and use your knowledge of the geometry of solid objects.)

REFERENCES

[1] Morgans, W.M., *Outlines of Paint Technology*, 3rd ed., Wiley, New York, 1990.

[2] McGinnis, V.D. and G.W. Gruber, Curing methods for coating, Chapter 35 in *Applied Polymer Science*, 2nd ed., R.W. Tess and G. W. Poehlein (eds), American Chemical Society, Washington, DC, 1985.

[3] Randell, D.R. (ed.), *Radiation Curing of Polymers*, The Royal Society of Chemistry, London, 1987.

[4] Athey, R.D., Jr., *Prog. Org. Coatings* **7**, 289 (1979).

[5] Beck, F., *Prog. Org. Coatings* **4**, 1 (1976).

[6] Brewer, G.E.F., Electrodeposition of paint, Chapter 34 in *Applied Polymer Science*, 2nd ed., R.W. Tess and G. W. Poehlein (eds.), American Chemical Society, Washington, DC, 1985.

[7] Deasy, P.B., *Microencapsulation and Related Drug Processes*, Marcel Dekker, New York, 1984.

CHAPTER 22

POLYMER APPLICATIONS: ADHESIVES

22.1 ADHESIVES

Adhesives have been a technologically important application of polymers for thousands of years. Many of the early natural adhesives are still used. These include starch and protein-based formulations such as hydrolyzed collagen from animal hides, hooves, and bones, and casein from milk. As new adhesive formulations based on synthetic polymers (often the same polymers used in other applications) continue to be developed, the range of applications for adhesives has expanded dramatically [1–7].

An adhesive has been defined as a substance capable of holding materials (adherends) together by surface attachment. Adhesives offer a number of significant advantages as a means of bonding. (1) They are often the only practical means available, particularly in the case of small adherends. For example, it is hard to imagine welding abrasive grains to a paper backing to make sandpaper or bolting the grains together to make a grinding wheel. (2) In the adhesive joining of large adherends, forces are fairly uniformly distributed over large areas of the adherend, resulting in low stresses, and holes (necessary for riveting or bolting), which invariably act as stress concentrators in the adherends, are eliminated, thus lowering the possibility of adherend failure. (3) In addition to joining, adhesives may also act as seals against the penetration of fluids. In the case of corrosive fluids, this, coupled with the absence of holes, where corrosion usually gains an initial foothold, can minimize corrosion problems. (4) In terms of weight, it does not take much adhesive to join much larger adherends. Usually, a molecular thickness of the glue is adequate. Hence, it is not surprising that many high-performance adhesives were originally developed for aerospace applications. (5) Adhesive joining may offer economic advantages, often by reducing the hand labor necessary for other bonding techniques.

Fundamental Principles of Polymeric Materials, Third Edition. Christopher S. Brazel and Stephen L. Rosen.
© 2012 John Wiley & Sons, Inc. Published 2012 by John Wiley & Sons, Inc.

A detailed treatment of the science of adhesion is beyond the scope of this chapter. Nevertheless, some important generalizations will be drawn [8,9]. Adhesion results from (1) mechanical bonding between the adhesive and adherend, and (2) chemical forces—either primary covalent bonds or polar secondary bonds—between the two surfaces. The latter are thought to be the more important, and this explains in part why inert nonpolar polymeric substrates, such as polyethylene and polytetrafluoroethylene (PTFE), present challenges to forming adhesive bonds (the process for bonding PTFE to cookware is a closely guarded trade secret of DuPont). To adhere these kinds of polymers to a surface, they must first be chemically treated to introduce polar sites on the surface. To promote mechanical bonding, adherend surfaces are often roughened before joining, but this is sometimes counterproductive. It can trap air bubbles at the bottom of crevices that act as stress concentrators and promote failure in rigid adhesives.

The strength of adhesion can be measured using an automated materials testing system, such as the one described in Chapter 16. By pulling the bond apart (e.g., by applying axial tension), the work of adhesion can be measured. Peel tests are also commonly done, and these largely resemble pulling a piece of tape (or a Band-Aid®) off a surface.

With good bonding between adhesive and adherend, joint failure is cohesive (the adhesive itself or the substrate fails). Where the adhesive is weaker than the substrate, to a good approximation, the properties of the adhesive polymer determine the properties of the adhesive joint, that is, the bond can be no stronger than the glue line. Brittle polymers give brittle joints, polymers with high shear strengths give bonds of high shear strength, heat-resistant polymers produce bonds with good heat resistance, and so on.

To form a successful joint, the adhesive must intimately contact the adherend surface. First this requires that it wet the surface. The subject of wetting is considered in detail in a number of sources on surface chemistry [10,11]. In general, wetting is promoted by polar secondary forces between adhesive and substrate. This is another reason why low-polarity polymeric adherends such as polyethylene and PTFE are difficult to bond with adhesives. To insure proper wetting and interfacial bonding, it is often necessary to clean the adherend surfaces carefully before joining [12]. Good contact also requires a viscosity low enough under conditions of application to allow the adhesive to flow over the surface and into its nooks and crannies. Once contact has been established, the adhesive must harden to provide necessary joint strength. There are five general categories of organic adhesive that accomplish these objectives in different ways.

22.1.1 Solvent-Based Adhesives

Here the adhesive polymer is made to flow by dissolving it in an appropriate solvent. Thus, the polymers used must be linear or branched to allow solution, and the joints formed will not be resistant to solvents of the type used initially to dissolve the polymer. To get a good bond, it helps if the solvent attacks the adherend also. In fact, solvent alone is often used to "solvent weld" polymers, dissolving some of the adherend surface to make it sticky and form an adhesive on application.

One of the drawbacks to solvent-based adhesives based on rigid polymers is the shrinkage that results when the solvent evaporates. This can set up stresses that weaken the joint. An example of this type of adhesive is the familiar model airplane cement, basically a cellulose nitrate solution, with perhaps some plasticizer. Rubber cements, of course, maintain their flexibility, but cannot support as great a stress. Commercial rubber cements are based on natural, SBR (poly(butadiene-*co*-styrene)), nitrile (poly(butadiene-*co*-acrylonitrile)), chloroprene (poly(2-chlorobutadiene)), and reclaimed (devulcanized) rubbers. Examples are household rubber cement and Pliobond®. Rubber cements may also incorporate a curing agent to crosslink the polymer after application and evaporation of the solvent. This greatly increases solvent resistance and strength.

22.1.2 Latex Adhesives

These materials are based on polymer latexes made by emulsion polymerization. They flow easily while the continuous water phase is present and dry by evaporation of the water, leaving behind a layer of polymer. In order that the polymer particles coalesce to form a continuous joint and be able to flow to contact the adherend surfaces, the polymers used must be above their glass transition temperature at use temperature. These requirements are similar to those for latex paints, so it is not surprising that some of the same polymers are used in both applications, for example, styrene-butadiene copolymers and poly(vinyl acetate). Nitrile and neoprene rubbers are used for increased polarity. A familiar example of a latex adhesive is "white glue," basically a plasticized poly(vinyl acetate) latex. Latex adhesives have displaced solvent-based adhesives in many applications because of their reduced pollution and fire hazards. They are used extensively for bonding pile and backing in carpets.

22.1.3 Pressure-Sensitive Adhesives [13]

These are really viscous polymer that melts at room temperature, so the polymers used must be above their glass transitions. They are caused to flow and contact the adherends by applied pressure, and when the pressure is released, the viscosity is high enough to withstand the stresses produced by the adherends, which obviously cannot be very great. The key property for a polymer used in this application is tack; thus low molecular weight additives, which can compose up to 40 wt% of the adhesive, that increase tack are called tackifiers. Tack basically means that the polymer has a low enough viscosity to permit good surface contact, yet is high enough to resist separation under stress, something on the order of 10^4–10^6 cp [14], although elasticity probably also plays a role. Natural, SBR, and reclaimed rubbers are common in this application, although weaker adhesives have found one important use: Post-it® notes. The many varieties of pressure-sensitive tape are faced with this type of adhesive.

Contact cements are a variation in which the rubbery polymer is applied to each adherend surface in the form of a solution or, increasingly, a latex. Evaporation of the solvent or water leaves a polymer film with the tack necessary to grab and hold the adherends when they are pressed together.

22.1.4 Hot-Melt Adhesives

Thermoplastics often form good adhesives simply by being melted to cause flow and then solidifying on cooling after contacting the surfaces under moderate pressure. Polyamides and poly(ethylene-*co*-vinyl acetate) are used frequently as hot-melt adhesives. Electric "glue guns," which operate on this principle, are readily available in the consumer market (through hardware and craft stores).

22.1.5 Reactive Adhesives

These compounds are either monomers that are low molecular weight polymers, which solidify by a polymerization and/or crosslinking reaction after application. They can develop tremendous bond strengths and have good solvent resistance and good (for polymers, anyhow) high-temperature properties. The most familiar example of reactive adhesives are the epoxies (Example 2.4O) generally cured by multifunctional amines. These are often sold as two-part epoxies, where the two components are kept in physically separated chambers and contact each other to start the reaction when applied to a surface. Polyurethanes (Example 2.4Q) also make excellent reactive adhesives.

The α-alkyl cyanoacrylate "super glues" ("one drop holds 5000 lbs") are now a familiar part of the consumer market. Originally, the monomers had extremely low viscosities and so could crawl into narrow crevices and wet the adherend surfaces rapidly. On the other hand, they would not fill gaps and were absorbed into porous adherends, giving poor bonds. Newer versions are available with higher viscosities to overcome these drawbacks. Cyanoacrylates can polymerize in seconds by an anionic addition reaction believed initiated by hydroxyl ions from water adsorbed on the adherend surfaces:

$$x\ H_2C{=}C\text{-}\overset{\displaystyle O}{\overset{\|}{C}}\text{-}OR \quad \xrightarrow{OH^-} \quad HO\!\left[\!\!\begin{array}{c} OR \\ H\ C{=}O \\ C\text{-}C \\ H\ CN \end{array}\!\!\right]_{x}\!\!H$$

(with CN on the monomer)

Unfortunately (or fortunately, if you stick your fingers together), being linear and polar, they have poor resistance to polar solvents (acetone is a good solvent), and they are subject to hydrolysis and so have poor environmental stability. Octyl cyanoacrylate, where the R group is C_8H_{17}, is better known as Dermabond®, and has been used in medicine to seal wounds with a strong bond while also creating a barrier to bacteria that may infect a wound during the healing process. Here, the adhesion bonds are meant to slowly dissolve or wear away, replaced by the healed tissue. Other reactive surgical glues, such as fibrin glues, are based on natural proteins that are part of the blood coagulation cascade, and can be used to seal internal wounds.

Phenolic and other formaldehyde condensation polymers are also important reactive adhesives. Powdered phenolic resin is mixed with abrasive grains and the mixture is compression molded to form grinding wheels. A B-stage phenolic (see Chapter 18.4 on polymer additives, curing agents) in a solvent is used to impregnate tissue paper. The solvent is evaporated and the dry sheets are placed between layers of wood in a heated press, where the resin first melts and then cures, bonding the wood to form plywood. Similarly, sheets of paper impregnated with a B-stage melamine–formaldehyde resin are laminated and cured to form the familiar Formica countertops.

Unlike the previous examples of reactive adhesives, the phenolics and other formaldehyde condensation polymers evolve water as they cure. If trapped in the joint, this can result in serious weakness, which limits their adhesive applications.

Note that all these examples of reactive adhesives are highly polar polymers that often possess side groups with strong dipoles. It is largely the polarity and secondary bonds that account for their good bonding capabilities.

REFERENCES

[1] Temin, S.C., Adhesive compositions, p. 547 in *Encyclopedia of Polymer Science and Engineering*, 2nd ed., Vol. 1, J. Kroschwitz (ed.), Wiley, New York, 1985.

[2] Patrick, R.L., Chemistry and technology of adhesives, Chapter 34 in *Applied Polymer Science*, J.K. Craver and R.W. Tess (eds), American Chemical Society, Washington, DC, 1975.

[3] Shields, J., *Adhesives Handbook*, 3rd ed., Butterworths, London, 1984.

[4] Hartshorne, S.R. (ed.), *Structural Adhesives: Chemistry and Technology*, Plenum, New York, 1986.

[5] Wake, W.C., *Adhesion and the Formulation of Adhesives*, 2nd ed., Applied Science, London, 1982.

[6] Wake, W.C. (ed.), *Synthetic Adhesives and Sealants*, Wiley, New York, 1987.

[7] Landrock, A.H., *Adhesives Technology Handbook*, Noyes, Park Ridge, NJ, 1985.

[8] Gent, A.N. and G. R. Hamed, Adhesion, p. 476 in *Encyclopedia of Polymer Science and Engineering*, 2nd ed., Vol. 1, J. Kroschwitz (ed.), Wiley, New York, 1985.

[9] Meyer, F.J., Bonding, p. 518 in *Encyclopedia of Polymer Science and Engineering*, 2nd ed., Vol. 1, J. Kroschwitz (ed.), Wiley, New York, 1985.

[10] Adamson, A.W., *Physical Chemistry of Surfaces*, 5th ed., Wiley, New York, 1990.

[11] Hiemenz, P.C., *Principles of Colloid and Surface Chemistry*, 2nd ed., Dekker, New York, 1986.

[12] Wegman, R.F., *Surface Preparation Techniques for Adhesive Bonding*, Noyes, Park Ridge, NJ, 1989.

[13] Sates, D.(ed.), *Handbook of Pressure-Sensitive Adhesives*, 2nd ed., Van Nostrand Reinhold, New York, 1987.

[14] Skeist, I., Adhesive compositions, p. 482 in *Encyclopedia of Polymer Science and Technology*, Vol. 1, N. Bikales (ed.), Wiley, New York, 1971.

INDEX

ABS, 364

Accelerators, 378

Acetal, 25

Acid. *See also* names of acids, 6, 13

Acrilan®, 354

Acrylic acid, 18

Acrylic fibers, 354, 381

Acrylic monomers, 385, 386

Acrylic polymers, 386

Activation energy, 92, 155–156, 182, 259–262

Activators, 378

Addition polymerization, 17–19, 131, 146–180
  anionic, 186–193
    kinetics, 192–193
    atom transfer radical, 195
  cationic, 185–186
  free-radical, 146–180
    average chain lengths, 153–173
    chain transfer, 157–160
    distributions, 160–165
    kinetics of, 149–153, 176–180
    mechanism of, 147–148
  gelation, 148–149
  group-transfer, 194–195
  heterogeneous stereospecific, 196–202

Addition polymers, 17–19

Additives, 363

Adherends, 390

Adhesives, 2, 390–393
  hot-melt, 392
  latex, 392
  pressure-sensitive, 392
  reactive, 393
  solvent-based, 391–392

Alcohol, 6, 13

Alkali-metal salt, 186

Alkyl vinyl ethers, 186

Aluminum triethyl, 196

Amines, 6, 14

Amino acid, 16

Ammonium persulfate, 146, 174, 175

Amorphous polymers, 46, 51–53, 91, 375
  molecular motions in, 92

Anionic addition polymerization, 186–193

Anionic addition polymers, 63, 186–193

Anionic, kinetics, 191–193

Antiblock agent, 371

Antioxidants, 378

Aramid, 57, 363

Arrhenius equation, 142, 155

A-stage resin, 369

Atactic polypropylene, 41

Atactic structures, 40–42

Autoacceleration, 220

Fundamental Principles of Polymeric Materials, Third Edition. Christopher S. Brazel and Stephen L. Rosen.
© 2012 John Wiley & Sons, Inc. Published 2012 by John Wiley & Sons, Inc.

Average molecular weights, 62–65, 76
  cumulative, 166–169
  determination of, 66–75, 79–85
  number, 62
  viscosity, 72–73
  weight, 63
Average velocity, 267
Azobisisobutyronitrile, 146

Banbury mixer, 356
Benzoyl peroxide, 146
Berry number, 118
Biocide, 371
Bisulfite, 175
Block copolymerization, 187, 194
Block copolymers, 21, 53, 187, 194, 376
Blow extrusion, 338, 346
Blowing agents, 370–371
Blow molding, 347–348
Boiling-point elevation, 67
Boltzmann distribution, 92
Boltzmann superposition principle, 293–297
Bonding, 35–40
  action of solvents, 38
  dipole interaction, 37
  distances, 37
  hydrogen, 36, 37, 40, 99, 111
  ionic, 36–37
  primary covalent, 36–37
  strengths, 36–37
  temperature response, 37–38
  types of, 35
  van der Waals, 35
Bonds, 4
  double, 6, 17–19, 23, 42, 146, 147, 367
Branched polymers, 21–22, 50
  as thermoplastic, 37
Branching coefficient, 138–140
Branch unit, 138–140
B-stage resin, 369
Bulk molding compound, 373
Bulk polymerization, 220–225
Butadiene, 43
n-Butyllithium, 186, 189

Calendering, 349–350
Caliber®, 25
Carbon black, 366, 368, 376–378
Carbon-carbon backbone, 4, 40
Carboxylic acid, 6, 13
Carothers' equation, 134
Carreau model, 258

Casting, 351–352
  solution, 351
Catalyst, 185, 196–202
  deactivation, 200–201
  Phillips, 196
  single-site, 202
  Ziegler-Natta, 196–197
Cationic polymerization, 185–186
Cavity, 340–341
Celcon®, 25
Cellophane, 69
Cellulose, 38
Cellulose esters, 38, 362
Chain-growth polymerizations, 146
Chain length, 13, 63
  average, 63, 75–79
  cumulative, 166–173
  distribution of, 75–78, 83–85, 133–137,
    160–164
  instantaneous, 153–155, 165–166
  kinetic, 154
  number-average, 62
  weight-average, 62–63
Chains, 20–21, 40, 46, 136–137
  distribution of lengths, 75–78, 83–85,
    133–137, 160–164
  internal mobility of, 95
  network, 246
  stiffness of, 96
Chain transfer, 157–160, 196
Chain-transfer agent, 158–160, 173, 196, 220,
    225, 230
Chain-transfer constant, 158–159
Characteristic time, 288–289
Charge-transfer complexes, 216–217
Chlorinated waxes, 123
Chromatography
  gel permeation, 79–85
  size exclusion, 79–85
Cohesive energy density, 112
Colbalt naphthenate, 384
Colligative property measurements, 67–69
Combination, 148, 163–165
Complex modulus, 298
Complex viscosity, 298
Compliance, 281
  creep, 281
Compounding, 355–357, 379
Compression molding, 342
Condensation, 12
Condensation polymers, 13–17, 131–144
θ-Condition, 119

Cone-and-plate viscometers, 313–314
Consistency, 256
Constitutive relations, 256
Contact cements, 392
Continuous distribution
    of chain lengths, 75–79, 83–85, 133–137,
        160–164, 189, 190
    of molecular weights, 75–79
    of relaxation times, 290
    of retardation times, 290
Continuous stirred tank reactors, 171, 193,
        214–215, 232–233
Conversion, 134, 151
    composition and, 210–216
    gel point, 138–141
Copolymerization, 207–217
    block, 187
    effect on T_g, 97
    ideal, 209–210
    influence on properties, 101–102
    kinetics, 216–217
    mechanism, 213
    reactivity ratio significance, 209
    variation of composition with conversion,
        210–216
Copolymers, 20–22
    block, 21, 53, 187, 367, 375
    graft, 21–22, 367
    random, 20, 53, 97, 101–102
    statistical, 20
    styrene-butadiene, 20, 175, 190–191, 375
Couette flow, 311–312
Couette viscometer, 311–313
Counter ions, 186
Coupling agents, 366–367
Creep
    compliance, 281, 306
    recovery, 280, 290, 305
    response, 284, 286, 291
    tests, 280, 290
Creslan®, 354
Critical micelle concentration, 174
Crosslinked polymers, 19, 22–23, 27–29, 37, 38,
        108, 352, 369–370, 375
Crosslinking, 23, 108, 137–142, 342, 369–370
    effects of, 103
    solubility and, 108
Crystalline melting point, 46, 93, 97–103, 361
Crystalline polymers, 46, 47
Crystallinity, 46–50, 56–58
    density as measure of, 49
    effects of, 48–53

extended-chain crystals, 56–57
folded-chain crystallites, 55–56
fringed micelle model, 53–54
liquid-crystal polymers, 57–59
requirements for, 46–47
spherulites, 55–56
Crystallites, 47
    extended-chain, 56–57
    folded-chain, 55–56
Cumulative average chain lengths, 166–173
Cumulative copolymer composition, 210–216
Curing, 12, 22, 339, 352, 385–387
    electron-beam, 386
    radiation, 386
    ultraviolet, 386
Curing agents, 369–370
Curing systems, 377–378
Cyanoacrylate adhesives, 393

Dacron®, 24, 100
Damping, 316, 322
Dead-stop polymerization, 152
Deborah number, 270, 288–289, 306, 323
Decomposition, 147
Degree of polymerization, 14
Degree of unsaturation, 23
Delrin®, 25
Denier, 380
Density
    cohesive energy, 112
    as measure of crystallinity, 49
    of polyethylene, 50
        high, 50, 225
        linear-low, 50, 225
        low, 50, 223
        medium, 50
Diacids, 14, 20, 23, 135, 188
Dialcohol, 14
Diamines, 14, 20, 135
Dicumyl peroxide, 370
Diene polymers, 18–19, 42–44
Dienes, 18
1,4-Dienes, 18, 42
Dies, 344–345, 353
Die swell, 345
Differential refractometers, 80
Differential scanning colorimetry, 93–94
Difunctional monomers, 13, 15, 132
Diisocyanate, 15, 25, 31
Dilatant, 253
Dilute solutions, 118
2,6-Dimethylphenol, 25

Dioctyl phthalate, 103, 123, 124
Diol, 14–15, 188
Dipole, 6
Dipole interaction bonding, 36–38
Dipole moments, 113
Disk-plate viscometer, 314–316
Disproportionation, 148, 154, 161, 164, 186
Dissociation energy, 36, 38
Distribution
  Boltzmann, 92
  most probable, 133, 136, 161, 189
  Poisson, 189
Distributions of molecular weight, 75–78,
    83–85, 132–133, 136–137, 160–165,
    189–190, 201–202
Double bonds, 17, 18, 23, 42, 141, 144, 146, 185,
    367, 369
Drag reduction, 269–271
Drying oil, 144, 384
Dry spinning, 353
Ductile deformation, 362
Dyeing, 381
Dyes, 368, 381
Dynamic mechanical testing, 95, 297–304,
    316–323

Einstein equation, 120
Elasticity, 239–248
  energy, 241–245, 286
  entropy, 241–245, 286
  statistics, 246–248
  thermodynamics, 239–245
  types of, 241–242
Elastic recovery, 285
Elastomers, 2, 375
  thermoplastic, 375–376
Electrodeposition, 387
Electron-beam curing, 386–387
Electrostatic spraying, 385–386
Elution volume, 80, 81
Emulsion, 174
Emulsion polymerization, 173–180,
    232–234, 384
  kinetics of, 175–180
Enamel, 384
End-group analysis, 66–67
Energy
  activation, 92, 155–156, 182, 259–262
  cohesive density, 112
  dissociation, 36, 37
  elasticity, 98, 241, 286
  Gibbs free, 67, 111, 118

  internal, 112, 239, 243
  kinetic, 75
  potential, 37
  thermal, 36, 47, 92
  viscous dissipation, 251, 264,
    337, 346
Engineering plastics, 362
Engineering tensile stress, 246
Enthalpy, 97–99, 111–113, 117–118, 345
Entropy, 69, 97–98, 110, 116, 119, 240–242, 286
  elasticity, 241–242, 286
Epoxies, 25, 28, 344, 352, 363
  plasticizers, 124
Equations
  Arrhenius, 155
  Carothers', 134
  Carreau, 258
  Einstein, 120
  Huggins, 71
  Mark-Houwink-Sakurada, 72
  Modified Cross, 258
  Rayleigh, 212
  Williams-Landel-Ferry, 260, 326
Esterification reaction, 13
Esters, 6, 13
  cellulose, 38, 362
Ethylene, 17, 24, 196, 223, 224
Ethylene glycol, 22, 24, 67, 141, 143
Extended-chain crystals, 56–57
Extending oils, 377, 378
External lubricants, 368–369, 371
External plasticizer, 122
Extrusion, 344–346
Extrusion, blow, 345–346
Extrusion blow molding, 347, 348,

Fiberglass reinforced plastics, 23, 352–353, 363
Fibers, 1, 380
  spinning, 353–355
  synthetic, 380–382
  dyeing, 381
  heat effects, 122, 381–382
  moisture effect, 122, 381–382
  processing, 380–381. *See also* names of fibers
Filament winding, 353
Fillers, 52, 364, 366, 367, 371, 372, 373,
    376–377
First-order thermodynamic transition, 93
Flame retardants, 371
Flory-Huggins interaction parameter, 117–118
Flory-Huggins theory, 117–118
Flory temperature, 119

Flow
  Couette, 311–314, 330–331
  Elongational, 270
  Poisueille, 265–268
  tube, 265–269
  turbulent, 268–269
  viscometric, 251–252, 265, 308, 311–313, 315
  viscous, 250–255
    basic definitions, 251–252
    curves, 252–255
    influence of molecular weight, 262–263
    polymer melts, 253, 255–256
    polymer solutions, 253, 255–256
    pressure effects, 263–264
    quantitative representation of behavior, 256–259
    temperature dependence, 259–262
    time-dependent behavior, 254–255
Flow curves, 252–259
Flow index, 256
Fluidized-bed coating, 349, 385, 388
Fluids
  dilatant, 253, 257
  Maxwell element as, 279
  Newtonian, 256, 262, 268–269, 277
  pseudoplastic, 253, 256, 257
  rheopectic, 254
  thixotropic, 254
Folded-chain crystallites, 55
Formica®, 24, 27, 393
Four-parameter model, 258, 273, 277, 285–288, 305–307, 322, 341
Free-radical addition polymerization, 146–180
  average chain lengths, 153–155, 158, 166–169
  chain transfer, 157–160
  distribution of chain lengths, 160–165
  gelation in, 148–149
  kinetics of, 148–153
  mechanism of, 147–148
Free-radical initiators, 146–147, 149, 166, 351, 369
Free volume, 95
Freezing-point depression, 67
Fringed-micelle model, 53–54
Functionality, 13, 15, 18, 23, 137

Gate, 274, 339, 340, 341
Gaussian distribution, 247
Gegen ion, 186, 187
Gel, 79, 109, 167, 376
Gel coat, 352

Gel effect, 220
Gel permeation chromatography, 66, 79–85
Gel point, 137–141
Gibbs free energy, 67, 98, 110
Glass capillary viscometers, 74, 264, 311
Glass transition, 91–92
Glass transition temperature, 91–97, 109, 242, 260, 322, 324, 325, 361, 368, 375, 376, 384
Glassy modulus, 322–323, 327
Glue, super, 393
Glue, white, 233, 392
Glue, surgical, 393
Glycerin, 15, 32, 138, 140, 144, 384
Glycol, 22, 24, 25, 30, 134, 142
Glyptal, 24
Graft copolymers, 21, 22, 202–203, 207, 364, 367
Grafted polymer surfaces, 202–203
Group-transfer polymerization, 194–195
Gutta-percha, 42

Hand layup, 352
Heat effect on fibers, 122, 381–382
Heat of vaporization, 67
Heterogeneous stereospecific polymerization, 196–202, 227
Hexamethylenetetramine, 369
Homopolymers, 20–21, 40, 50, 52, 97, 101, 123, 189, 207
Hopper, 337, 339–340, 344, 373
Hot-runner molds, 340
Huggins equation, 71
Hydrocarbon oils, 377
Hydrogen bonding, 35–40, 47, 54, 99, 113, 121, 380
Hydrosol, 124
Hydroxy acid, 16, 18, 134

Ideal copolymerization, 210
Ideal rubber, 242–244, 325
  statistics, 246–248
Ideal solutions, 67, 119, 210
Impact modifiers, 364
Impact strength, 225, 367
Infrared measurements, 67, 80
Inherent viscosity, 71–72
Inhibitors, 141, 153–154, 157–159
Initiators, 23, 67, 146–148, 149–154, 157, 166–167, 174–176, 179, 186–187, 191, 193, 195, 209, 222, 230, 352, 369, 372

Injection blow molding, 338, 347–348
Injection molding, 105, 264, 337–344
Instantaneous chain length, 153–155,
    169–174, 209
Instantaneous copolymer composition, 209
Interatomic distances, 36–37
Internal energy, 112, 114, 241, 248
Internal lubricant, 369
Internal plasticizer, 123
Intrinsic viscosity, 70–75, 83, 118, 262
Ion-exchange resins, 230
Ionic bonding, 35–37
Ionic polymerizations, 185–193
Ions, counter or gegen, 186, 187
Isobutylene, 186
Isocyanate, 6, 15, 342
Isotactic polypropylene, 41, 381
Isotactic structures, 41–42
Isotropic compression, 240, 249
Isotropic pressure, 241, 242

Kevlar®, 57–58, 355, 363
Kinematic viscosity, 75
Kinetic chain length, 154, 165
Kinetic energy, 75

Lacquers. 227, 383–384
Lactam, 16
Lactic acid, 18
Lamellae, 54–56, 57
Latex, 30, 105, 178–179, 232–233,
    384–385, 392
Latex adhesives, 392
Latex paints, 30, 384–385
Leathery region, 327
Lexan®, 20, 25, 28
Light scattering, 66, 69–70
Limiting viscosity number, 72
Linear polyethylene, 57, 93,
    196, 348
Linear polymer chain, 20
Linear polymers, 20–21, 28–29
  as thermoplastic, 37
Linear viscoelasticity, 276–304
  Boltzmann superposition principle, 293–296,
    301
  Deborah number, 270, 288, 306, 323–324
  dynamic mechanical testing, 297–304,
    316–323
  four-parameter model, 289, 305–307,
    322, 341
  limit of, 279

mechanical models, 276–284
quantitative approaches, 293–297
time-temperature superposition, 323–329
Liquid-crystal polymers, 57–59, 111
  lyotropic, 57, 111
  thermotropic, 57–59
Liquid-liquid transition, 104
Lithography, 358
Living polymers, 187–189, 191, 195, 202
Logarithmic viscosity number, 72
Loss tangent, 300
Lost-core molding, 342
Lower critical solution temperature,
    110, 121
Lower-Newtonian, viscosity, 74, 255
Low-shrink additive, 372–373
Lubricants, 363, 368–369, 372
Lucite®, 91, 102

Macromolecular hypothesis, 3
Makrolon®, 25, 28
Maleic acid, 22–23
Maleic anhydride, 22–23, 210–211, 217
Maltese Cross appearance, 55
Mandrel, 346
Mass polymerization, 220
Master batching, 232–233
Maxwell element, 279–282, 289, 294–295, 296,
    301, 303
Maxwell orthogonal rheometer, 319
Maxwell relation, 244
Mechanical models for linear viscoelastic
    response, 276–285
Melmac®, 24, 27
Melt fracture, 345
Melting, thermodynamics of, 97–100
Melting point, crystalline, 50, 58, 91, 93,
    97–100, 361
Melt spinning, 353
Membrane osmometry, 68–69
Memory function, 294, 301
Mercaptans, 158–159
Metastable amorphous state, 100
2-Methyl-1,3 butadiene, 19
Methyl methacrylate, 31, 45, 64, 194,
    220, 231
α-Methyl styrene, 96, 186
Micelles, 21, 174–178, 387
  critical concentration, 174
  fringed model, 53–54
Microencapsulation, 387–388
Modified Cross model, 258–260, 266

Modulus, 50–51, 239, 247, 277–278, 281–282, 287–288, 291, 297–298, 301, 308–309, 322, 325, 361, 363, 366, 368, 377
  complex, 298
  glassy, 287, 322–323, 327
  Hooke's, 277, 308
  loss, 298, 308
  relaxation, 281, 289, 290, 292–293, 294, 301, 328
  rubbery, 322, 327
  shear, 277
  storage, 297–298, 301
  tensile, 50–51, 57, 362
  Young's, 247, 277, 308–309
Moisture, effect on fibers, 122, 381–382
Molding, 337–344
  blow, 338, 347–348
  compression, 342
  fluidized-bed, 349
  foam, 338
  injection, 337–341, 343–344
  lost-core, 342
  reaction injection, 342–343
  reinforced thermoset, 352–353
  resin-transfer, 343–344
  rotational, 338, 348–349
  slush, 348–349
  transfer, 342
Molecular motions in an amorphous polymer, 92
Molecular response, 97, 285–288
Molecular structure, 1, 3–4, 11, 19, 35–44
Molecular weight, 61–85
  average, 62–75
    defined, 62–65
    determination of, 66–75, 79–85
    number, 61–62, 75–76
    viscosity, 72–73
    weight, 62–63, 69
  defined, 61
  distributions, 75–79, 202
  influence on flow properties, 262–263
Monodisperse polymers, 63–64, 72, 80–82, 189–194
Monomers, 5, 12, 15–16, 20–21, 24
  difunctional, 15, 20
  vinyl, 17–18
Most-probable distribution, 133, 161, 190, 201
Motionless mixers, 343
Multiviscosity motor oils, 120
Mylar®, 24, 26, 100
Natta, G., 41–42, 196

Natural rubber, 12, 19, 22, 42, 243, 245, 376, 378–379
Network chains, 246–247
Network polymers, 37, 148
Newtonian fluids, 253–254, 256, 262, 268–270, 277
Newtonian viscosity, 74, 291
Non radical addition polymerization, 185–202
  anionic, 186–194
  cationic, 185–186
  group transfer, 194
  heterogeneous stereospecific, 196–202
Nozzle, 342, 357
Number-average chain lengths, 133–136, 155–156, 168, 169, 189, 209, 213
Number-average molecular weight, 63–64, 66, 67, 69, 246
Nylons, 2, 5, 14, 16, 20, 21, 24, 26, 57–58, 98, 102, 134, 228–229, 362, 381

Oil paints, 383–384
Oils
  drying, 144, 384
  extending, 248, 377, 378
  motor, 120, 123
  polyunsaturated vegetable, 124
Oil varnish, 384
Olefins, 50, 196, 225–226
Oligomers, 12, 61, 123, 133, 202, 385–386
Optical properties, 51–53, 100, 224
Organic functional groups, 14
Organosol, 124
Orientation blow molding, 347
Orlon®, 354
Orthogonal rheometer, 319–320
Osmometers, 68–69
Osmometry
  membrane, 68–69
  vapor-pressure, 69
Osmotic pressure, 67–68
Ostwald viscometers, 74
Oxidative coupling, 28
Oxirane rings, 124

Paints, 383–387
  latex, 30, 384–385
  oil, 384
Parison, 338, 347–348
Parison programmers, 347
Penultimate effects, 216–217
Persulfate, 146, 174
Phase angle, 297

Phenolics, 12, 366, 369, 393
Phenoxy, 25, 28
Phillips catalysts, 196
Photoinitiators, 146–147, 358, 386
Phthalic anhydride, 23, 24, 140–141
Pigments, 115, 124, 351, 355, 363, 368, 369,
    378, 381, 383
Plasticizers, 102, 103, 122–124, 363, 368
    efficiency, 123
    epoxy, 124
    external, 122–123
    internal, 123
    permanence, 123
    polymeric, 123
Plastics, 1–2, 361–373
    contents of compounds, 363–370
    mechanical properties, 362–363
    rubber toughened, 364
Platens, 342
Plexiglas®, 91, 102
Plywood, 393
Poisson distribution, 190
Poisueille flow, 265–267, 308
Polyacrylonitrile, 17, 38, 95, 223, 364, 380
Polyamides, 14, 16, 99, 392
Polybutadiene, 52, 91, 364, 367, 376
Poly (butylene terephthalate), 24, 99
Polycarbonates, 24, 363, 365
Polychlorotrifluoroethylene, 103
Polycondensation, 14, 131–142
    gel formation, 137–142
    interfacial, 228–229
    kinetics of, 142–143
    statistics of, 132–133
Poly (diallyl phthalate), 24, 27
Polydispersity index, 63, 65, 80, 82, 86, 91, 137,
    170, 172, 190, 201–202
Polyesters, 2, 5, 14–16, 20, 23, 58, 66, 140, 143,
    352, 355, 363, 372, 386
    unsaturated, 22–23, 140–141, 344, 351, 355,
        367, 369, 372
Polyethyl acrylate, 91
Polyethylene, 11, 24, 26, 40, 41, 49, 50,
    51, 93, 97, 99–100, 102–103, 108,
    158, 196, 223–226, 228, 256, 348,
    372, 373, 391
    density of, 49–50
        high, 49–50, 225
        low, 49–50, 99, 223
        medium, 50
        ultra-low, 50
        very-low, 50

linear, 24, 26, 40, 57, 93, 196, 348
linear low density, 50, 225
properties of, 49–50
Poly (ethylene terephthalate) 24, 26, 100, 373
Polyimide, 14–15, 25, 29, 96
Polyisoprene, 42–43, 91, 376
1,2-Polyisoprene, 43
1,4-Polyisoprene, 42–43, 376
    cis, 42–43
    trans, 42–43
3,4-Polyisoprene, 43
Polymeric plasticizers, 123
Polymerization, 12–18, 131–234
    anionic, 186–192
    bulk, 220–225
    cationic, 185–186
    chain growth, 146–180
    condensation, 12–14, 131–143
    degree of, 14
    emulsion, 173–180, 232–234
        kinetics of, 176–180
    free-radical addition, 146–180
        average chain lengths, 155–156, 159–173,
            179, 186, 189–190
        chain transfer, 157–160
        kinetics of homogeneous, 149–153
        mechanism of, 146–148
    gas-phase olefin, 225–226
    group-transfer, 194–195
    heterogeneous bulk, 220
    heterogeneous stereospecific, 196–202
    interfacial, 228
    ionic, 227
    mass, 220–225
    oxidative coupling, 28
    precipitation, 223
    ring scission, 16, 28, 42
    solution, 226–228
    step-growth, 131–143
    suspension, 229–232
    vinyl, 17–18
    Ziegler-Natta, 227–234
Polymers, 1–6
    addition, 14, 17
    amorphous, 47–48, 91, 322, 327
        molecular motions in, 92
    branched, 19, 21–22, 37, 50, 103, 137
    condensation, 13–17, 131
    crosslinked, 19, 22, 37, 38, 53, 79, 113, 137,
        140, 149, 224, 239, 284, 290, 328, 352,
        361, 369, 370, 375, 386
    crystalline, 47–59, 93–104

diene, 6, 18–19, 42–44, 196, 376
linear, 15, 18, 20–21, 46, 80, 134, 282
living, 187–189, 195, 202,
monodisperse, 63, 72, 80, 81, 189, 194
network, 22, 37, 148
stereoregular, 41, 47, 49, 194
structure of, 19–30
thermoplastic, 11–12, 37–38, 58, 222,
 337–339, 342–344, 350, 356, 363,
 369–370, 375
thermosetting, 12, 37–38, 137, 227, 327, 342,
 351, 352, 355, 362, 369–370, 373, 385
types of, 11–30
unsaturated, 18, 23, 377–378
Polymethyl methacrylate, 31, 49, 71, 75, 91, 194,
 220, 222, 231, 322–323, 353, 362
Poly (4-methyl-1-pentene), 50, 52
Polyphenylene oxide, 25, 117
Polypropylene, 5, 11, 25, 41, 50–53, 75, 196,
 225, 228, 346, 363, 366, 381
Polystyrene, 11, 20, 26, 41, 52, 75, 80–83, 96,
 108, 191, 230–231, 328, 340, 362–365,
 367, 369, 373, 376
Polysulfone, 25
Polytetrafluorethylene, 25, 38, 391
Polyureas, 31, 99, 342, 372
Polyurethane, 15, 25, 99, 338, 342–344, 372,
 385, 393
Polyvinyl chloride, 11, 17, 20, 75, 123, 223, 231,
 367, 368, 371, 375
Polyvinylidene chloride, 103
Potential energy, 36, 37, 95–96, 250, 288
Potting, 222
Power law, 256–258, 266–269, 315–316
Preform, 343, 347
Pressure
 hydrostatic, 241
 isotropic, 241–242
 osmotic, 66–69
 T_g and, 95
 viscosity and, 263–264
Pressure-sensitive adhesives, 392
Primary covalent bonding, 35–38
Procedure X, 292
Processing, 337–359
 calendering, 338, 349–350
 casting, 338, 351–352
  solution, 351
 extrusion, 58, 338, 344–346, 354
 fiber spinning, 57, 338, 353–355, 381
 molding, 337–344
  blow, 338, 347–348

  compression, 342
  fluidized-bed, 348–349
  injection, 337–341, 342
  lost-core, 342
  reaction injection, 342
  reinforced thermoset, 352–353
  resin transfer, 343
  rotational, 348
  slush, 348
  transfer, 342
 sheet forming, 350
 stamping, 351
 thermoforming, 350–351
Processing aids, 363, 369, 378
Promoters, 378
Propagation, 148–150, 157–158, 160, 162, 176,
 186–187, 192, 194, 196–199, 208, 216
Propylene, 17, 25, 53, 225
Propylene oxide, 42
Protective coatings, 3
Protective colloids, 230
Prototyping, 358
Pseudoplastic fluids, 253, 256–257, 272
Pseudoplasticity, 267
Pultrusion, 353
Pyramidal polymer crystals, 54

Radiation curing, 386
Radioactively tagged initiators, 67
Random copolymers, 20, 53, 97, 101–102,
 207, 212
Rate of strain, 281–283, 288, 297, 309
Rayleigh equation, 212
Reaction injection molding, 338, 342
Reactive adhesives, 393
Reactivity ratios, 208–209, 217
Reactors continuously stirred tank, 171, 193,
 215, 220, 223, 233
 batch, 134, 152, 167, 169–173, 190, 192, 210,
 213, 232
 plug-flow, 171, 193, 215, 224
 semi-batch, 215, 220
 tubular, 171, 193, 224
Recycling, 2, 104, 121, 373
Redox systems, 146, 175
Reduced viscosity, 70–72
Reference temperature, 94, 260, 325
Refractive index, 52, 69, 80
Refractometers, 80
Regrind, 339–340
Reinforced thermoset molding, 338,
 352–353

Reinforcing agents, 222, 363–364, 366, 367, 369, 376, 377
Relative viscosity, 70–72
Relaxation
  modulus, 281, 289, 292, 293
  stress, 276, 281–282, 284, 289, 292, 293, 295–297, 301, 303–304, 324–325, 327–328
  time, 281, 288, 290, 325
Repeating unit, 14, 16, 20–23, 43, 58, 63, 101, 103, 148
Resins, 12, 100, 124, 141, 338, 343, 344, 348, 350, 372, 393
  ion-exchange, 230
Resin-transfer molding, 343
Resoles, 369
Retardation times, 290
Retarders, 153–154
Retention volume, 80
Reynolds number, 268–270
Rheology, 250–271
Rheopectic fluids, 254–255
Ring-scission polymerization, 16, 28, 42
Rotational molding, 338, 348–349
Rubber Reserve Program, 232
Rubbers, 1, 3, 12, 19, 20, 22, 25, 38, 42, 43, 52, 91, 107, 173, 187, 225, 227–228, 232, 233, 239, 242–244, 246–247, 287, 301, 318, 351, 364, 375–379
  antioxidants, 378
  contents of compounds, 376–378
  curing systems, 377–378
  elasticity, 239–248
    statistics, 246–248
    thermodynamics. 239–246
  extending oils, 248, 377
  fillers, 376–377
  ideal, 242–243, 244, 246–248
    statistics, 246–248
  natural, 12, 19, 22, 42, 243, 245, 376, 378, 379
  pigments, 376, 378
  reinforcing fillers, 376
  silicone, 3, 25, 95, 250, 351, 371, 376
  stabilizers, 378
  styrene-butadiene, 20, 232, 367, 385, 392
  thermoplastic elastomers, 375–376
  vulcanizing systems, 377–378
Rubber-toughened plastics, 364
Rubbery modulus, 287, 322, 327
Rubbery plateau, 309, 322, 327
Runners, 339, 340, 342

Salts
  alkali metal, 186
  ionic, 67
  organic acid, 173
Saran®, 103
Secondary bonds, 35, 37–38, 40, 53, 122, 367, 381, 391, 393
Second-order thermodynamic transition, 93
Semi-permeable membranes, 68–69
Shear rate, 74, 120, 251–259, 261–264, 266, 304, 314, 356
Shear strain, 240, 251–252, 277–282, 284, 295, 302
Shear stress, 240–241, 247, 251–254, 256, 259–261, 264–266, 277–282, 284, 287, 291, 293, 301, 313, 321
Shear-thickening, 253
Shear-thinning, 253
Shear viscosity, 255, 258–263, 270, 277
Sheet forming, 350–351
Sheet molding compound (SMC), 351, 371–373
Silane coupling agents, 367
Silicone rubbers, 25, 351, 371, 376
Silly putty, 3, 250, 270, 288
Sirup, 222
Size exclusion chromatography (SEC), 79–83
Slip agent, 371
Slush molding, 348–349
Soap, 174–175, 232, 367
Solubility, 107–124
  crosslinking and, 108
  general rules for, 107–108
  parameters, 95, 112–114
    three dimensional, 114–115
  thermodynamic basis of, 110–111
Solution casting, 338, 351
Solution polymerization, 226–228
Solutions, 48, 54, 57, 67, 70, 73–75, 107–124, 140, 192, 210, 221, 250–251, 253–255, 258, 262, 268, 270, 292, 308, 311, 313, 314, 349, 386
  concentrated-plasticizers, 122–124
  dilute, properties of, 48, 57, 69–71, 75, 80, 109–110, 118–121
  flow properties, 255–256
Solution viscosity, 70–72, 74, 120, 262, 269
Solvent-based adhesives, 391–392
Solventless coatings, 385–387
Solvent power, 119–120

Solvents
  action of, 38
  polymer-polymer-common solvent
    system, 121
Specific viscosity, 70, 72
Specific volume, 92–93, 100, 101
Spherulites, 55–56
Spinnerettes, 274, 369
Spinning
  dry, 353–354
  fiber, 338, 353–358
  melt, 353
  wet, 57, 354–355
Spirit varnish, 384
Stabilizers, 367–368, 378
Standard linear solid, 306
Staple, 380
Statistical copolymers, 20
Staudinger, Dr. Herman, 3
Step-growth polymerizations, 131–143
Stereoisomerism, 40–44
  in diene polymers, 42–44
  in vinyl polymers, 40–42
Stereoisomers, 41, 43–44
Stereoregular polymers, 41, 47, 185, 194
Stereoregular polypropylenes, 50
Stereospecific polymerization, 196–202
Storage modulus, 297–298, 301, 308
Strain, 240, 247, 252, 276–284, 287, 288,
    293–295, 297, 300, 302, 309–310,
    317–322, 324, 341
  rate of, 252, 281–283, 288, 297, 310
  shear, 277–278, 280, 282, 284, 295
  tensile, 247, 277–278, 280, 282–284, 318
Stress, 239–240, 251–254, 256, 260–261,
    264–266, 277–287, 313
  shear, 240, 251–254, 256, 261, 264–266, 277–
    282, 284, 313
  tensile, 240, 242–243, 246, 277–282,
    284–285, 287, 291, 293, 324
  engineering, 246, 276
  true, 247
  ultimate, 308, 311
  yield, 308, 311
Stress cracking, 342, 348–349, 370
Stress relaxation, 276, 281–282, 284, 290, 292,
    295, 303–304, 324–325, 327–328
Stretch blow molding, 348
Structural unit, 132, 244
Styrene, 17, 20, 22–23, 41, 52, 96, 111, 117, 180,
    186, 189, 195, 211, 214, 217, 223, 367,
    369, 378

Styrene-Butadiene Rubber, 20, 232, 376, 392
Super glue, 393
Surface finishes, 3, 341, 361, 369, 383–387
  enamel, 384
  lacquers, 227, 383–384
  latex paints, 384–385
  oil paints, 383–384
  solventless, 385–387
  varnish, 384
Suspension polymerization, 229–232
Syndiotactic polypropylene, 41
Syndiotactic structures, 41–42
Synthesis, chemistry of, 12–19
Synthetic fibers, 380–382
  dyeing, 381
  heat effects, 381–382
  moisture effects, 381–382
  processing, 380–381
  spinning, 353–355

Tangent, loss, 300
Temperature
  bonding response, 37–38
  and chain length, 155–156
  at constant force, 243–244
  at constant length, 244–245
  Flory, 119
  flow properties dependence, 259–262
  glass transition, 91–97, 109, 242, 260, 322,
    324–325, 361, 368, 375–376, 384
  lower critical solution, 110, 121
  and rate of polymerization, 155–156
  reaction to, 11–12
  reference, 94, 260, 325–326
  shift factor, 325–326
  time superposition, 292–297, 308,
    323–325, 328
  upper critical solution, 109–111
Tenacity, 380
Tensile stress, 240, 242–243, 246, 277–282, 284,
    285, 287, 291, 293, 324
  engineering, 246, 276
  true, 247
Termination, 148, 150, 153–155, 163–165, 176,
    179–180, 186–187, 195, 216, 221
Thermal degradation, 103
Thermal energy, 36, 47, 95, 120, 242, 256,
    260, 324
Thermodynamics
  of elasticity, 239–245
  of melting, 97–100
  of solubility, 110–118

Thermodynamic transition
  first-order, 93
  second-order, 93
Thermoplastic elastomers, 375–379
Thermoplastic polymers, 11, 12, 37–38, 58,
  337–339, 342
Thermosetting polymers, 12, 30, 37, 38,
  137, 351
Theta condition, 119
  solvent, 119
  temperature, 119
Thixotropic fluids, 254–255
Thixotropy, 385
Three-parameter model, 258, 284
Time-dependent behavior, 254–255
Times
  characteristic, 258, 288, 324
  relaxation, 270, 281, 288–290, 325
  retardation, 290
  temperature superposition, 292–297,
    308, 323–325
Titanate coupling agents, 367
Titanium tetrachloride, 196
Titration, 66
Torsional flow, 315–317
Transfer molding, 342
Transient time-dependent effects,
  254–255
Transitions, 91–104
  alpha, 104
  beta, 104
  first-order, 93
  glass, 91–97, 101–103
  liquid-liquid, 103
  melting, 93–94, 97–100
  second-order, 93
Tricresyl phosphate, 123
Tromsdorff effect, 155, 220, 234
Trouton's viscosity, 277
True tensile stress, 247
Tube flow, 250–271
Turbulent flow, 268–270
Two-roll mills, 356–357, 379

Ubbelohde viscometers, 74–75
Ultraviolet curing, 386
Unsaturated polyesters, 22, 23, 140, 344, 351,
  355, 367, 369, 372
Unsaturated polymers, 18
Unsaturation, degree of, 23
Upper critical solution temperature,
  109–111

Valve gate, 340
Van der Waals bonding, 35–37, 114
Vaporization, 91, 112
  heat of, 67
Vaporization temperatures, 112
Vapor-pressure osmometry, 69
Varnish, 383–384
Vectra A®, 58
Velocity, average, 267
Velocity gradient, 251–252, 265–266, 312
Vinyl monomers, 17–20, 23, 41, 50, 153, 221,
  369, 370
Vinyl polymerization, 17–18
Vinyl polymers, stereoisomerism in, 40–42
Viscoelastic solids, 284, 290
Viscoelasticity, linear, 276–304
  Boltzmann Superposition, 293–297, 301
  Deborah number, 270, 288–289, 323–324
  dynamic mechanical testing, 297–304,
    316–323
  four-parameter model, 258, 277, 285–288,
    322, 341
  limit of, 279
  mechanical models, 276–285
  molecular response, 285–288
  procedure X, 292
  quantitative approaches, 289–293
  time-temperature superposition, 292–297,
    308, 323–325
Viscoelastic response, 250, 276–288
Viscometers
  capillary, 74, 264, 311
  cone-and-plate, 313–314
  Couette, 311–313
  disk-plate, 314–316
  glass capillary, 74
  Ostwald, 74
  Ubbelohde, 74–75
Viscometric flows, 251–252, 265, 308,
  311–313, 315
Viscometry, 66, 120, 264, 311–312, 316, 318
Viscosity, 66, 70–75, 85, 118, 120–121, 123,
  167, 221, 223, 226, 230, 233, 250–260,
  262–264, 268–271, 277–278, 291–292,
  298, 303–304, 311–316, 341, 343–344,
  363, 367, 369, 383–387, 391
  average molecular weight, 66, 252
  complex, 298
  inherent, 71–72
  intrinsic, 70–75, 83
  kinematic, 75
  limiting number, 72

logarithmic number, 72
lower-Newtonian, 74, 255–256
multi-, 120, 123
Newtonian, 291
pressure and, 263–264
reduced, 70–72
relative, 70, 72
specific, 70, 72
tensile, 305
terminology, 72
Trouton's, 277
upper Newtonian, 256, 258
zero-shear, 74, 255, 258–260, 263–264,
    292, 303
Viscosity number, 72
Viscosity ratio, 72
Viscous energy dissipation, 251, 256, 264–265,
    301, 339, 345–346
Viscous flow, 250–271
  basic definitions, 250–252
  curves, 252–254
  influence of molecular weight,
    262–263
  polymer melts, 250–251, 253, 255–256
  polymer solutions, 255–256
  pressure effects, 263–264
  quantitative representation of behavior,
    256–259
  region, 325, 327
  temperature dependence, 259–262
  time-dependent behavior, 254–255
Voigt-Kelvin model, 283–285, 288, 290–291

Volume
  elution, 80–81, 83
  free, 95, 97
  molar, 112–114, 118
  specific, 92–93, 100, 101
Vulcanization, 12, 22, 239, 376–378
Vulcanizing systems, 377–378

Weight, molecular, 61–86
  average, 62–75
  defined, 61
  distributions, 75–79, 85, 202
  number-average, 62–68, 246
  viscosity-average, 72–73
  viscous flow influence, 262–263
  weight-average, 63–64
Weissenberg effect, 3, 287
Wet spinning, 57, 354
White glue, 233, 392
Williams-Landel-Ferry equation, 260, 326

X-ray studies, 48, 55
Xydar®, 58, 99

Young's modulus, 247, 277, 308–309

Zero-shear viscosity, 74, 255, 258–260,
    263–264, 292, 303
Ziegler-Natta catalysis, 185, 196–197, 199, 202,
    214, 225, 228, 232, 234
Zimm plot, 69
Zinc oxide, 378